T0229870

GENE THERAPY FOR
DISEASES OF THE LUNG

LUNG BIOLOGY IN HEALTH AND DISEASE

Executive Editor

Claude Lenfant
Director, National Heart, Lung and Blood Institute
National Institutes of Health
Bethesda, Maryland

ADDITIONAL VOLUMES IN PREPARATION

The opinions expressed in these volumes do not necessarily represent the views of the National Institutes of Health.

GENE THERAPY FOR DISEASES OF THE LUNG

Edited by

Kenneth L. Brigham

Vanderbilt University School of Medicine
Nashville, Tennessee

CRC Press
Taylor & Francis Group
Boca Raton London New York

CRC Press is an imprint of the
Taylor & Francis Group, an **informa** business

CRC Press
Taylor & Francis Group
6000 Broken Sound Parkway NW, Suite 300
Boca Raton, FL 33487-2742

© 1997 by Taylor & Francis Group, LLC
CRC Press is an imprint of Taylor & Francis Group, an Informa business

No claim to original U.S. Government works

ISBN-13: 978-0-8247-0060-7 (hbk)

Visit the Taylor & Francis Web site at
http://www.taylorandfrancis.com

and the CRC Press Web site at
http://www.crcpress.com

INTRODUCTION

In a 1791 letter to the National Assembly, Edmund Burke said, "You can never plan the future by the past." Much later, Winston Churchill stated, "A pessimist sees the difficulty in every opportunity; an optimist sees the opportunity in every difficulty." Of course, since these were comments by statesmen and politicians, one could ask what this has to do with scientific achievements. Quite a lot, I should say!

In science, as in politics, one cannot really "plan the future by the past"; nonetheless, the past undeniably provides the foundation for future opportunities.

The field of genetic therapy is a creation of contemporary visionaries who not only acknowledged the difficulties but also saw the opportunities. In the first chapter of this volume, Dr. W. French Anderson describes vividly the steps that led to gene transfer—hopefully for therapeutic purposes—as we know it today. All this happened within the last 30 years, but it was not easy, as illustrated by Anderson's own story. Indeed, in 1968, his manuscript "Current Potential for Modification of Genetic

Defects" was rejected because one reviewer scorned it as "medical prediction."

Although gene therapy applications are in clinical trials today, no one would yet venture to say that gene transfer will become a routine, or even common, therapeutic approach. Fortunately, researchers working in this area are good—in fact, extremely good—and they are also optimists after the manner of Churchill. For these reasons, the scientific community can be—must be—as enthusiastic about the prospects for gene therapy as it may be cautious in its approaches.

Gene Therapy for Diseases of the Lung reports on the state of gene therapy with regard to lung diseases. The optimism and enthusiasm it conveys are infectious. Dr. Kenneth Brigham, in the very bold undertaking of editing this volume, has assembled a cast of superb scientists whose contributions in the laboratories and in this volume are paving the way for great hope. These contributions add a new milestone to the Lung Biology in Health and Disease series. For this I am very grateful to Dr. Brigham and all his outstanding colleagues.

Claude Lenfant, M.D.
Bethesda, Maryland

PREFACE

If there are three phases to scientific discovery—*gee whiz, aw shucks*, and *yes, but*—we are well into the *yes, but* phase of gene therapy. That is particularly true for such therapy directed at lung diseases.

Gee whiz. DNA—the stuff of genes, the fundamental unit of life—can be isolated, sequenced, rearranged, and domesticated. (As W. French Anderson points out in Chapter 1, quite early on, the real visionaries saw the potential for that technology as therapy.) And, human diseases can be directly attributed to specific pieces of DNA, the exact location and structure of which are known. Devilishly (or divinely) clever strategies exist for linking disease to gene, even when one is starting essentially blind (see Francis S. Collins' Chapter 2). Such strategies, in a gesture of grandly elegant science and mind-boggling bravado, explained the fundamental pathobiology of one of the most common and previously baffling inherited diseases of the lungs, cystic fibrosis. Other, more complex, explanations await only sufficient resources.

Cystic fibrosis is one of the two major inherited diseases of the lungs (the other is alpha-1 antitrypsin deficiency). With gene in hand and the rationale transparently simple, rapid attempts at gene therapy were inevitable. Development of the replication-deficient adenovirus vector and demonstrations of its utility in delivering genes to respiratory epithelium in animals forced the issue. If cystic fibrosis could be cured, the same basic technology could be used to wipe out other diseases, perhaps even acquired ones. Promises were made and embellished by the popular press. Industrial-academic partnerships flourished. Books were written (there are six books about gene therapy written for the lay public on my shelf); stars were born; a new therapeutic millennium seemed at hand.

Aw shucks. Alas, in science as well as in the rest of life, things that look too good to be true usually are. We have learned, and continue to learn, much from attempts to cure cystic fibrosis with gene therapy, but the disease is not cured. The adenoviral vector is not so perfect. The inflammatory and immune responses that it triggers are serious, perhaps fatal, flaws in the treatment of a disease that is largely a consequence of inflammation gone awry. Too, clinical studies raised questions about the efficiency of the vector. The linear reasoning of simply delivering the normal gene to the airway epithelium and curing the disease is probably not flawed, but the devil is in the details. How can enough of the normal gene be delivered to the critical target population of cells without doing more harm than good? At some point it became clear that rumors of the imminent dawn of a new therapeutic millennium were gross exaggerations.

Yes, but. The beauty of science is that new knowledge often comes in the disguise of unanticipated results. True, curing cystic fibrosis with gene therapy was not as simple as originally thought, but that fact has uncovered a host of new possibilities. Better ways to get DNA where you want it, in some cases with surprising precision, are being developed and tested, even in humans. Different viral vectors, plasmid DNA escorted by lipids or other molecules, and combined viral and lipid delivery strategies, are all discussed in Part Three of this volume. *Yes*, it is possible with existing technology to deliver foreign DNA to the lungs via either the airways or the circulation, *but* better and safer delivery systems are needed to make the therapy work.

This volume is organized according to the sequence of scientific discovery discussed above—Part One is *gee whiz*, Part Two is *aw shucks* and Part Three is *yes, but*. We have added a fourth phase of discovery which seems imperative when considering the wonderful potential of the technology of molecular biology to benefit human beings—*what if*.

What if. If it were possible to develop safe and efficient gene delivery systems with all of the characteristics required of any therapy for human use, the potential disease targets would be myriad. The simple linear reasoning that may apply to cystic fibrosis need not be limiting. Diseases such as cancer, which may be a complex of genetic and environmental insults, or even acquired diseases, which are not a result of any genetic abnormality, could be vulnerable to either transient or prolonged genetic manipulation. The possibility of a completely new category of therapies based on the most fundamental mechanisms in biology is too pregnant with potential and simply too seductive for either the lay public or science to ignore. Several of the possibilities are discussed in Part Four.

The final section of this book is presented as an Epilogue. If all of this is not just pie-in-the-sky, ivory-tower science, if we really mean to develop gene-based drugs to be used in humans, then there are regulatory issues to deal with. Suzanne Epstein discusses such issues in Chapter 17, pointing out some issues that are unique to DNA. Finally, Theodore Friedmann admonishes us in Chapter 18 to keep our feet on the ground. We best do that. The risks are too high and the potential too great to queer the deal by being reckless.

I am sincerely honored that Claude Lenfant asked me to edit this volume. Claude deserves much of the credit for recruiting such an outstanding cast of authors. As a result, this is a remarkable book for the community of scholars seeking to cure lung diseases. Gene therapy will come into common use. Some of us will live to see it. The authors of this book will play major roles in making it happen.

Kenneth L. Brigham

Susanna Chiocca, Ph.D.   Staff Scientist, Institute for Molecular Pathology, Vienna, Austria

Francis S. Collins, M.D., Ph.D.   Director, National Center for Human Genome Research, National Institutes of Health, Bethesda, Maryland

Matthew Cotten, Ph.D.   Group Leader, Institute for Molecular Pathology, Vienna, Austria

David T. Curiel, M.D.   Professor of Medicine, Division of Pulmonary and Critical Care Medicine; Director, Gene Therapy Program, University of Alabama at Birmingham, Birmingham, Alabama

Joanne T. Douglas, Ph.D.   Postdoctoral Fellow, Gene Therapy Program, University of Alabama at Birmingham, Birmingham, Alabama

Randy C. Eisensmith, Ph.D.   Associate Professor, Institute for Gene Therapy and Molecular Medicine and Department of Human Genetics, Mount Sinai School of Medicine, New York, New York

Suzanne Epstein, Ph.D.   Chief, Molecular Immunology Laboratory, Division of Cellular and Gene Therapies, Center for Biologics Evaluation and Research, Food and Drug Administration, Bethesda, Maryland

Theodore Friedmann, M.D., M.A. (Oxon.)   Professor, Pediatrics, University of California at San Diego, San Diego, California

Xiang Gao, Ph.D.   Assistant Professor of Medicine, Division of Allergy, Pulmonary and Critical Care Medicine, Vanderbilt University School of Medicine, Nashville, Tennessee

Robert I. Garver, Jr., M.D.   Associate Professor of Medicine, Division of Pulmonary and Critical Care Medicine, University of Alabama at Birmingham, Birmingham, Alabama

Stephan W. Glasser, Ph.D.   Assistant Professor of Pediatrics, Pulmonary Biology, Children's Hospital Medical Center, Cincinnati, Ohio

Larry G. Johnson, M.D.   Associate Professor of Medicine, Cystic Fibrosis/Pulmonary Research and Treatment Center, The University of North Carolina at Chapel Hill, Chapel Hill, North Carolina

Choon-Taek Lee, M.D., Ph.D. Chief, Section of Pulmonology, Department of Internal Medicine, Korea Cancer Center Hospital, Seoul, Korea

Kathleen E. B. Meyer, Ph.D. Postdoctoral Fellow, Biopharmaceutical Sciences, University of California at San Francisco, San Francisco, California

Elizabeth G. Nabel, M.D. Professor of Internal Medicine and Physiology; Director, Cardiovascular Research Center, Internal Medicine, University of Michigan, Ann Arbor, Michigan

David M. Rodman, M.D. Associate Professor of Internal Medicine, University of Colorado Health Sciences Center, Denver, Colorado

Arlene A. Stecenko, M.D. Associate Professor of Medicine, Division of Allergy, Pulmonary and Critical Care Medicine, Vanderbilt University School of Medicine, Nashville, Tennessee

Francis C. Szoka, Jr., Ph.D. Professor, Biopharmaceutical Sciences, University of California at San Francisco, San Francisco, California

Lisa S. Uyechi, M.S. Graduate Student, Biopharmaceutical Sciences, University of California at San Francisco, San Francisco, California

Jeffrey A. Whitsett, M.D. Director, Division of Neonatology & Pulmonary Biology, Pediatrics, Children's Hospital Medical Center, Cincinnati, Ohio

Savio L. C. Woo, Ph.D. Professor and Director, Institute for Gene Therapy and Molecular Medicine, Mount Sinai School of Medicine, New York, New York

Pamela L. Zeitlin, M.D., Ph.D. Associate Professor of Pediatrics, Johns Hopkins University School of Medicine, Baltimore, Maryland

CONTENTS

Part One

OVERVIEW

1

Human Gene Therapy: The Initial Concepts

W. FRENCH ANDERSON

University of Southern California School of Medicine
Los Angeles, California

I. Introduction

A number of reviews and histories of gene therapy have been published (1–5). Rather than repeat the well-worn story of 1980 to the present, this review examines the earliest writings that envisioned the concept of gene therapy in human beings. Because some of the sources are a bit difficult to obtain, more extensive quotations than usual will be used to make clear the thinking of the authors. The reader is also referred to the excellent article by Jon Wolff and Joshua Lederberg entitled "An Early History of Gene Transfer and Therapy" (5).

II. The 1960s

The earliest references that specifically address a scientific approach to carry out human gene therapy are in 1966; the original thinkers were Edward Tatum and Joshua Lederberg. However, as early as 1963 Lederberg wrote in an article "Biological Future of Man" (6):

We might anticipate the *in vitro* culture of germ cells and such manipulations as the interchange of chromosomes and segments. The ultimate application of molecular biology would be the direct control of nucleotide sequences in human chromosomes, coupled with recognition, selection and integration of the desired genes....

On May 26, 1966, a symposium took place at Columbia University College of Physicians and Surgeons in New York City entitled "Reflections on Research and the Future of Medicine." The symposium was published in book form, and the contribution by Tatum was also published separately (see Ref. 7). Tatum addressed human genetic engineering directly:

> Finally we can anticipate that viruses will be used effectively for man's benefit, in theoretical studies concerning somatic cell genetics and possibly in genetic therapy.... We even can be somewhat optimistic about the long-range possibility of therapy based on the isolation or design, synthesis, and introduction of new genes into defective cells of particular organs. (p. 33)

> I would define *genetic engineering* as the alteration of existing genes in an individual. This could be accomplished by directed mutation, or by the replacement of existing genes by others. . . . Precedents for the introduction or transfer of genes from one cell to another exist in microbial systems, and now are being tried with mammalian cells in culture. . . .
> Hence, it can be suggested that the first successful genetic engineering will be done with the patient's own cells, for example liver cells, grown in culture. The desired new gene will be introduced, by directed mutation, from normal cells of another donor by transduction, or by direct DNA transfer. The rare cell with the desired change will then be selected, grown into a mass culture, and reimplanted in the patient's liver. (p. 43)

Severo Ochoa, in a lengthy comment after Tatum's presentation, ended with the following statement (only reproduced in the book version of the symposium):

> I have been impressed by Dr. Tatum's speculations, shall we say, on genetic engineering. I must confess that I have often been embarrassed by the question: What will be the practical implications of the remarkable developments in molecular biology? Will we be able to fix genes? Will we be able to make genes to order?
> There can be no doubt that a wide gap between basic knowledge and possible applications still exists. However, some of the remarks made by Dr. Tatum may not be as farfetched as they might have sounded some time ago—in particular, as regards the possibility of introducing genic material with the required characteristics: either by direct transfer of the DNA or through transduction. Further, one can envisage the possibility of utilizing nonpathogenic viruses for transfer of fragments of the genome, containing the desired traits, from cells in culture to the organism.

In the September-October 1966 issue of *The American Naturalist*, Lederberg addresses the concept of engineering human cells in an article entitled "Experimental Genetics and Human Evolution" (8):

> The cultural revolution has begun its most critical impact on human evolution, having generated technical power which now feeds back to biological nature. The last decade of molecular biology has given us a mechanistic understanding of heredity, and an entry to the same for development. These are just as applicable to human nature as they are to microbial physiology. Some themes of biological engineering are already an inevitable accompaniment of scientific and medical progress over the next five to 20 years.
>
> The sharpest challenges to our pretensions about human nature are already in view—and may be overlooked by too farsighted focussing on more sophisticated possibilities, like "chemical control of genotype". (To save repeating a phrase, let me call this genetic alchemy, or *algeny*). (p. 521)

> This leads us finally to algeny. Man is indeed on the brink of a major evolutionary perturbation, but this is not algeny, but *vegetative propagation*. (No one will be surprised that Haldane had anticipated this reasoning years ago.) [Note: see Ref. 9.]
>
> For the sake of argument, suppose we could mimic with human cells what we know in bacteria, the useful transfer of DNA extracted from one cell line to the chromosomes of another cell. Suppose we could even go one step further and sprinkle some specified changes of genotype over that DNA. What use could we make of this technology in the production as opposed to the experimental phase?
>
> Repair genetic-metabolic disease? . . .
>
> To recapitulate, if the desired effect is achieved by modifying some somatic cells, the same end is available by transplanting cells already known to have these properties. In general this should be much easier than systematically changing the existing ones. . . .
>
> If we have efficacious methods for testing and selecting new genotypes, do we have much need for algeny? Would not recombination and mutation give ample material for test? Perhaps for some time. But I would credit the possibility of designing a useful protein from first premises, replacing evolution by art. It would then be requisite to implant a specified nucleotide sequence into a chromosome. (pp. 526–527)

Thus, in 1966, the concept of genetic engineering of human beings was clearly delineated. In a January 8, 1967, column in *The Washington Post* (10), Lederberg states:

> Dr. Rogers points out how to cure a genetic disease, like phenylketonuria, which causes severe mental retardation. We would find a silent virus that encodes for the missing enzyme, phenylalanine hydroxylase. We would then infect the infant with this virus. Our present knowledge of virus genetics and biochemistry

supports even grander expectations—to breed viruses for calculated virogenic effects.

And again in a January 13, 1968, column (11) in the same paper:

> It is a scheme that we might call "virogenic therapy". This is an extension of the already well-founded use of tempered live viruses as vaccines to stimulate immunity. . . . The infection of a cell by a virus is therefore tantamount to adding some new genes to that cell. . . .
>
> We can, however, think of extracting the DNA molecules that code, say, for insulin and chemically grafting these to the DNA of an existing tempered virus. These new hybrid viruses would then have to be very carefully studied and perhaps modified even further, to select those appropriate for virogenic therapy in man.

The first discussion of the social/ethical implications of human genetic engineering was in the August 11, 1967, issue of *Science*, where Marshall Nirenberg wrote a thoughtful editorial entitled "Will Society Be Prepared?" (12). In it he stated:

> My guess is that cells will be programmed with synthetic messages within 25 years. The point which deserves special emphasis is that man may be able to program his own cells with synthetic information long before he will be able to assess adequately the long-term consequences of such alterations. . . . When man becomes capable of instructing his own cells, he must refrain from doing so until he has sufficient wisdom to use this knowledge for the benefit of mankind.

Two other scientists discussed human gene therapy at this time. In December, 1967, following the announcement of the first in vitro synthesis of DNA, Arthur Kornberg said (13):

> It may be possible then to attach a gene to a harmless viral DNA and use this virus as a vehicle for delivering a gene to the cells of a patient suffering from a hereditary defect and thereby cure him.

In 1968, French Anderson wrote an article for *New England Journal of Medicine* in which he postulated how gene therapy could one day be accomplished. The editors rejected the paper as "very erudite and fascinating . . . but too speculative [for the *Journal*]" (14). Although the manuscript is lost, a talk at the Kennedy Foundation Symposium on Mental Retardation based on the paper was reported in the June 1968 issue of *Pediatric News* (15, pp. 1, 29).

> In order to insert a correct gene into cells containing a mutation, it will first be necessary to isolate the desired gene from a normal chromosome. Then this gene will probably have to be duplicated to provide many copies. And, finally, it will be necessary to incorporate the correct copy into the genome of the defective cell, Dr. Anderson explained.

While any discussion of potential approaches to this problem is based sole-ly on speculation, one of the most promising methods for accomplishing this sequence would be the development of nonpathogenic viruses capable of trans-ferring genetic material from one DNA genome to another, he said.

This virus could be used to infect tissue culture cells containing the normal gene. Theoretically the virus would incorporate the desired DNA sequence. Then the virus could be reisolated, multiplied in mass culture, harvested, purified, and stored for administration to a patient suffering a genetic defect. . . .

Only an enlightened public and scientific community will assure that this power will be used effectively for the benefit of mankind, [said] Dr. Anderson.

The theme of caution and the importance of public understanding of the implications of human genetic engineering, first expounded by Nirenberg, be-came widely discussed by the end of the decade. Robert Sinsheimer published a most provocative article entitled "The Prospect of Designed Genetic Change" in the April 1969 issue of *Engineering and Science* (16). He declared:

A new eugenics has arisen, based upon the dramatic increase in our understand-ing of the biochemistry of heredity and our comprehension of the craft and means of evolution. . . . The old eugenics would have required a continual selection for breeding of the fit, and a culling of the unfit. The new eugenics would permit in principle the conversion of the unfit to the highest genetic level. The old eugenics was limited to a numerical enhancement of the best of our existing gene pool. The horizons of the new eugenics are in principle boundless—for we should have the potential to create new genes and new qualities yet undreamed. . . . Indeed, this concept marks a turning point in the whole evolution of life. For the first time in all time, a living creature understands its origin and can undertake to design its future. . . . Today we can envision that chance—and its dark companion of awesome choice and responsibility.

The first isolation of a gene, by Jonathan Beckwith and his colleagues at Harvard, was described in an article on the front page of the November 23, 1969, *New York Times*. In an accompanying news conference as well as in fol-low-up articles, Beckwith declared that scientists would one day treat patients suffering with genetic diseases with viruses carrying therapeutic genes. He strongly emphasized, however, that the power of genetic manipulation could be used for evil purposes and, in fact, *would* be misused if society was not constantly vigilant (17).

This concern about the potential misuse of human genetic engineering was also reflected in books written with explosive titles: *The Biological Time Bomb*, by Gordon Rattray Taylor (1968), and *The Second Genesis: The Coming Control of Life*, by Albert Rosenfeld (1969).

But up to this point (1969), all discussions were theoretical. In that year, however, Lederberg read an article in *Lancet* (18) describing two German sisters who suffered from a genetic defect in which toxic levels of arginine built up in their bloodstream. Lederberg had been interacting with Stanfield Rogers for several years. Rogers had been studying the Shope rabbit papilloma virus and had reported that animals infected with this virus had, among other characteristics, a decreased level of blood arginine. In fact, in a 1966 paper in *Nature* (19), Rogers had pointed out that many lab scientists who worked with the Shope virus had reduced blood arginine levels but, apparently, no other side effects from exposure to the virus. Lederberg told Rogers about the *Lancet* paper, and Rogers contacted the authors with the suggestion that the Shope virus be tried as a treatment for the girls.

III. The 1970s

In 1970, Rogers carried out the first human genetic engineering experiment when he administered the Shope virus to the two German girls suffering from argininemia. He described these experiments in a paper presented at a New York Academy of Sciences Conference entitled "Ethical and Scientific Issues Posed by Human Uses of Molecular Genetics" on May 15-16, 1975, in New York City (20):

> These children were epileptic, spastic, grossly retarded, and progressively becoming worse. In view of laboratory experience that went back almost 40 years, it seemed worthwhile to take the risk (which we had no reason to believe existed anyway) of administering the virus to the children in the hope of replacing their genetically deficient enzyme. . . . Since the use of virus genetic information to replace that lost because of a deficiency disease had not been attempted before, extreme caution was used. The children with argininemia were given a dose of virus that had been purified in cesium chloride and was shown by electron microscope scanning to contain only the Shope virus. Immunological studies and blind passage in tissue-cultured cells revealed no other virus or harmful effects. The dose of virus first used was about 1/20th of that which we had previously found harmless to mice.

For various technical and logistic reasons, a "therapeutic" dose of virus was never administered. Considerable controversy erupted about this experiment the result of which was that Rogers never engaged in another human experiment although he worked extensively in the genetic engineering of plants. He drew the following conclusion concerning his 1970–1972 human genetic engineering experiments with Shope virus (20):

> Although these results were at best disappointing, it still seems to us that the chance to prevent progressive deterioration in these children's condition was the only ethical route to take and that, should other children be found to have this disease, they should be so treated. (p. 67)

Anderson supported the Rogers' experiment in a 1971 lecture, later published (21):

> In this case, the experiment appears justified by the tragic circumstances of the disease involved. Nonetheless, success here might encourage less justified attempts at premature gene therapy. . . . At this time the virus has been given to the youngsters, but no word has yet been released as to whether the therapy is successful. If nothing happens, or if the patients grow worse, then the treatment will have failed. But what if their blood arginine levels return to normal? Might these children have had a completely normal life if only they had received the virus right after birth? If you were the parent of a newborn baby who was found to have argininemia and therefore to be doomed to suffer and die, would you allow your baby to be exposed to this relatively unproven technique of gene therapy? Many parents would answer, I believe, that they would not only allow it but would urge its use. In this case, with several decades of experience indicating the safety of this particular virus and absolute certainty of suffering and death without it, there seems to me to be little question about moral justification.
>
> But what about the larger question? What about other genes, other viruses? Where does one draw the line? . . . Who is to decide what are "good" uses and what are "bad" uses?

On the other hand, in March 1972, Theodore Friedmann and Richard Roblin, in an excellent article in *Science* entitled "Gene Therapy for Human Genetic Disease?" (22), expressed concern that the Rogers experiment was premature. They delineated a number of problems that they felt should have been addressed before carrying out the human experiment and then went on to say:

> But we are concerned that this first attempt at gene therapy, which we believe to have been premature, will serve as an impetus for other attempts in the near future. For this reason, we offer the following considerations as a starting point for what we hope will become a widespread discussion of appropriate criteria for the use of genetic manipulative techniques in humans.

They concluded by stating:

> For the foreseeable future, however, we oppose any further attempts at gene therapy in human patients.

Robert M. Veatch, an ethicist, placed the Rogers experiment in a broader ethical framework in an article entitled "Ethical Issues in Genetics" published in 1974 (23). Concurrent with the controversy over the Rogers experiment, a number of other papers were published. In the December 18, 1970, issue of *Science*, Bernard Davis published a careful analysis of "Prospects for Genetic Intervention in Man" (24). Michael Hamilton, Canon of the Washington Cathedral, convened a conference in 1971, which was published in 1972 in a book entitled *The New Genetics and the Future of Man*. Following an overview of "Genetic Therapy" by Anderson (21), experts in three separate areas responded: Arno Motulsky, clinical geneticist; Alexander Capron, lawyer; and Paul Ramsey, theologian. Ernst Freese convened a conference entitled "The Prospects of Gene Therapy" in 1971 at the NIH. Lederberg followed up his 1960s articles with a paper entitled "Tomorrow's Babies" in 1970 (25) and "Genetic Engineering, or the Amelioration of Genetic Defect" in January 1971 (26):

> Another approach to constructive therapy, which may mitigate a variety of diseases, is an extension of the existing uses of specific virus strains. At present, their role in medicine is confined to their use as vaccines ... We can visualize the engineering of other viruses so that they will introduce compensatory genetic information into appropriate somatic cells, to restore functions that are blanked out in a given genetic defect. As with vaccine viruses, this presumably will leave the germ cell DNA unaltered, and therefore does not attack the defective gene as such. If we can cope with the disease, should we bother about the gene? Or may we not leave that problem to another generation?

In March 1972, Lederberg published another paper, entitled "Options for Genetic Therapy" (27). Thus there was much activity at least in terms of meetings and papers in the early 1970s.

However, not everyone was enthusiastic about the technical capability of carrying out genetic engineering of human beings. Sir MacFarlane Burnet was so upset by all the discussions that he wrote a whole book in opposition to the idea. It was published in 1971 and entitled *Genes Dreams and Realities* (28). The opening sentence of the book reads:

> The stimulus to write this book came from a suggestion that too much sensational material was being written about the future significance for medicine of discoveries in molecular biology. . . . Are we going to see 'genetic engineering' applied to the cure of genetic disease. . . ?

After chapters explaining genes, molecular biology, and cell biology, Burnet launches into his main theme in Chapter 4, "Human Applications of the 'New' Biology":

> If rocks from the moon can be brought to Houston, why should we not be able to apply molecular biology, cytology, and the rest, to cure what is now incurable? In this chapter I want to look primarily at the potential applications of molecular biology to human affairs. (p. 69)

> If we can synthesize a gene of the 'right' type, why not use it to replace the bad gene more or less the way that we can use an accident victim's healthy kidney to replace a diseased one? (p. 70)

> The next step would be the crucial and probably impossible one, to incorporate the gene into the genetic mechanism of a suitable virus vehicle in such a fashion

that the virus in turn will transfer the gene it is carrying to cells throughout the body and in the process precisely replace the faulty gene with the right one. I should be willing to state in any company that the chance of doing this will remain infinitely small to the last syllable of recorded time. (p. 72)

[My objective] is to indicate how far beyond the bounds of the practical it is to look for deliberate chemical rearrangement of the nucleotide units in DNA or RNA to produce a predictable result even with the smallest 'organism' of practical importance. (p. 81)

In the Kornberg approach the synthetic gene is implanted in one of the 'cancer viruses' and ferried to all relevant cells of the patient by injection of large amounts of treated virus into the blood circulation. Tatum's suggestion was to treat a tissue culture of the patient's cells with the 'good' gene, sort out the tiny proportion that would be 'cured' and grow them up to the required bulk for implantation into the patient. . . . Neither approach has yet been tried and I doubt whether they ever *will* be tried by responsible investigators. (p. 85)

Burnet returned to emphasize his contention in the latter parts of his book:

In Chapter 4 I discussed in general terms the widely accepted suggestion that 'genetic engineering' was a possibility of the future—that some day a faulty gene could be replaced by a normal one. I discussed the possibility, perhaps in too cavalier a fashion, as something which would remain for ever impossible to apply to actual human individuals with genetic disease. To wind up this account of some of the genetic diseases to which it might be applied, I felt that I should restate the difficulties in more specific fashion. (p. 119)

At every step, one can with some hesitation say that with the rather restricted knowledge we have, this may be possible in principle but the practical problems would take an army of first-rate scientists and technicians to overcome them. There could never, in my opinion, be adequate motivation for any government to finance such a research or for any individual scientist to undertake the responsibility of its technical direction. (p. 121)

Thus Burnet makes amply clear that he does not believe that genetic engineering of human beings would ever be possible, or at least not for many decades. This general scepticism about just the *ability* to do gene therapy (never mind the safety and ethical aspects) was still very much present when we were attempting to get approval for our initial gene therapy protocol (for ADA deficiency) and accounts, in part, for why it took us 3½ years to get final approval.

One reason Burnet was so adamant was that he felt that all the major scientific discoveries had been made and that there would therefore be no new tools that could be used:

It may be that most of the discoveries needed for practical medicine have been made. (p. 3)

Rightly or wrongly I think that with those two discoveries [chemical structure of an RNA bacteriophage and 'adequate description of the ribosome and the precise mechanism of translation'], Nobel prizes in molecular biology will stop. (p. 42)

After working for a year on the present book I cannot avoid the conclusion that we have reached the stage in 1971 when little further advance can be expected from laboratory science in the handling of the 'intrinsic' types of disability and disease. There will always be possibilities of improvement in detail but I am specially impressed by the fact that since 1957 there has been no new thought on the handling of cancer, of old age, or of autoimmune disease. (p. 217)

None of my juniors seems worried as I am, that the contribution of laboratory science to medicine has virtually come to an end. . . . Almost none of modern basic research in the medical sciences has any direct or indirect bearing on the prevention of disease or on the improvement of the medical care. . . . Gunther Stent and Niels Jerne, both of them [are] distinguished experimentalists. They feel that both biological and physical science may have passed the era of great discoveries. I have to agree with them and can quote a statement I made elsewhere, that future historians may speak of an age of scientific discovery that started with Galileo in 1586 and ended something less that four hundred years later. (p. 218)

I believe that in all the major sciences the general picture has been competently and in broad outline completely delineated by 1970. The task now is to fill in the detail. (p. 219)

We shall have to forego that special place we used to claim amongst scholars that there was special virtue in our work because of its potentiality for saving life. (p. 226)

Burnet left a legacy of doubt that was very difficult to overcome. Even as Burnet was working on his book, Smith and Nathans were publishing their papers in 1970–1971 (29) describing sequence-specific restriction endonucleases, thereby laying the cornerstone for the gene revolution to come.

In 1971, Paul Berg had made sufficient progress in splicing genes into the mammalian DNA virus SV-40 to cause an uproar at the Cell Culture Course at Cold Spring Harbor in June 1971. Robert Pollack, who taught the course, called Berg and expressed his concerns about the biosafety of these experiments. This led to the first Asilomar Conference, January 22–24, 1973, that assessed the biohazard issues associated with recombinant DNA research. (For details of the social history of recombinant DNA research during these years, see Refs. 30–33.)

By March 1973, Cohen and Boyer had succeeded in creating "molecular cloning." When Boyer presented the work at the June 1973 Gordon Conference on Nucleic Acids, the conferees recognized the enormous power and potential danger of this new technology. The conference leaders published a letter in *Science* (34) expressing concern that not enough was known about the potential consequences of such experiments. The National Academy of Sciences established a committee chaired by Paul Berg which sent a second letter to *Science* (35) suggesting a voluntary moratorium on some types of recombinant DNA research. At this point, the press and the public really got involved.

The definitive Asilomar Conference took place February 24–26, 1975. This was a "defining moment" for the whole field of gene research, including the gene therapy research of the future. Emotional discussion of potential biohazards took place in front of the world. The day after the Conference ended, the first Recombinant DNA Advisory Committee (RAC) meeting was convened. Guidelines for carrying out recombinant DNA research were hammered out over the next year. The first official NIH Guidelines were published June 23, 1976.

On that same day, Mayor Alfred Vellucci held a formal and very emotional hearing in Cambridge, Massachusetts, into Harvard University's request to build a P3 facility on its campus in Cambridge. Several months later, Sen. Edward Kennedy held congressional hearings on the potential hazards associated with recombinant DNA. Clearly, the safety of recombinant DNA, and the bacterial organisms that might emerge from recombinant DNA research, had become a topic of enormous public interest and concern.

Not just politicians, but many highly respected scientists had real worries. Two particularly well articulated articles were published in 1976–1977. Erwin Chargaff wrote in *Science* (36):

> Have we the right to counteract, irreversibly, the evolutionary wisdom of millions of years? . . . I am one of the few people old enough to remember that the extermination camps in Nazi Germany began as an experiment in genetics.

And Robert Sinsheimer followed up on his 1969 article by expressing his fear that, if nature has provided a natural barrier between prokaryotes and eukaryotes, breeching that barrier could be disastrous. He wrote in the *New Scientist* (37):

> They regard our ecological niche as wholly secure, deeply insulated from potential onslaught, with no chinks or unguarded section or perimeter. I cannot be so sanguine. In simple truth, just one—just one—penetration of our niche could be sufficient to produce a catastrophe.

In 1977 the National Academy of Sciences held a symposium that was invaded by activists that forcefully took over the microphone and marched through the auditorium carrying dramatic anti-science placards. More hearings and many bills/amendments were introduced in Congress proposing to severely regulate DNA research. By the end of 1978, most of the legislative/media frenzy had subsided without the passage of restrictive legislation. Science's successful transit through these treacherous waters can be attributed, I believe, to two factors. First was the acquisition by scientists of considerable data demonstrating the safety of recombinant DNA and recombinant DNA research when carried out according to the NIH Guidelines. Second was the superb job done by Donald Fredrickson, the NIH Director, in steering the ship of science through all the reefs and barriers to a safe exit. Those of us who lived through this extraordinary exposure to democracy in action can never forget it, nor can we adequately thank Dr. Fredrickson for his wisdom and leadership.

By 1979, the RAC and its Guidelines had become highly respected and the public's fear of an Andromeda strain of bacteria escaping from a recombinant DNA research laboratory had greatly subsided.

Within this background, questions about applying recombinant DNA research to human beings became dominant as the 1980s began. The first event was a letter signed by the Secretary General of each of the three major religious organizations in the United States (the National Council of Churches, the Synagogue Council of America, and the United States Catholic Conference) asking President Carter to establish a forum for addressing the fundamental ethical questions relating to genetic engineering. They wrote in a letter dated June 20, 1980 (38):

> History has shown us that there will always be those who believe it appropriate to "correct" our mental and social structures by genetic means, so as to fit their vision of humanity. This becomes more dangerous when the basic tools to do so are finally at hand. Those who would play God will be tempted as never before.

This was the social setting in which the human gene therapy research of the next decade took place.

Bibliography

1. Anderson WF. Prospects for human gene therapy. Science 1984; 226:401–409.
2. Freidmann T. Progress toward human gene therapy. Science 1989; 244:1275–1281.
3. Miller AD. Human gene therapy comes of age. Nature 1992; 357:455–460.
4. Anderson WF. Human gene therapy. Science 1992; 256:808–813.
5. Wolff JA, Lederberg J. An early history of gene transfer and therapy. Hum Gene Ther 1994; 5:469–480.
6. Lederberg J. Biological future of man. In: Wolstenholme G, ed. Man and His Future. London: Churchill, 1963:265.
7. Tatum EL. Molecular biology, nucleic acids, and the future of medicine. In: Lyght CE, ed. Reflections on Research and the Future of Medicine. New York: McGraw-Hill, 1967: 31–49. [Also in Perspect Biol Med 1966; 10:19–32.]
8. Lederberg J. Experimental genetics and human evolution. Am Nat 1966; 100:519–531.

9. Dronamraju KR. J.B.S. Haldane's (1892–1964) biological speculations. Hum Gen Ther 1993; 4:303–306.
10. Lederberg J. Dangerous delinquents: the most lethal viruses can disappear quickly, like that of the 1918 influenza epidemic. Washington Post, Jan. 8, 1967.
11. Lederberg J. DNA breakthrough points way to therapy by virus. Washington Post, Jan. 13, 1968.
12. Nirenberg MW. Will society be prepared? Science 1967; 157:633.
13. Kornberg A. As quoted in Ref. 28, p. 71.
14. Culliton BJ. French Anderson's 20-year crusade. Science 1989; 246:748.
15. Anon. Repair of genetic defects predicted. Pediatr News, June, pp. 1, 29, 1968.
16. Sinsheimer R, The prospect of designed genetic change. Engineering and Science 1969; 32:8–13.
17. Beckwith J. Social and political uses of genetics in the United States: past and present. In: Lappe M, Morison RS, eds. Ethical and Scientific Issues Posed by Human Uses of Molecular Genetics. New York: New York Academy of Sciences, 1976:46–58.
18. Terheggen HG, Schwenk A, Van Sande M, Lowenthal A, Columbo JP. Argininemia with arginase deficiency. Lancet 1969; 2:748–749.
19. Rogers S. Shope papilloma virus: a passenger in man and its significance to the potential control of the host genome. Nature 1966; 212:1220–1222.
20. Rogers S. Reflections on issues posed by recombinant DNA molecule technology. II. In: Lappe M, Morison RS, eds. Ethical and Scientific Issues Posed by Human Uses of Molecular Genetics. New York: New York Academy of Sciences, 1976:66–70.
21. Anderson WF. Genetic therapy. In: Hamilton M, ed. The New Genetics and the Future of Man. Grand Rapids: Eerdmans, 1972:109–124.
22. Friedmann T, Roblin R. Gene therapy for human genetic disease? Science 1972; 175:949–955.
23. Veatch RM. Ethical issues in genetics. In: Steinberg AG, Bearn AG, eds. Prog Med Genet 1974; 10:223–264.
24. Davis BD. Prospects for genetic intervention in man. Science 1970; 170:1279–1283.
25. Lederberg J. Tomorrow's babies. Proceedings of the Sixth World Congress on Fertility and Sterility, Tel Aviv, 1968:18–23.
26. Lederberg J. Genetic engineering, or the amelioration of genetic defect. Pharos Jan. 1971: 9–12.
27. Lederberg J. Options for genetic therapy. Medical Dimensions, March 1972: 16, 17,63.
28. Burnet M. Genes, Dreams and Reality. New York: Basic Books, 1971.
29. Nathans D, Smith HO. Restriction endonucleases in the analysis and restructuring of DNA molecules. Ann Rev Biochem 1975; 44:273–293.
30. Watson JD, Tooze J. The DNA Story: A Documentary History of Gene Cloning. San Francisco: Freeman, 1981.
31. Krimsky S. Genetic Alchemy: The Social History of the Recombinant DNA Controversy. Cambridge: MIT Press, 1982.
32. Wright S. Molecular Politics: Developing American and British Regulatory Policy for Genetic Engineering, 1972–1982. Chicago: University of Chicago Press, 1994.

33. Thompson L. Correcting the Code: Inventing the Genetic Cure for the Human Body. New York: Simon & Schuster, 1994.
34. Singer M, Soll D. Guidelines for DNA hybrid molecules. Science Sept. 21, 1973; 181:1114. Letter.
35. Berg P, Baltimore D, Boyer HW, et al. Potential biohazards of recombinant DNA molecules. Science July 26, 1974; 185:303. Letter.
36. Chargaff E. On the dangers of genetic meddling. Science 1976; 192:938–940.
37. Sinsheimer R. An evolutionary perspective for genetic engineering. New Scientist, Jan. 20, 1977.
38. Capron A, ed. Splicing Life: A Report on the Social and Ethical Issues of Genetic Engineering with Human Beings. Washington, DC: President's Commission for the Study of Ethical Problems in Medicine and Biomedical and Behavioral Research, Nov. 1982:95–96.

2

Discovering Genes That Cause Disease

FRANCIS S. COLLINS

National Center for Human Genome Research
National Institutes of Health
Bethesda, Maryland

It should be inherently obvious that specific gene therapy requires identification of the appropriate gene to be delivered. In the case of an autosomal-recessive genetic disease (e.g., cystic fibrosis; see Chap. 13), the ideal gene to be delivered is a normal copy of the gene which is homozygously altered in affected individuals. On the other hand, in dominant conditions where the disease arises as a result of a gain of function of the protein product (e.g., the expanded glutamine repeat in Huntington's disease), addition of an extra normal copy will be unlikely to achieve benefit, and a successful gene therapy strategy will need to be targeted at turning off the mutant allele (e.g., by an antisense strategy) or blocking its effects by other manipulations of the pathogenic pathway. For complex, polygenic, or acquired disorders (e.g., cancer, coronary artery disease) the choice of therapeutic gene will depend upon an understanding of the vulnerable points in pathogenesis, and in many instances

---

Adapted from Collins FS. Positional cloning moves from perditional to traditional. Nature Genet 1995; 9:347–350.

17

may utilize genes whose endogenous copies are entirely normal in the affected patients, by expressing these in non-physiologic locations or amounts.

A revolution in gene discovery is currently underway. The Human Genome Project, initiated in 1990, aims to determine the complete sequence of the human genome, including the estimated 100,000 genes, by the year 2005. Progress has been gratifying, and the Project is currently running ahead of schedule and under budget (1). Approximately half of these gene sequences will reveal homologies that suggest their function. But how does one determine that hereditary susceptibility to a particular disorder is due to alterations in a particular one of these 100,000 genes? This needle-in-a-haystack problem presents a major challenge for genetic medicine but is being increasingly successfully addressed by a process called "positional cloning," which is the topic of this chapter.

Two methods define the extremes of disease gene cloning methodology (though hybrid methods, as we shall see, are becoming common): functional cloning and positional cloning (Fig. 1).

Functional cloning refers to identification of the gene causing a human disease based on fundamental information about the basic biochemical defect, without reference to chromosomal map position (e.g., phenylketonuria, sickle cell anemia). Unfortunately, our level of ignorance about the fundamental underpinnings of human disease is sufficiently profound that this route to gene discovery is usually not available—most disease genes where such information exists have already been cloned. At the opposite extreme, pure "positional cloning" assumes *no* functional information and must locate the responsible gene solely on the basis of map position. Fifteen years ago it would have been difficult to find many human geneticists who thought it would be realistically possible to clone a specific disease gene solely on the basis of its map position in the three gigabase human genome. Ten years ago, Frank Ruddle suggested that methods might eventually be developed to accomplish such daunting tasks

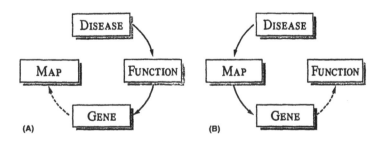

Figure 1   Methods for disease gene identification—the two extremes. (A) Functional cloning. (B) Positional cloning.

(2), but it was only with the advent of discussions about the possibility of a Human Genome Project that such scenarios began to be seriously considered as having general potential. Linkage analysis of multiple affected families is usually the first step, although the identification of patients who have visible cytogenetic rearrangements can greatly assist the low and high resolution mapping of the responsible gene.

Progress in positional cloning (originally dubbed reverse genetics) moved relatively slowly at first. The first success was reported in 1986 with the cloning of the gene for chronic granulomatous disease (3), providing a crucial proof of principle. By 1992, the positional cloning of thirteen inherited human disease genes had been reported (4). A year later, there were nineteen (5). A review in 1995 catalogued a total of 42. As of this writing, there are 55.

The majority of positional cloning successes have relied on cytogenetic rearrangements. A special case of DNA rearrangement, the expanded triplet repeat (6), has also greatly aided those gene searches where it proved to represent the mutational basis of the disease. Positional cloning in the absence of one or the other of these clues is still a laborious business, but the recent growth in this subroster (which included only cystic fibrosis until 1993) testifies to the progress in genetic mapping, physical mapping, and gene hunting technologies arising from the Human Genome Project.

The summary in Table 1 is actually a considerable underrepresentation of the impact of positional cloning, as it omits a long list of stunning successes in the cloning of genes with somatically acquired mutations, most of which were identified because of their role in cancer (7). Most of the common characteristic cytogenetic rearrangements in lymphoma and leukemia have by now had their breakpoints cloned and transcript participants identified using the positional cloning strategy, and many solid tumor genetic alterations have also been revealed.

As predicted, however (5,8), the positional cloning strategy in its purest form is already beginning to give way (much to the relief of those who have labored mightily to generate this list) to a streamlined and accelerated version, which can be conveniently referred to as the "positional candidate" approach (Fig. 2; Table 2). This strategy relies on a combination of mapping to the correct chromosomal subregion (generally using linkage analysis) followed by a survey of the interval to see if attractive candidates reside there. Marfan syndrome, for example, was mapped to 15q by standard linkage (9). At almost that same moment, the fibrillin gene (an attractive biochemical candidate) was mapped to 15q (10). In a matter of weeks mutations were found in the fibrillin coding region of patients with Marfan syndrome (11). Similarly, multiple endocrine neoplasia type II was mapped to chromosome 10 by linkage (12). Perusal of the local map identified the *ret* proto-oncogene lying suspiciously in the correct area, eventually leading to the identification of point mutations in

Table 1 Disease Genes Identified by Positional Cloning

1986
Chronic granulomatous disease
Duchenne muscular dystrophy
Retinoblastoma

1989
Cystic fibrosis

1990
Wilms tumor
Neurofibromatosis type 1
Testis determining factor
Choroideremia

1991
Fragile X syndrome
Familial polyposis coli
Kallmann syndrome
Aniridia

1992
Myotonic dystrophy
Lowe syndrome
Norrie syndrome

1993
Menkes disease
X-linked agammaglobulinemia
Glycerol kinase deficiency
Adrenoleukodystrophy
Neurofibromatosis type 2
Huntington's disease
Von Hippel–Lindau disease
Spinocerebellar ataxia 1
Lissencephaly
Wilson disease
Tuberous sclerosis

1994
McLeod syndrome
Polycystic kidney disease
Dentatorubral pallidoluysian atrophy
Fragile X "E"
Achondroplasia
Wiskott-Aldrich syndrome
Early-onset breast/ovarian cancer (BRCA1)
Diastrophic dysplasia
Aarskog-Scott syndrome
Congenital adrenal hypoplasia
Emery-Dreifuss muscular dystrophy
Machado-Joseph disease

1995
Spinal muscular atrophy
Chondrodysplasia punctata
Limb-girdle muscular dystrophy
Ocular albinism
Ataxia telangiectasia
Alzheimer's disease (chromosome 14)
Alzheimer's disease (chromosome 1)
Hypophosphatemic rickets
Hereditary multiple exostoses
Bloom syndrome
Early-onset breast cancer (BRCA2)

1996
Friedreich's ataxia
Progressive myoclonic epilepsy
Treacher-Collins syndrome
Long QT syndrome (chromosome 11)
Barth syndrome
Simpson-Golabi-Behmel syndrome

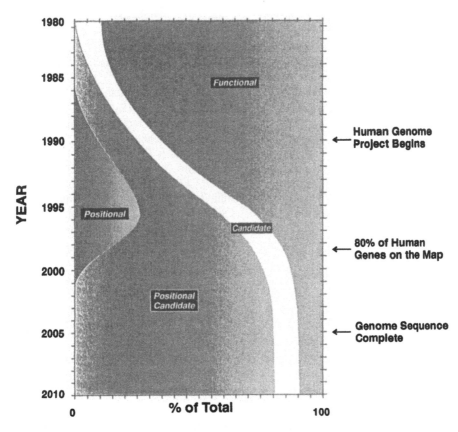

Figure 2   Trends in methods for cloning human disease genes, 1980–2010. Exact quantitation is not intended; trends after 1995 are highly speculative.

affected patients (13). The truly surprising finding that other mutations in this same gene are capable of causing a completely different phenotype (Hirsch-sprung's disease) (14). Perhaps the most medically significant positional candidate gene successes have been the genes for hereditary nonpolyposis colon cancer (HNPCC). In that instance, linkage information pointing to chromosome 2p and phenotypic information about DNA instability in colon cancers from affected patients came together in a stunningly productive positional candidate confluence. When the dust settled, no fewer than four genes involved in DNA mismatch repair had been implicated in HNPCC (15–17).

   The future expanding success of the positional candidate approach is predicated on an increasingly dense transcript map. Large databases of cDNA sequences may assist the pure candidate gene approach, but without associated

Table 2   Representative Inherited Disease Genes Identified by Positional Candidate Methods

Disease	Affected protein
Alzheimer's disease	β-Amyloid protein precursor, apolipoprotein E
Amyotrophic lateral sclerosis	Superoxide dismutase
Charcot-Marie-Tooth disease type 1A	Peripheral myelin protein 22
Charcot-Marie-Tooth disease type 1B	Myelin protein zero (P_0)
Crouzon syndrome	Fibroblast growth factor receptor 2
Familial hypertrophic cardiomyopathy	Cardiac myosin heavy chain
Familial melanoma	p16
Hereditary hemorrhagic telangiectasia type 1	Endoglin
Hereditary nonpolyposis colon cancer	*hMSH2, hMLH1, hPMS1, hPMS2*
Hyperekplexia	Inhibitory glycine receptor, α-subunit
Jackson-Weiss syndrome	Fibroblast growth factor receptor 2
Long QT syndrome	SCN5A, HERG cardiac ion channels
Malignant hyperthermia	Ryanodine receptor
Marfan syndrome	Fibrillin
Multiple endocrine neoplasia type 2a	Proto-oncogene *RET*
Pfeiffer syndrome	Fibroblast growth factor receptor 1
Supravalvular aortic stenosis	Elastin
Retinitis pigmentosa	Peripherin, rhodopsin
Waardenburg syndrome	Human homolog of the mouse homeobox gene *Pax-3*

mapping information their usefulness is greatly blunted. Since linkage analysis of affected families usually results in candidate intervals of 0.5 to 5 Mb, ideally the transcript map should also have this degree of resolution. Mapping of cDNAs to traditional somatic cell hybrids or by fluorescent in situ hybridization (FISH) will not usually achieve this; utilizing large insert clone libraries (such as yeast artificial chromosomes, or YACs) or radiation hybrids is more appropriate.

It is likely that the human transcript map will improve dramatically in the coming months. Partial cDNA sequences (expressed sequence tags, or ESTs) have been accumulating in public databases at an impressive rate, and a recent Merck-funded cDNA sequencing initiative at Washington University in St. Louis has produced 5' and 3' sequences for 140,000 cDNAs, and will reach 280,000 by the end of 1996. In a testimonial to international cooperation, an EST mapping consortium (with major genome center participants at Stanford, MIT, Oxford, Cambridge, and Paris) plans to map some 70,000 ESTs to 0.5 Mb intervals (perhaps 0.1 Mb intervals in the near future), utilizing the whole genome radiation hybrid and YAC panel approaches. Some 7000 have already been mapped. Recent announcements by the Institute for Genomic Research (TIGR) and Sandoz of their willingness to donate primers for this mapping effort are most welcome. With all of this activity, it seems likely that more than half of the human transcripts will be placed on the map in the next 12 months. The effect on the success rate of the positional candidate approach should be profound.

The major future challenge to positional cloning (and its rapidly growing offspring the positional candidate approach) will be the elucidation of genes responsible for predisposition to common oligogenic disorders such as diabetes, asthma, hypertension, many forms of cancer, and the major mental illnesses. Indeed, it can be argued that virtually any disease (except possibly trauma) has some genetic component, and the determination of these genetic influences is a high priority for modern medicine.

As described in a recent review of the genetics of complex traits (18), methods now exist which should be capable of mapping predisposing loci which contribute a relative risk to siblings of 1.5 or greater for a particular disease. Such an analysis has recently been reported for type I diabetes (IDDM), for instance, resulting in the mapping of five contributing loci (including HLA and the insulin gene) (19). But for complex traits the difficulty of the task of going from a blurred chromosomal location (rarely better than 2 to 5 Mb) to the responsible gene can hardly be overstated. No hard and fast "recombinants" can be counted on to delimit the boundaries. It is likely that the sequence variation of the contributing loci, like their quantitative contribution to the phenotype, will be subtle—few chromosomal rearrangements or large deletions can be expected. Even worse, some of the responsible sequence changes are likely to lie *outside* the coding region, as seems to be the case for the insulin gene in IDDM (20).

So are such efforts doomed? In the spirit of recent experience, apparently insurmountable obstacles are still likely to yield to advancing genome technology. For the case of common diseases, the solution will likely be found in one of three approaches: (1) Identification of a prevalent predisposing haplotype (linkage disequilibrium) in a genetically homogeneous population. The recent use of this strategy to clone the gene for diastrophic dysplasia in Finland (21) lends strong support to the concept that such populations allow tracking of a much more refined predisposing region. Counting on such linkage disequilibrium to come to the rescue in a more outbred population may be risky, however. (2) Identification of a syntenic animal model. Animal models of human disease may not necessarily reflect the same predisposing loci, of course; but if genetic linkage analysis indicates that a small syntenic region is contributing relative risk in animals and humans, shifting the analysis to the animal species will be a highly attractive option. The ability to perform selective cross-breeding allows much more precise mapping of a polygene in animal experiments than in humans. (3) If all else fails, the availability of a highly dense transcript map (reducing the hunt for candidate genes to a computer exercise), coupled with powerful methods to search for sequence variation over large numbers of candidate DNA segments (where DNA chip hybridization (22) may greatly speed throughput), may still prevail.

Though challenging, the effort to expand the successes of positional cloning is clearly a high scientific priority. Indeed, it could be argued that this is the major medical justification for the Human Genome Project. Successful identification of a gene predisposing to human illness represents the traversal of a major bottleneck in understanding of that disease (Fig. 3), and potentially allows the development of diagnostic and therapeutic advances of immense medical benefit.

In many instances, however, the diagnostic consequences of gene discovery (prediction of susceptibility to disease) will be in hand well prior to the development of curative therapies. This will stretch the abilities of medicine and society to be sure that such diagnostic abilities are used to benefit individuals, rather than expose them to genetic discrimination. But the ultimate

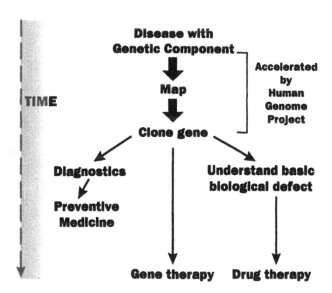

Figure 3   Progress in molecular medicine. The time needed to clone a disease gene has been rapidly shortening, based on maps and technologies emanating from the Human Genome Project. After gene cloning, diagnostic capabilities often arrive quickly, in some instances allowing preventive strategies to be initiated (colonoscopy for individuals at genetically high risk for colon cancer, for example). The timetable for development of effective gene or drug therapies is much less predictable, however.

solution to that dilemma will come from the successful development of simple and cost-effective gene or drug therapies. The timeline for such breakthroughs cannot currently be predicted; however, even though we can state with some confidence that all of the human genes will be identified by 2005, and their sequences will be publicly available, making predictions about therapeutic applications for any given disorder is much more difficult.

References

1. Collins FS. Ahead of schedule and under budget: the Genome Project passes its fifth birthday. Proc Natl Acad Sci USA 1995; 92:10841–10848.
2. Ruddle FH. The William Allan Memorial Award Address: reverse genetics and beyond. Am J Hum Genet 1984; 36:944–953.
3. Royer-Pokora B, Kunkel LM, Monaco AP, et al. Cloning the gene for an inherited human disorder-chronic granulomatous disease-on the basis of its chromosomal location. Nature 1986; 322:32–38.
4. Collins FS. Positional cloning: let's not call it reverse anymore. Nature Genet 1992; 1:3–6.
5. Ballabio A. The rise and fall of positional cloning? Nature Genet 1993; 3:277–279.
6. Bates G, Lehrach H. Trinucleotide repeat expansions and human genetic disease. Bioessays 1994; 16:277–284.
7. Stanbridge EJ. Functional evidence for human tumor suppressor genes: chromosome and molecular studies. Cancer Surveys 1992; 12:5–24.
8. Collins FS. Positional cloning moves from perditional to traditional. Nature Genet 1995; 9:347–350.
9. Kainulainen K, Pulkkinen L, Savolainen A, Kaitila I, Peltonen L. Location on chromosome 15 of the gene defect causing Marfan syndrome. N Engl J Med 1990; 323:935–939.
10. Magenis RE, Maslen CL, Smith L, Allen L, Sakai LY. Localization of the fibrillin (FBN) gene to chromosome 15, band q21.1. Genomics 1991; 11:346–351.
11. Dietz HC, Cutting GR, Pyeritz RE, Maslen CL. Marfan syndrome caused by a recurrent de novo missense mutation in the fibrillin gene. Nature 1991; 352:337–339.
12. Larimore TC, Howe JR, Korte JA, et al. Familial medullary thyroid carcinoma and multiple endocrine neoplasia type 2B map to the same region of chromosome 10 as multiple endocrine neoplasia type 2A. Genomics 1991; 9:181–192.
13. Mulligan LM, Kwok JBJ, Healey CS, et al. Germ-line mutations of the RET proto-oncogene in multiple endocrine neoplasia type 2A. Nature 1993; 363:458–460.
14. Edery P, Lyonnet S, Mulligan LM, et al. Mutations of the *RET* proto-oncogene in Hirschsprung's disease. Nature 1994; 367:378–380.
15. Leach FS, Nicolaides NC, Papadopoulos N, et al. Mutations of a mutS homolog in hereditary nonpolyposis colorectal cancer. Cell 1993; 75:1215–1225.
16. Papadopoulos N, Nicolaides NC, Wei Y-F, et al. Mutation of a mutL homolog in hereditary colon cancer. Science 1994; 263:1625–1629.

17. Nicolaides NC, Papadopoulos N, Liu B, et al. Mutations of two PMS homologues in hereditary nonpolyposis colon cancer. Nature 1994; 371:75–80.
18. Lander ES, Schork NJ. Genetic dissection of complex traits Science. 1994; 265: 2037–2048.
19. Todd JA. Genetic analysis of type 1 diabetes using whole genome approaches. Proc Natl Acad Sci USA 1995; 92:8560–8565.
20. Bennett ST, Lucassen AM, Gough SC, Powell EE. Susceptibility to human type 1 diabetes at IDDM2 is determined by tandem repeat variation at the insulin gene minisatellite locus. Nature Genet 1995; 9:284–292.
21. Hästbacka J, de la Chapelle A, Mahtani MM, et al. The diastrophic dysplasia gene encodes a novel sulfate transporter: positional cloning by fine-structure linkage disequilibrium mapping. Cell 1994; 78:1073–1087.
22. Pease AC, Solas D, Sullivan EJ, Cronin MT, Holmes CP, Fodor SPA. Light-generated oligonucleotide arrays for rapid DNA sequence analysis. Proc Natl Acad Sci. USA 1994; 91:5022–5026.

Part Two

DELIVERING GENES TO HUMAN LUNGS

3

Adenoviral Vectors for Gene Therapy of Inherited and Acquired Disorders of the Lung

DAVID T. CURIEL and ROBERT I. GARVER, JR.

University of Alabama at Birmingham
Birmingham, Alabama

I. Overview of Adenovirus Biology

Human adenoviruses are comprised of 47 to 49 serotypes that have been grouped into six subgenera, designated A through F, on the basis of genotypic and phenotypic characteristics (1). Recombinant adenoviruses, defined here as adenoviruses modified to contain a nonviral protein coding sequence, are derived from human serotypes 2 and 5 of the group C adenovirus family. These two serotypes share an identical genomic organization with a nearly identical nucleotide sequence so that several common vector backbones employed today are hybrids of serotypes 2 and 5. The viral replication is comprised of early and late phases; genes bidirectionally transcribed during the two phases are designated either "E" or "L" followed by a number that corresponds to a cluster of genes that generally share coding and regulatory sequences. The rationale for using these two serotypes is primarily based on the knowledge that the group C adenoviruses are not oncogenic and that much of the biology of these two serotypes has been extensively delineated over the past two decades.

Although it is beyond the scope of this review to comprehensively evaluate adenoviral biology, several key aspects in this regard are helpful in terms of understanding both strengths and deficiencies of this vector system. First, like most viruses, the adenovirus gains entry to the host cell by binding to a specific cell surface receptor that subsequently undergoes internalization. The critical determinants of the receptor binding and internalization have been identified recently and will be discussed in depth elsewhere in this monograph (see Chap. 11). Although serotypes 2 and 5 are primarily respiratory pathogens, it appears that almost all epithelial cells efficiently bind and internalize recombinant adenoviruses. However, it is important to note that the human serotypes have a much lower infection efficiency for non-primate cells, a fact that is important in designing and interpreting animal experiments with these vectors. Second, adenoviruses contain a double stranded DNA genome of approximately 36 kb. Several studies have carefully demonstrated that the native genomic size can only be varied in the range of ±5% in order to achieve acceptable levels of viral production and stability (2). This size requirement has obvious impact on the design of adenoviral vector backbones and the protein coding sequences that are inserted within them. Third, several of the viral genes necessary for viral replication can be deleted from the viral genome and supplied in *trans* as a means of replication-enabling the defective virus (e.g., 3). This principle has been applied to the design of replication-defective vectors that are produced by passaging the defective virus through cell lines that supply the products of the deleted viral genes.

II. Conventional Recombinant Adenovirus Vectors

A. Adenovirus Vectors Are Replication-Defective

Recombinant adenovirus is rendered replication-defective while retaining the capacity for replication under controlled circumstances by exploiting the knowledge that some viral genes necessary for new viral synthesis can be deleted from the viral genome and supplied in *trans*. The result is the ability to produce a large amount of virus that is replication-defective except in those cells containing the deleted viral gene products. In practical terms, this is accomplished by established cell lines that contain stably integrated adenoviral genes so that when infected with the otherwise defective virus, a normal infectious life cycle occurs with the generation of new, defective viruses.

The most widely used adenoviral vectors contain a deletion within the E1 region that inactivates both of the two regions, E1A and E1B. The complex biology of these two genes and protein products have been comprehensively reviewed (see 4–6), but key features relevant to adenoviral vectors will be briefly described here. The two E1 regions have separate transcriptional regulatory signals, and each region actually codes for multiple proteins generated by differential splicing of the transcripts. The function of the three major E1A proteins are multifaceted and complex, but for the purpose of this discussion, the most relevant activity is the promiscuous transactivation of viral gene promoters by one of the E1A proteins. This transactivation is considered a primary initiating event of viral replication, and in its absence, viral replication occurs at negligible levels except under very specific conditions, such as a high multiplicity of infection (7–10). The E1A proteins by themselves appear to be strong inducers of apoptosis (11–13). In the context of adenoviral replication, one of the most important functions of the E1B proteins is to prevent the onset of apoptosis in the presence of E1A (12,14). Both the E1A and E1B proteins can be supplied in *trans* to replication-enable E1-defective adenoviruses. This has been most accomplished by the use of the 293 cell line, which is human embryonic kidney cells containing the E1A and E1B gene region of human adenovirus serotype 5 (3,15).

Other defective virus/packaging cell line systems have been developed for gene transfer, although they are much less commonly used than the E1-defective vectors. One example is E4 defective viruses for use with the W162 packaging cell line containing stably integrated E4 viral sequences (16).

B. Common Methods for Creating Recombinant Adenoviruses

Recombinant adenovirus genomes are constructed from plasmid intermediates containing portions of the wild-type adenovirus genome. Berkner and Sharp first reported that large portions of the adenoviral genome could be propagated as a plasmid insert and later rejoined to produce a functional adenoviral genome (17). These initial strategies involved heterologous recombination between linear DNA segments. This initial observation was followed by subsequent refinements in the vector intermediates. In this regard, Graham and colleagues have produced the most widely used adenoviral vector system which is based on plasmid vectors which undergo recombination and that have convenient cloning sites for new coding sequences (as reviewed in 18).

The details of plasmid design and common methods of use have been described in recent reviews (18,19) and therefore will not be reiterated in detail here (Fig. 1). In brief, the most widely used method involves three major steps. First, the new coding sequence with appropriate transcriptional start and stop regulatory sequences is added to a multiple cloning site within the deleted E1 region of a plasmid containing a portion of the left-hand (5') end of the adenovirus genome. Second, this plasmid vector containing the new coding sequence is cotransfected into 293 cells with a second plasmid that contains the entire adenovirus genome with an E1 deletion modified to contain a "stuffer fragment" of plasmid DNA. The stuffer fragment not only contains the plasmid origin of replication and antibiotic resistance gene for bacterial propagation, but it is sufficiently large to prevent that adenoviral DNA from being packaged into a stable viral particle. Homologous recombination occurs be-

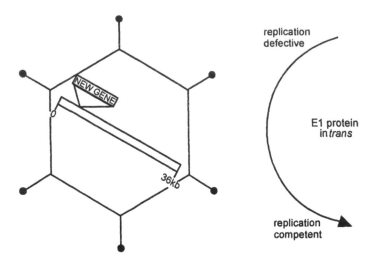

Figure 1 Recombinant adenovirus paradigm. On left is a representation of the defective virus containing a genome represented by the open bar. The genome has a deletion in the E1 region at the left-hand end into which is inserted a new gene. The recombinant virus is replication-defective, but can be replication-enabled within cells that supply the E1 proteins.

tween the two plasmids so that the E1 region containing the coding sequence of interest replaces the plasmid stuffer within the otherwise intact genome, and the E1 proteins made by the 293 cells activate the recombinant genome replication with the result that recombinant virus is made. The third step is a series of plaque purifications with screening assays at each step to eliminate undesired wild type virus that is generated by homologous recombination between the viral sequences within 293 cells and the adenoviral plasmid with the stuffer fragment.

III. Adenoviral Vectors for Gene Delivery to the Lung

A. The Lung as a Target for Genetic Intervention

Despite the variety of potential disease states approachable by gene therapy methodologies, diseases of the lung have to date represented a disproportionately large fraction of the approved human gene therapy protocols (20). This reflects the fact that the lung is a particularly attractive target for gene therapy interventions for a number of reasons. First, the genetic basis for a number of pulmonary diseases has been elucidated. The genes for two inherited pulmonary disorders, cystic fibrosis (CF) and α1-antitrypsin (α1AT) deficiency, have been identified. Second, these disorders are relatively common, representing the two most common fatal inherited disorders of Caucasians in North America. Thus, from the standpoint of inherited disorders, these two pulmonary diseases are particularly attractive targets. Some of these considerations pertain to lung cancer, as this is a common disease for which genetic mechanisms are being elucidated and for which conventional therapy is ineffective. Lung cancer is the major lethal cancer in both men and women; its incidence is increasing (21), and over the past two decades overall mortality from this disease has only minimally improved (21). Similarly, there is no effective therapy for malignant mesothelioma, which represents a relatively unique model of localized intrathoracic malignancy without a major metastatic component.

In addition, there are several practical considerations that make the lung an attractive gene therapy target. As a general principle, delivery of the therapeutic gene to the target cell represents a significant limitation to many gene therapy applications. This problem has been overcome for a subset of diseases by genetically modifying cells ex vivo and then reimplanting the corrected cells to the target organ. This approach, however, has been limited to contexts in which the target cells can be propagated and maintained in tissue culture for prolonged periods of time and then reinfused after gene transfer. In practice, therefore, the ex vivo technique limits gene therapy approaches to a relatively small subset of diseases. For the lung, where the ex vivo strategy is not feasible, direct in vivo gene delivery is therefore required. This is a realistic goal, however, because unlike other solid organs that are involved by genetic diseases, the lung is accessible for gene delivery. Airway access offers a means to directly approach both conducting airway and lung parenchyma, and the dual vascular supply of the lung offers unique opportunities to achieve gene delivery via the vascular route. These practical aspects have facilitated the development of gene delivery strategies for lung diseases.

B. Adenoviral Vectors for In Vivo Gene Transfer to the Lung

As gene therapy strategies for lung disorders require in vivo gene delivery, the subset of vectors appropriate for this employment is limited. In this regard, one of the unique attributes of recombinant adenoviral vectors is their capacity to accomplish in situ gene delivery to differentiated parenchyma in vivo. This fact was first noted by Siegfried, whereby ornithine transcarbamylose (OTC) deficient mice could be phenotypically corrected by in vivo gene delivery of adenoviral vectors (22). Based on this finding, as well as the concept that the adenovirus possessed a natural tropism for airway epithelia, Crystal proposed the use of these vector agents for direct in situ transduction of airway epithelium (23). In this regard, Rosenfeld et al. delivered an $\alpha 1$ antitrypsin ($\alpha 1AT$) encoding recombinant adenovirus via the airway luminal route to rodents and reported successful in situ transduction of airway epithelium. This was manifest in that airway epithelial explants possessed the capacity to achieve biosynthesis of heterologous, human α1AT, based upon airway epithelial synthesis of the protein. In addition, epithelial lining fluid in vector transduced animals possessed augmented levels of human α1AT. Thus, this finding established the concept that the natural tropism of adenoviral vectors, as well as their in vivo stability, allowed phenotypically relevant levels of airway epithelial transduction.

With the recognition that airway epithelial could be genetically modified in this manner, a variety of additional gene therapy strategies for pulmonary disorders was suggested. In this regard, the ability to accomplish in vivo gene transfer to airway epithelium allowed the development of gene therapy approaches for both inherited and acquired disorders of the lung (24). Discussion of these approaches is extended in this chapter, as well as in other sections of this monograph. For the present context, however, a number of lung biology issues will be considered which extend the initial recognition of targetable lung cells. Each of these studies also extends the possible range of gene therapy approaches which may thereby be conceptualized for lung disorders employing adenoviral vectors.

With the recognition that airway epithelium could be transduced by this route, a relevant question was what cell types were capable of being modified. To this end, Mastrangeli et al. explored airway epithelial cell subsets modified after airway delivery of the vector (25). Their study demonstrated that vector efficacy was relevant for both ciliated and non-ciliated airway epithelium. This fact has been noted for rodent and primate model systems. Thus, it appeared that all airway epithelial cells accessible via the luminal rate were successfully transduced via adenoviral vectors.

Because of their relevance to the pathogenesis of CF, attempts have also been made to achieve transduction of bronchial submucosal gland cells. In this regard, initial attempts to achieve genetic transduction of this cell subset via luminal delivery did not show the high levels of transduction (26). Studies have shown that this may largely reflect issues related to mechanical access to these cells via the luminal route (27,28). Based on this, attempts have been made to achieve transduction of this cellular target viral alternate means. These strategies have included mechanical abrasion to enhance vector access to target cells. In addition, it has been recognized that non-luminal routes may offer attractive alternatives for accessing these cells. In this regard, the desire to transduce the bronchial submucosa gland cells for CF gene therapy thus led to the exploration of other routing of adenoviral vector delivery to the lung. To this end, Crystal and colleagues delivered the adenovirus via the pulmonary vascular route. Their study demonstrated that vascular administration of the vector could accomplish high levels of transduction of pulmonary vascular cells (28). In these studies, some level of transduction of bronchial submucosal gland cells was also achieved. The ability to modify pulmonary vascular epithelium has also been noted by Dichek et al. They showed higher transduction levels with local delivery via a cellular-lumen system than with simple vascular delivery (29). Whereas these various studies have not demonstrated an effective means to accomplish adenoviral vector modified submucosal gland gene transfer, they have related the capacity to transduce other pulmonary cellular subsets by nonluminal routing strategies. This recognition, however, may provide the means to develop gene therapy strategies based upon the capacity to transduce these pulmonary cells in vivo.

Another issue relevant to the employment of this vector for gene therapy strategies is the issue of longevity of gene expression. In this regard, the initial studies of the Crystal group in the context of delivery of the α1AT and CF genes in rodent models demonstrated transient levels of gene expression (30). Two factors would have predicted such a finding. First, the adenovirus vector is nonintegrative; delivered DNA is not incorporated into the host genome and thus gene expression would be transient. Second, the turnover of airway epithelium would have diluted the proportion of transduced cells. Thus, for gene therapy strategies as related to sustained phenotypic correction, this aspect of adenoviral vector biology was problematic.

In addition to these factors, however, other biological variables appeared to limit adenoviral vector mediated transgene expression. In this regard, vector administration to the lung was associated with a prominent inflammatory cell infiltration (31–34). Analysis of this phenomenon revealed that transduced cells were recognized as foreign and cleared by immunological mechanisms (31–34). This subject is treated more extensively elsewhere in this monograph. For the present context, however, the issue of transient gene expression led to the recognition that gene therapy strategies for inherited genetic disorders such as CF and α1AT would require reputative delivery of adenoviral vectors to achieve sustained phenotypic correction. Studies by Trapnell et al., however, revealed that primary immunologic recognition of the vector prevented this mode of repetitive vector delivery of transgenes to the lung (35). These studies were extended by McCoy et al. who showed that even devitalized vector particles could induce a humoral response which prevented repetitive vector delivery (36). Thus, issues of vector and transduced cell immunogenicity have important consequences with respect to longevity of transgene expression, as well as vector employment schemas.

IV. Gene Therapy Approaches Based on Adenovirus Vectors for Inherited Pulmonary Disorders

A. α1-Antitrypsin Deficiency

α1-Antitrypsin deficiency is an autosomal recessive disease characterized by the development of panacinar emphysema and cirrhosis of the liver.

Based on the recognition that α1AT lung disease results from deficient serum and lung levels of the α1AT antiprotease, it was logical to attempt to correct this abnormality by restoration of the deficient lung levels of α1AT, either indirectly via serum augmentation or directly via protein delivery to the lower respiratory tract. Trials employing serum or airway delivery of purified human α1AT, or alternatively a distinct antiprotease in the form of recombinant secretory leukoproteinase inhibitor (SLIPI), have been developed (37). Although these trials have demonstrated that direct protein replacement may reverse the biochemical abnormalities characterizing this disorder, it is still premature to determine the clinical efficacy of this therapy. In addition, practical issues, such as the cost and inconvenience of weekly or monthly infusion of the purified α1AT for the remainder of the patient's life, make this therapy suboptimal. The chronic administration of human derived plasma also raises risks related to transmission of blood-born infectious agents.

As α1AT deficiency meets the criteria to develop gene therapy approaches, a number of strategies have been undertaken. The α1AT gene is available, and disease pathogenesis is well characterized (38,39). From the practical standpoint, α1AT derived by cellular biosynthesis can access the plasma compartment from multiple potential sites and will localize to the lower respiratory tract. In addition, α1AT is a serum glycoprotein that does not require specialized processing and has been demonstrated to be successfully produced by a variety of heterologous cells. Thus, the cellular site of α1AT synthesis is not crucial. Additionally, a clear threshold exists whereby there is no disease associated with levels above this threshold; hence, highly regulated gene expression does not appear imperative.

Based upon this concept, methods have been developed to accomplish direct *in situ* modification of hepatocytes, the normal cell responsible for α1AT biosynthesis. Jaffee et al. employed vascular delivery of an α1AT-encoding recombinant adenovirus to achieve high-efficiency transduction of hepatocytes and augmented serum and lung levels of α1AT in a rodent model (23). However, observed levels were subtherapeutic and expression lasted only several days. Given the problems with vector-induced inflammation, it is unlikely that repetitive delivery will be a feasible method to address the problem of transient expression. In this regard, in an analogous model, repetitive delivery to augment hepatocyte LDL-R was not feasible due to the induced antivector immunity (40). Thus, unless methods can be developed to allow prolonged and elevated gene expression in transduced cells and/or reduce the host antivector immune response, this approach will not likely be of clinical value.

With the recognized difficulties of genetically inducing hepatocytes to produce sufficient α1AT levels, alternative approaches have been developed to target the lower respiratory tract directly. In this regard, epithelial lining fluid (ELF) levels of α1AT required to maintain effective anti-elastase activity are of a much lower magnitude than the required serum levels. Therefore, it has been logical to attempt to directly augment ELF α1AT levels through genetic modification of airway epithelial cells. Rosenfeld et al. have employed this strategy utilizing a recombinant adenovirus encoding the α1AT cDNA (23). Direct intratracheal delivery of the vector allowed airway epithelial expression of human α1AT in a rodent model. The transduced cells secreted a human α1AT of normal size that functioned as an inhibitor of its natural substrate, elastase. Importantly, vector transduction by this route augmented lung ELF levels of α1AT. In practical terms, this strategy may be susceptible to the same limitations as adenoviral vector transduction of the liver.

B. Cystic Fibrosis

Cystic fibrosis (CF) is an autosomal-recessive disease caused by mutations in the gene encoding for the cystic fibrosis transmembrane conductance regulator (CFTR) protein. Using positional cloning, the gene was isolated to chromosome 7 and found to consist of approximately 250,000 base pairs that encode an mRNA of 6.5 kb (41,42) (reviewed in 43). Over 350 mutations have been identified, resulting in aberrant transcription, translation, cellular trafficking, or membrane function (reviewed in 44).

Isolation of the CF gene has led to the development of gene therapy strategies to replace the defective gene (45). As the pulmonary disease in CF is the major cause of morbidity and mortality, numerous investigators have focused on the lung as the first target organ for gene replacement.

Rapid progress towards gene replacement has been made since the identification of the CF gene (43), but the underlying pathophysiology of CF, and hence the necessary target cell(s) for gene therapy, remain unclear (reviewed in 43). At the cellular level, there is abundant evidence that CFTR is an apical membrane protein which serves as a regulated chloride channel (46). Through unknown mechanisms, the absence of functional CFTR in CF epithelial cells results in both sodium hyper absorption and lack of cAMP-mediated chloride secretion (reviewed in 47). Moreover, there is accumulating evidence that CFTR is important for other cellular functions, including sulfation of high-molecular-weight glycoconjugates (48), acidification of intracellular organelles (49), sialylation of cell surface receptors (50), regulation of membrane trafficking (51), and secretion of mucus (52).

At the organ level, CFTR has been localized to both the surface epithelium (53,54) and submucosal glands (55,56) in the lung, as well as to the ductal epithelium in the biliary tree and pancreas. It is unknown how the cellular defect in these organs translates to the abnormally viscous secretions seen in patients with CF (reviewed in 57), but current gene therapy approaches are based on the hypothesis that the sodium hyperabsorption and lack of cAMP-mediated chloride transport result in a dehydrated epithelial lining fluid and reduced mucociliary clearance. If this is correct, delivery of a normal copy of the CF gene to the surface epithelium of conducting airways is likely to correct the epithelial cell ion transport abnormalities, which will in turn reduce mucus viscoelasticity and restore normal mucociliary clearance. As previously discussed, adenoviral and other candidate vectors being used in human trials efficiently achieves gene transfer to submucosal glands. Thus, if CFTR function in submucosal glands is necessary for normal airway clearance, current approaches are not likely to impact on disease pathogenesis in the cartilaginous airways. However, efficient gene transfer to surface epithelium in noncartilaginous airway could ameliorate distal airway disease and thereby affect the clinical course of the disease without necessarily achieving a cure.

Several in vitro studies have demonstrated that delivery of a normal CFTR gene to CF epithelial cells restores cAMP-mediated chloride transport, but it remains unknown what percentage of epithelial cells in vivo must express CFTR to correct ion transport. Using recombinant retroviruses and vaccinia virus, Drumm and colleagues (58) and Rich and colleagues (46) showed that expression of CFTR restored cAMP-mediated chloride secretion in a CF pancreatic cell line and primary CF epithelial cells. Moreover, in a polarized epithelial sheet, Johnson et al. demonstrated that expression of CFTR in as few as 6% to 10% of CF airway epithelial cells restored normal chloride transport properties (59). In more recent studies, however, the same investigators demonstrated that close to 100% of CF epithelial cells must be transduced to reduce the sodium hyperabsorption characteristic of CF (60). These data indicate that if sodium hyperabsorption across surface epithelia plays a major role in the pathogenesis of CF lung disease, very efficient gene delivery will be required to restore normal airway function.

The initial complementation experiments provided encouragement to investigators in search of gene transfer vectors to deliver the CF gene to intact airways. As discussed above, the ideal vector would efficiently deliver the gene to the appropriate target cells (including the columnar epithelial cells and/or submucosal gland serous cells in the conducting airway) without causing toxicity or inflammation. To provide for indefinite expression, the vector would need to integrate the gene into a population of stem cells since the airway surface epithelium regenerates slowly over time. Although none of the available vectors meets these criteria, studies in a number of animal models demonstrated gene transfer to the lung and provided impetus for the human trials in progress.

Several groups of investigators have demonstrated that recombinant adenoviruses can be used to deliver the CF gene to airway epithelium, but recent work has revealed problems with gene transfer efficiency and vector-related toxicity. In bronchial xenografts (26) and cotton rats (30), recombinant adenoviruses mediate transgene expression to approximately 5–20% of airway epithelial cells with most of the expression in ciliated cells and minimal transfer to the basal cells of a pseudostratified columnar epithelium (25). In contrast to cotton rats, when these vectors are instilled in the airways of large animals such as primates (61), in human airway xenografts (27) or in the trachea of CFTR-deficient mice, gene transfer to conducting airway is much less efficient, with <1% of airway epithelial cells expressing the transgene. In addition, there is preferential gene transfer to undifferentiated and injured epithelium (62), a finding that appears to be related to the expression of specific integrin cell adhesion receptors (63,64). This data suggests that very high titers (a multiplicity of infection or number of infectious particles per cell greater than 10^3/ cell) will be necessary to achieve physiologically meaningful gene transfer to normal differentiated airway epithelium.

Thus, some of the preclinical data suggests that both the efficiency and toxicity profile of recombinant adenoviruses must be improved for these vectors to be clinically useful for CF gene therapy. It remains to be seen whether alternative delivery methods, such as instillation of highly concentrated viral solutions or aerosolization, can improve gene transfer efficiency, and whether further modifications of the adenovirus genome will abrogate the host inflammatory response. Finally, the development of neutralizing antibodies (65,66) in the lung may preclude repeated administration, which will be necessary since the transferred gene does not integrate and transgene expression decreases within several weeks. These issues have had important implications with respect to human clinical trails for cystic fibrosis. This topic is covered elsewhere in this monograph.

V. Potential Utility of Adenoviral Vectors for Treatment of Acquired Lung Conditions

A. Non-Small-Cell Lung Cancer

Lung cancer is the best example of an acquired disease where preliminary results have suggested that adenoviral-mediated gene therapy has the potential for clinical utility. As a background to adenoviral-mediated gene therapy of lung cancer, it is useful to broadly consider that gene therapy approaches to cancer can be divided into three categories: immunopotentiation, mutation compensation, and molecular chemotherapy. Immunopotentiation is defined here as the introduction of genetic modifications in host cells to augment immunologically-mediated destruction of the tumor cells. Examples of immunopotentiation include the overexpression of cytokines by malignant cells (e.g., 67) and the use of "therapeutic" polynucleotide vaccines (e.g., 68). Mutation compensation refers to the addition of genes to malignant cells intended to dominantly compensate for the effects of mutations strongly associated with neoplasia. Examples of mutation compensation include the addition of genes encoding wild type anti-oncogenes to cell lines having a mutated antioncogene (e.g., 69) or genes encoding intracellular antibodies that disrupt production of oncogene products (70). Molecular chemotherapy involves the addition of genes that will lead to the death of the malignant cells within the host (e.g., 71) in which the therapeutic gene functions analogously to conventional chemotherapeutic agents.

The mutation compensation strategy has been employed in preclinical, and recently clinical, investigational therapies for non-small-cell lung cancer (NSCLC). NSCLC, like the majority of cancers, has been found to have mutations in both proto-oncogenes and antioncogenes (see 72 for review). Among antioncogenes, a.k.a. tumor suppressor genes, the p53 gene is deleted or mutated in approximately 50% of NSCLC (e.g., 73,74). Two reports in 1990 demonstrated with in vitro studies that introduction of a wild-type p53 coding sequence into sarcoma cells lacking endogenous p53 protein (75), or colon carcinoma cells with mutant p53 protein (76), suppressed the neoplastic phenotype of those cells. In 1992, these earlier reports were extended to lung cancer by Unger et al. by cotransfecting wild-type and mutant p53 to show that the presence of wild-type p53 restored the normal transcriptional activity to the host cells (77). The same year, Takahashi et al. showed that transfection of wild type p53 into lung cancer cell lines with mutant or deleted p53 suppressed both in vitro and in vivo growth of the lung cancer cell lines (78). In 1994, Fujiwara et al. used retroviral-mediated transfer of wild-type p53 to demonstrate a significant reduction in the growth of a human lung cancer cell line orthotopically engrafted into nude mice (79).

More recently, Jack Roth's group has used recombinant adenovirus carrying the wild type p53 gene to "treat" preclinical models of NSCLC. Spheroids comprised of human lung cancer cells with a homozygous deletion of p53 exposed to the p53-adenovirus were found to undergo apoptosis (80). This same study showed that the same adenovirus could be directly administered to subcutaneously engrafted tumor nodules of the same cell line in combination with cisplatin therapy to produce a marked reduction in tumor nodule size (80). A subsequent report by the same group showed that the same p53 adenovirus prevented the orthotopic growth of a human lung cancer cell line when the virus was administered three days after intratracheal introduction of the cells (69). These promising preclinical results have provided the rationale for the Roth group to develop a Phase I human lung cancer protocol that will test the coadministration of the p53-adenovirus in conjunction with cisplatin to individuals with unresectable NSCLC.

Although the preclinical studies with either retroviral or adenoviral transfer of wild-type p53-engrafted human lung cancers have demonstrated significant, local tumor reduction, the adenovirus has become the preferred vector for this lung cancer strategy. The stated reason for this preference is based on the ability to achieve higher titers with adenovirus; hence, more virus can be administered in vivo. However, there have not been any published studies of head-to-head comparisons of retrovirus vs. adenovirus for preclinical p53-based therapies.

Recently, the p16^{INK4} gene has been shown in one study to mitigate the malignant phenotype in human lung cancer cell lines not expressing the normal gene product (81). In this study by Jin et al., an adenoviral vector containing the p16^{INK4} coding sequence administered in vitro to several human NSCLC cell lines resulted in markedly diminished in vitro growth. One of the cell lines was found to have moderately reduced in vivo growth when the cells were engrafted into immunosuppressed mice following the adenovirus treatment. These results suggest that the p16^{INK4} gene is a candidate tumor suppressor gene that could be administered in a therapeutic context to NSCLC, although the data are more preliminary than those obtained with p53.

B. Malignant Mesothelioma

Recombinant adenoviruses have also been used recently for molecular chemotherapy strategies directed toward another intrathoracic malignancy, malignant mesothelioma. This particular neoplasm has been chosen by Albelda and Kaiser and colleagues since it is generally localized to the chest cavity but ineffectively treated by conventional therapies. An initial report from this group used an adenovirus containing the Lac Z marker gene to define the transduction characteristics in vitro and in vivo (82). Two human mesothelioma cell lines were transducible in vitro by the adenovirus, although to a lesser degree than a NSCLC line used as a control. The recombinant adenovirus was shown to efficiently transduce one of the cell lines engrafted to the peritoneal cavity of immunosuppressed mice with no significant marker gene transfer detected in any of the intraperitoneal organs. This same group reported that a recombinant adenovirus containing the herpes simplex virus thymidine kinase (HSVTK) gene conferred ganciclovir sensitivity to human mesothelioma cell lines in vitro (83). The use of the HSVTK-adenovirus was extended to the peritoneal in vivo models where it was shown to produce a significant reduction in tumor burden in combination with ganciclovir therapy (84). These studies have formed the basis for an ongoing Phase I clinical trial of adenoviral-mediated HSVTK + ganciclovir therapy for intrapleural mesothelioma.

VI. Strategies to Improve Adenoviral Vectors from Lung Disorder Gene Therapy Approaches

A. Ongoing Improvements in Recombinant Adenoviruses

The preclinical and clinical investigations of adenoviral-mediated gene transfer for treatment of various pulmonary conditions were performed with E1-

defective adenoviruses, some of which also had inactivated E3 genes and will be referred to as "initial adenovirus vectors." These investigations with the initial adenovirus vectors, as outlined in the previous sections, have served to define both the strengths and limitations of recombinant adenoviruses for gene therapy of lung conditions. A number of groups are working to make improvements in the initial adenovirus vector system that will be discussed in the following sections.

B. Self-Limited Replication Enablement

One area of improvement concerns the development of strategies for amplifying adenoviral-mediated gene transfer by inducing self-limited, replication enablement of the replication-defective virus. The rationale for replication-enabling a replication-defective virus is based on some earlier work with recombinant retroviruses that demonstrated a marked increase in intratumoral gene transfer by introducing recombinant retroviral producer cells into a solid tumor nodule (71). The retroviral studies used stable retrovirus-producing cell lines, an approach that could not be extrapolated to the adenoviral system since it is a lytic virus that destroys the cell making the virus; hence, stable adenoviral producing lines are not possible. The adenoviral replication-enablement strategy capitalized on the knowledge that E1-defective adenoviruses are replication-enabled in cell lines that express the proteins from the E1 genes deleted from the virus (Fig. 2). The initial report by Goldsmith et al. demonstrated the concept of the self-limited replication enablement by codelivery of an E1-defective adenovirus and a plasmid encoding the deleted E1 proteins to a carcinoma cell line in vitro (85). It was shown that codelivery of the virus and the replication-enabling plasmid resulted in the production of new, E1-defective adenovirus. A subsequent study by the same group has shown that the codelivery strategy was operative in several different cell lines, including one derived from NSCLC, and that cells exposed to the virus and the replication-enabling plasmid produced more than 500 new replication-defective adenoviruses for every adenovirus initially input into the system (86). Importantly, this same study showed that cells of a tumor cell line that had been cotransduced with the replication-enabling plasmid and a defective adenovirus carrying a marker gene led to markedly increased viral transgene expression within solid tumor nodules in vivo compared to tumor nodules containing cells only transduced with the virus alone.

One important concern of this system is the production of small amounts of replication-competent virus (i.e., wild type) that appears to arise by homologous recombination between the recombinant adenoviral genome and the replication-enabling plasmid. The conventional method for producing recombinant adenovirus by passaging virus through the 293 cell line has also been associated with wild type virus production (87), but the cotransduction method

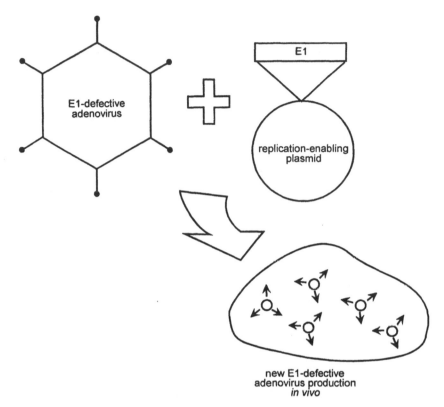

Figure 2 Self-limited replication-enablement strategy. The E1-defective virus and a plasmid encoding the deleted E1 functions are both introduced into cells so that new recombinant adenovirus is produced. Introduction of both the plasmid and the virus into sites in vivo results in the establishment of recombinant adenovirus producer cells (circles) that subsequently release additional virus locally so as to infect adjacent cells (arrows). The result is an amplification of the viral-mediated gene transfer in situ.

has been shown to have a higher frequency of wild type production (86). This deficiency has recently been overcome by the development of new replication-enabling plasmids that have greatly reduced the opportunities for homologous recombination (88). It now appears that this self-limited replication enablement system has the potential to provide a safe means of amplifying adenoviral gene transfer in vivo.

C. Mitigating Immunological Destruction

Another aspect of recombinant adenovirus gene therapy that is leading to new adenovirus vector designs and/or changes in administration strategy is that of immunogenicity. It had been recognized by several investigators that adenoviral transgene expression decreased to nominal levels several weeks after administration in immunocompetent animals but persisted for months in immunosuppressed animals. Yang et al. showed that cell-mediated immunity directed against viral proteins played a significant role in the time dependent elimination of the adenoviral-transduced cells (89). The viral protein production occurred in cells containing the E1-defective viruses presumably because cellular factors were capable of transactivating low levels of viral gene transcription in the absence of E1 proteins. This hypothesis led Wilson and colleagues to develop an E1-defective adenovirus with a temperature sensitive mutation of the E2a gene with the intent of further reducing viral protein production in transduced cells, and thereby prolong the duration of viral transgene expression (90). This virus permits E2 protein expression at 32°C so that virus can be produced at the permissive temperature, but lower amounts of E2 expression occurred at 37°C (90). In mouse liver, nonhuman primate lung, and cotton rat lung, significantly prolonged levels of viral transgene expression were obtained with the E1 and E2-defective virus compared to conventional E1-defective virus (90–92). This refined adenoviral vector may hold promise for prolonging viral transgene expression from a single administration. However, the recombinant adenoviruses are immunogenic as a consequence of the viral coat proteins; hence, repeat administration of recombinant adenovirus—even the E2-defective virus—will remain problematic.

As an alternative to adenoviral modification as a means of mitigating immune-mediated elimination of transgene-expressing cells, several groups have shown that the administration of immunosuppressants to the host will also address this problem. Two groups have shown that FK506 prolonged adenoviral transgene expression in skeletal muscle in mice with a significantly reduced infiltration of muscle with immunomodulatory cells (93,94). Cyclosporine administration was shown to similarly prolong adenoviral transgene expression in the livers of immunocompetent mice, but not to the extent achieved with parallel investigations using the E2-defective vector described in the preceding paragraph (90). In another strategy, Kolls et al. have shown that administration of a monoclonal antibody directed against CD4 positive cells to achieve more than a doubling of adenoviral transgene expression in immunocompetent mice (95). None of the published studies with the immunosuppressive regimens have addressed the problem of repeat administration of the adenovirus.

D. Elimination of Wild Type Virus Production

A third broad area of adenoviral vector improvement is a variety of strategies with the common objective of further reducing the probability for wild type virus—replication-competent virus production. In the conventional adeno-virus virus, the defective E1 region can homologously recombine with E1 se-quences present in the 293 cell genome so as to contaminate viral stocks with small amounts of wild-type virus (87). This problem has been anecdotally re-ported to pose a significant problem in large-scale adenovirus preparations intended for clinical use (96). A second point at which wild-type virus may arise is within the host that may contract an adenovirus infection coincident with administration of the recombinant adenovirus, thereby providing opportuni-ties for homologous recombination.

Several modifications to the adenoviral genome have been made to greatly minimize chances for a single homologous recombinant event to pro-duce a replication-competent virus. Krougliak and Graham have described a 293 cell line modified to produce E4 and pIX proteins under control of induc-ible promoters that rescued E1- and E4-defective viruses (97). The result is a doubly defective virus that would require two separate homologous recombi-nant events in order to replace both defective regions with wild type sequences. Similarly, Yeh et al. have described a combined E1 and E4 defective virus that is produced by a 293 cell line modified to contain a dexamethasone inducible minimal E4 coding sequence (98). An analogous but different approach has been described in which a modified 293 cell line was made to also produce the adenovirus precursor terminal protein (pTP), an essential component of viral replication strand synthesis (99). It was shown that these cells could *trans* com-plement a defective adenovirus with a temperature sensitive mutation of the pTP-encoding region at the nonpermissive temperature. It would be expected that this cell line could be used to produce a virus doubly defective in E1 and terminal protein, although this report did not demonstrate this point. An even more preliminary, multiple deletion strategy was described in which more than 7 kb of essential viral sequences were deleted, including L1, L2, VAI, VAII, and pTP genes so that a helper virus was necessary to produce infectious par-ticles containing the multiply defective genome (100). However, it is difficult to completely separate the defective virus from the helper virus with this sys-tem, and in addition, wild-type recombinants were produced.

Recombinant adenoviral vectors have thus been shown to possess the unique capacity to accomplish extremely efficient in vivo gene delivery to a variety of pulmonary cellular targets. This ability has allowed the proposition of a number of gene therapy approaches for both inherited and acquired dis-orders of the lung. In experimental model systems, these approaches have demonstrated the potential effect by this gene delivery technology. In practice, in the context of human clinical gene therapy trials, however, a number of problems have arisen which will require addressing before clinical utility may be claimed. Nevertheless, in their present form, these vector reagents often represent an extremely powerful research tool for employment in the context of disorders of the lung. Future work directed towards presently recognized liabilities may ultimately allow the realization of the full potential of this vector system for the approach of human pulmonary disorders.

Acknowledgments

This work was supported in part by the National Institutes of Health R015025505, U.S. Army DAMD17-94-J-4398, and the Muscular Dystrophy Association to D.T.C., and a VA Merit Review grant to R.I.G.

References

1. Bailey A Mautner V. Phylogenetic relationships among adenovirus serotypes. Virology 1994; 205:438–452.
2. Bett AJ, Prevec L, Graham FL. Packaging capacity and stability of human adenovirus type 5 vectors. J Virol 1993; 67:5911–5921.
3. Graham FL, Smiley J, Russell WC, Nairn R. Characteristics of a human cell line transformed by DNA from human adenovirus type 5. J Gen Virol 1977; 36:59–74.
4. White E. Regulation of apoptosis by the transforming genes of the DNA tumor virus adenovirus. Proc Soc Exp Biol Med 1993; 204:30–39.
5. Nevins JR. Adenovirus E1A: transcription regulation and alteration of cell growth control. Curr Top Microbiol 1995; 199(pt 3):25–32.
6. Jones N. Transcription modulation by the adenovirus E1A gene. Curr Top Microbiol Immunol 1995; 199(pt 3):59–80.
7. Shenk T, Jones N, Colby W, Fowlkes D. Functional analysis of adenovirus type 5 host range deletion mutants defective for transformation of rat embryo cells. Cold Spring Harbor Symp Quant Biol 1979; 44:367–375.
8. Carlock LR, Jones NC. Transformation-defective mutant of adenovirus type 5 containing a single altered E1a mRNA species. J Virol 1981; 40:657–664.
9. Nevins JR. Mechanism of activation of early viral transcription by the adenovirus E1A gene product. Cell 1981; 26:213–220.
10. Gaynor RB, Berk AJ. Cis-acting induction of adenovirus transcription. Cell 1983; 33:683–693.
11. Frisch M. Antioncogenic effect of adenovirus E1A in human tumor cells. Proc Natl Acad Sci USA 1991; 88:9077–9081.
12. Rao L, Debbas M, Sabbatini P, Hockenbery D, Korsmeyer S, White E. The adenovirus E1A proteins induce apoptosis, which is inhibited by the E1B 19-kDa and Bcl-2 proteins. Proc Natl Acad Sci USA 1992; 89:7742–7746.
13. Frisch SM, Francis H. Disruption of the epithelial cell-matrix interactions induces apoptosis. J Cell Biol 1994; 124:619–626.
14. White E, Cipriani R, Sabbatini P, Denton A. Adenovirus E1B 19-kilodalton protein overcomes the cytotoxicity of E1A proteins. J Virol 1991; 65:2968–2978.
15. Aiello L, Guilfoyle R, Huebner K, Weinman R. Adenovirus 5 DNA sequences present and RNA sequences transcribed in transformed human embryo kidney cells (HEK-Ad-5 or 293). Virol 1979; 94:460–469.

16. Weinberg DH, Ketner G. A cell line that supports the growth of a defective early region 4 deletion mutant of human adenovirus type 2. Proc Natl Acad Sci USA 1983; 80:5383–5386.
17. Berkner KL, Sharp PA. Generation of adenovirus by transfection of plasmids. Nucleic Acids Res 1983; 11:6003–6020.
18. Graham FL, Prevec L. Manipulation of adenovirus vectors. In: Murray, EJ ed. Manipulation of Adenovirus Vectors. Clifton: Humana Press Inc. 1991:109–128.
19. Becker TC, Noel RJ, Coats WS, et al. Use of recombinant adenovirus for metabolic engineering of mammalian cells. In: Methods in Cell Biology. New York: Academic Press, 1994:161–189.
20. Sobol RE, Scanlon KJ. Clinical protocols. Cancer Gene Ther 1995; 2:137–145.
21. Cancer Statistics 1995. Cancer Statistics 1995—A Cancer Journal for Clinicians 1995; 45.
22. Siegfried W. Perspective in gene therapy with recombinant adenoviruses. Exp Clin Endocrinol 1993; 101:7–11.
23. Rosenfeld MA, Siegfried W, Yoshimura K, et al. Adenovirus-mediated transfer of a recombinant α1-antitrypsin gene to the lung epithelium in vivo. Science 1991; 252:431–434.
24. Curiel DT, Pilewski JM, Albelda SM. Gene therapy approaches for inherited and acquired lung diseases. Am J Respir Cell Mol Biol 1996; 14:1–18.
25. Mastrangeli A, Danel C, Rosenfeld MA, et al. Diversity of airway epithelial cell targets for in vivo recombinant adenovirus-mediated gene transfer. J Clin Invest 1993; 91:225–234.
26. Engelhardt JF, Yang Y, Stratford-Perricaudet LD, et al. Direct gene transfer of human CFTR into human bronchial epithelia of xenografts with E1-deleted adenoviruses. Nature Genet 1993; 4:27–34.
27. Pilewski JM, Engelhardt JF, Bavaria JE, Kaiser LR, Wilson JM, Albelda SM. Adenovirus-mediated gene transfer to human bronchial submucosal glands using xenografts. Am J Physiol Lung Cell Mol Physiol 1995; 12:L657–L665.
28. Lemarchand P, Jones M, Danel C, Yamada I, Mastrangeli A, Crystal RG. In vivo adenovirus-mediated gene transfer to lungs via pulmonary artery. J Appl Physiol 1994; 76:2840–2845.
29. Schachtner SK, Rome JJ, Hoyt RF, Newman KD, Virmani R, Dichek DA. In vivo adenovirus-mediated gene transfer via the pulmonary artery of rats. Circ Res 1995; 76:701–709.
30. Rosenfeld MA, Yoshimura K, Trapnell BC, et al. In vivo transfer of the human cystic fibrosis transmembrane conductance regulator gene to the airway epithelium. Cell 1992; 68:143–155.
31. Brody SL, Metzger M, Danel C, Rosenfeld MA, Crystal RG. Acute responses of non-human primates to airway delivery of an adenovirus vector containing the human cystic fibrosis transmembrane conductance regulator cDNA. Human Gene Ther 1994; 5:821–836.
32. Engelhardt JF, Litzky L, Wilson JM. Prolonged transgene expression in cotton rat lung with recombinant adenoviruses defective in E2a. Human Gene Ther 1994; 5:1217–1229.

33. Crystal RG, McElvaney NG, Rosenfeld MA, et al. Administration of an adenovirus containing the human *CFTR* cDNA to the respiratory tract of individuals with cystic fibrosis. Nature Genet 1994; 8:42–51.
34. Simon RH, Engelhardt JF, Yang Y, et al. Adenovirus-mediated transfer of the CFTR gene to lung of nonhuman primates: toxicity study. Human Gene Ther 1993; 4:771–780.
35. Yei S, Mittereder N, Wert S, Whitsett JA, Wilmott RW, Trapnell BC. In vivo evaluation of the safety of adenovirus-mediated transfer of the human cystic fibrosis transmembrane conductance regulator cDNA to the lung. Human Gene Ther 1994; 5:731–744.
36. McCoy RD, Davidson BL, Roessler BJ, et al. Pulmonary inflammation induced by incomplete or inactivated adenoviral particles. Human Gene Ther 1995; 6:1553–1560.
37. Miller N, Vile R. Targeted vectors for gene therapy. FASEB J 1995; 9:190–199.
38. Morse JO. Alpha1-antitrypsin deficiency. N Engl J Med 1978; 299:1099–2005.
39. Crystal RG, Brantly ML, Hubbard RC, Curiel DT, States MD, Holmes MD. The alpha1-antitrypsin gene and its mutations. Chest 1989; 95:196–208.
40. Kozarsky K, Grossman N, Wilson JM. Adenovirus-mediated correction of the genetic defect in hepatocytes from patients with familial hypercholesterolemia. Somat Cell Mol Genet 1993; 19:449–458.
41. Riordan JR, Rommens JM, Kerem B-S, et al. Identification of the cystic fibrosis gene: cloning and characterization of complementary DNA. Science 1989; 245:1059–1065.
42. Rommens JM, Iannuzzi MC, Kerem BS, et al. Identification of the cystic fibrosis gene: chromosome walking and jumping. Science 1989; 245:1059–1065.
43. Collins FS. Cystic fibrosis: molecular biology and therapeutic implications. Science 1992; 256:774–779.
44. Welsh MJ, Smith AE. Molecular mechanisms of CFTR chloride channel dysfunction in cystic fibrosis. Cell 1993; 73:1251–1254.
45. Mulligan RC. The basic science of gene therapy. Science 1993; 260:926–932.
46. Rich DP, Anderson MP, Gregory RJ, et al. Expression of cystic fibrosis transmembrane conductance regulator corrects defective chloride channel regulation in cystic fibrosis airway epithelial cells. Nature 1990; 347:358–363.
47. Boucher RC. Airway epithelial fluid transport. Am Rev Respir Dis 1994; 150:271–281.
48. Cheng PW, Boat TF, Cranfill K, Yankaskas JR, Boucher RC. Increased sulfation of glycoconjugates by cultured nasal epithelial cells from patients with cystic fibrosis. J Clin Invest 1989; 84:68–72.
49. Barasch J, Kiss B, Prince A, Saiman L, Gruenert D, Al-Awqati Q. Acidification of intracellular organelles is defective in cystic fibrosis. Nature 1992; 352:70–73.
50. Imundo L, Barasch J, Prince A, Al-Awqati Q. Cystic fibrosis epithelial cells have a receptor for pathogenic bacteria on their apical surface. Proc Natl Acad Sci USA 1995; 92:3019–3023.
51. Bradbury NA, Jilling T, Berta G, Sorscher EJ, Bridges RJ, Kirk KL. Regulation of plasma membrane recycling by CFTR. Science 1992; 256:530–532.

52. Lloyd Mills C, Pereira MMC, Dormer RL, McPherson MA. An antibody against a CFTR derived synthetic peptide inhibits beta-adrenergic stimulation of mucin secretion. Biochem Biophys Res Commun 1992; 188:1146–1152.
53. Trapnell BC, Chu C-S, Paako PK, et al. Expression of the cystic fibrosis transmembrane conductance regulator gene in the respiratory tract of normal individuals and individuals with cystic fibrosis. Proc Natl Acad Sci USA 1991; 88:6565–6569.
54. Engelhardt JF, Zepeda M, Cohn JA, Yankaskas JR, Wilson JM. Expression of the cystic fibrosis gene in adult human lung. J Clin Invest 1992; 93:737–749.
55. Engelhardt JF, Yankaskas JR, Ernst SA, et al. Submucosal glands are the predominant site of CFTR expression in the human bronchus. Nature Genet 1992; 2:240–248.
56. Jacquot J, Puchelle E, Hinnrasky J, et al. Localization of the cystic fibrosis transmembrane conductance regulator in airway secretory glands. Eur Respir J 1993; 6:169–176.
57. Richardson PS, Alton EWFW. Cystic fibrosis transmembrane conductance regulator protein: what is its role in cystic fibrosis? Eur Respir J 1993; 6:160–162.
58. Drumm ML, Pope HA, Cliff WH, et al. Correction of the cystic fibrosis defect in vitro by retrovirus-mediated gene transfer. Cell 1990; 62:1227–1233.
59. Johnson LG, Olsen JC, Sarkadi B, Moore KL, Swanstrom R, Boucher RC. Efficiency of gene transfer for restoration of normal airway epithelial function in cystic fibrosis. Nature Genet 1992; 2:21–25.
60. Johnson LG, Boyles SE, Wilson J, Boucher RC. Normalization of raised sodium absorption and raised calcium-mediated chloride secretion by adenovirus-mediated expression of cystic fibrosis transmembrane conductance regulator in primary human cystic fibrosis airway epithelial cells. J Clin Invest 1995; 95:1377–1382.
61. Engelhardt JF, Simon RH, Yang Y, et al. Adenovirus-mediated transfer of the CFTR gene to lung of nonhuman primates: biological efficacy study. Human Gene Ther 1993; 4:759–769.
62. Grubb BR, Pickles RJ, Ye H, Yankaskas JR, Vick RN, Englehardt JF. Inefficient gene transfer by adenovirus vector to cystic fibrosis airway epithelia of mice and humans. Nature 1994; 371:802–806.
63. Wickham TJ, Mathias P, Cheresh DA, Nemerow GR. Integrins αvß3 and αvß5 promote adenovirus internalization but not virus attachment. Cell 1993; 73:309–319.
64. Goldman MJ, Wilson JM. Expression of αvß5 integrin is necessary for efficient adenovirus-mediated gene transfer in human airway. J Virol 1995; 69:5951–5958.
65. Gomez-Foix AM, Coats WS, Baque S, Alam T, Gerard RD, Newgard CB. Adenovirus-mediated transfer of the muscle glycogen phosphorylase gene into hepatocytes confers altered regulation of glycogen metabolism. J Biol Chem 1992; 267:25129–25134.
66. Zabner J, Peterson DM, Puga AP, et al. Safety and efficacy of repetitive adenovirus-mediated transfer of CFTR cDNA to airway epithelia of primates and cotton rats. Nature Genet 1994; 6:75–83.
67. Dranoff G, Jaffee E, Lazenby P, et al. Vaccination with irradiated tumor cells engineered to secrete murine granulocyte-macrophage colony-stimulating factor

stimulates potent, specific, and long-lasting anti-tumor immunity. Proc Natl Acad Sci USA 1993; 90:3539–3543.

68. Conry RM, LoBuglio AF, Kantor J, et al. Immune response to a carcinoembryonic antigen polynucleotide vaccine. Cancer Res 1994; 54:1164–1168.

69. Zhang W-W, Fang X, Mazur W, French BA, Georges RN, Roth JA. High-efficiency gene transfer and high-level expression of wild-type p53 in human lung cancer cells mediated by recombinant adenovirus. Cancer Gene Ther 1994; 1:5–13.

70. Deshane J, Loechel F, Conry RM, Siegal GP, King CR, Curiel DT. Intracellular single-chain antibody directed against erbB2 down-regulates cell surface erbB2 and exhibits a selective anti-proliferative effect in erbB2 overexpressing cancer cell lines. Gene Ther 1994; 1:332–337.

71. Culver KW, Ram Z, Wallbridge S, Ishii H, Oldfield EH, Blaese RM. In vivo gene transfer with retroviral vector-producer cells for treatment of experimental brain tumors. Science 1992; 256:1550–1552.

72. Ginsberg RJ, Kris MG, Armstrong JG. Cancer of the Lung. In: DeVita VT, Hellman S, Rosenberg SA, eds. Cancer of the Lung. Philadelphia: Lippincott, 1993: 673–723.

73. Fontanini G, Bigini D, Vignati S, et al. p53 expression in non small cell lung cancer: clinical and biological correlations. Anticancer Res 1993; 13:737–742.

74. Levin WJ, Casey G, Ramos JC, Arboleda MJ, Reissman PT, Slamon DJ. Tumor suppressor and immediate early transcription factor genes in non-small cell lung cancer. Chest 1994; 106:372S–376S.

75. Chen P-L, Chen Y, Bookstein R, Lee W-H. Genetic mechanisms of tumor suppression by the human p53 gene. Science 1990; 250:1576–1580.

76. Baker SJ, Markowitz S, Fearon ER, Willson JKV, Vogelstein B. Suppression of human colorectal carcinoma cell growth by wild-type p53. Science 1990; 249:912–915.

77. Unger T, Nau MM, Segal S, Minna JD. p53: a transdominant regulator of transcription whose function is ablated by mutations occurring in human cancer. EMBO J 1992; 11:1383–1390.

78. Takahashi T, Carbone D, Takahashi T, et al. Wild-type but not mutant p53 suppresses the growth of human lung cancer cells bearing multiple genetic lesions. Cancer Res 1992; 52:2340–2343.

79. Fujiwara T, Cai D, Georges RN, Mukhopadhyay T, Grimm EA, Roth JA. Therapeutic effect of a retroviral wild-type p53 expression vector in an orthotopic lung cancer model. JNCI 1994; 86:1458–1462.

80. Fujiwara T, Grimm EA, Mukhopadhyay T, Zhang W-W, Owen-Schaub LB, Roth JA. Induction of chemosensitivity in human lung cancer cells in vivo by adenovirus-mediated transfer of the wild-type p53 gene. Cancer Res 1994; 54:2287–2291.

81. Jin X, Nguyen D, Zhang W-W, Kyritsis AP, Roth JA. Cell cycle arrest and inhibition of tumor cell proliferation by the p16^{INK4} gene mediated by an adenovirus vector. Cancer Res 1995; 55:3250–3253.

82. Smythe WR, Kaiser LR, Hwang HC, et al. Successful adenovirus-mediated gene transfer in an in vivo model of human malignant mesothelioma. Ann Thorac Surg 1994; 57:1395–1401.

83. Smythe WR, Hwang HC, Amin KM, et al. Use of recombinant adenovirus to transfer the herpes simplex virus thymidine kinase (HSVtk) gene to thoracic neoplasms: An effective in vitro drug sensitization system. Cancer Res 1994; 54:2055–2059.

84. Hwang HC, Smythe WR, Elshami AA, et al. Gene therapy using adenovirus carrying the herpes simplex–thymidine kinase gene to treat in vivo models of human malignant mesothelioma and lung cancer. Am J Respir Cell Mol Biol 1995; 13:7–16.

85. Goldsmith KT, Curiel DT, Engler JA, Garver RI Jr. Trans complementation of an E1A-deleted adenovirus with codelivered E1A sequences to make recombinant adenoviral producer cells. Human Gene Ther 1994; 5:1341–1348.

86. Dion LD, Goldsmith KT, Garver RI Jr. Quantitative and in vivo activity of adenoviral-producing cells made by cotransduction of a replication-defective adenovirus and a replication-enabling plasmid. Cancer Gene Ther 1996; 3:230–237.

87. Lochmuller H, Jani A, Huard J, et al. Emergence of early region 1–containing replication-competent adenovirus in stocks of replication-defective adenovirus recombinants (ΔE1 + ΔE3) during multiple passages in 293 cells. Human Gene Ther 1994; 5:1485–1491.

88. Dion LD, Goldsmith KT, Garver RI Jr. Novel strategies for self-limited replication-enablement of E1-defective adenoviruses that minimize wild type virus production. 1996. Submitted.

89. Yang Y, Nunes FA, Berencsi K, Furth EE, Gonczol E, Wilson JM. Cellular immunity to viral antigens limits E1-deleted adenoviruses for gene therapy. Proc Natl Acad Sci USA 1994; 91:4407–4411.

90. Engelhardt JF, Ye X, Doranz B, Wilson JM. Ablation of E2A in recombinant adenoviruses improves transgene persistence and decreases inflammatory response in mouse liver. Proc Natl Acad Sci USA 1994; 91:6196–6200.

91. Goldman MJ, Litzky LA, Engelhardt JF, Wilson JM. Transfer of the CFTR gene to the lung of nonhuman priomates with E1-deleted, E2a-defective recombinant adenoviruses: a preclinical study. Human Gene Ther 1995; 6:839–851.

92. Yang Y, Nunes FA, Berensci K, Gonczol E, Engelhardt JE, Wilson JM. Inactivation of E2a in recombinant adenoviruses improves the prospect for gene therapy for cystic fibrosis. Nature Genet 1994; 7:362–369.

93. Vilquin J-T, Guerette B, Kinoshita I, et al. FK506 immunosuppression to control the immune reactions triggered by first-generation adenovirus-mediated gene transfer. Human Gene Ther 1995; 6:1391–1401.

94. Lochmuller H, Petrof BJ, Allen C, Prescott S, Massie B, Karpati G. Immunosuppression by FK506 markedly prolongs expression of adenovirus-delivered transgene in skeletal muscles of adult dystrophic [mdx] mice. Biochem Biophys Res Commun 1995; 213:569–574.

95. Kolls JK, Lei D, Odom G, Shellito J. Pre-treatment with a depleting monoclonal anti-CD4 antibody prolongs in vivo transgene expression of a recombinant adenoviral vector. Am J Respir Crit Care Med 1995; 151(part 2):A670.

96. Armentano D, Sookdeo CC, Hehir KM, et al. Characterization of an adenovirus gene transfer vector containing an E4 deletion. Human Gene Ther 1995; 6:1343–1353.

97. Krougliak V, Graham FL. Development of cell lines capable of complementing E1, E4, and protein IX defective adenovirus type 5 mutants. Human Gene Ther 1995; 6:1575–1586.
98. Yeh P, Dedieu J-F, Orsini C, Vigne E, Denefle P, Perricaudet M. Efficient dual transcomplementation of adenovirus E1 and E4 regions from a 293-derived cell line expressing a minimal E4 functional unit. J Virol 1996; 70:559–565.
99. Schaack J, Guo X, Ho WY-W, Karlok M, Chen C, Ornelles D. Adenovirus type 5 precursor terminal protein-expressing 293 and HeLa cell lines. J Virol 1995; 69:4079–4085.
100. Mitani K, Graham FL, Caskey CT, Kochanek S. Rescue, propagation, and partial purification of a helper virus-dependent adenovirus vector. Proc Natl Acad Sci USA 1995; 92:3854–3858.

4

Adeno-Associated Virus-Based Delivery Systems

PAMELA L. ZEITLIN

Johns Hopkins University School of Medicine
Baltimore, Maryland

I. Introduction

The adeno-associated virus (AAV) has been the subject of intense study recently by investigators working with viral-based gene delivery systems. AAV is a member of the *Dependovirus* genus in the Parvoviridae family. Parvoviruses are small, single-stranded DNA viruses with a host range in both the vertebrates and invertebrates (1). AAV is often called a defective or dependent virus because it requires a helper virus to generate a productive infection. From 49% to 80% of human adults are seropositive for AAV type 2 (2–5). It has been speculated that AAV evolved from cellular "junk" DNA into a protective interfering transposon under evolutionary pressures which favored its transposonlike qualities (6). Transposons would interfere with invading pathogenic viruses (such as the parvovirus B19) by competing with the pathogen DNA for replicative enzymes. Thus the common ancestry for autonomous and defective parvoviruses may have been in cellular DNA which then developed under different evolutionary pressures to establish two genera. The features that make AAV an attractive potential gene therapy vector include nonpathogenicity, site-specific integration in the absence of a helper virus infection, and the ability to remove all of the viral genes without loss of infectivity.

II. AAV Structure

The AAV virion is a nonenveloped icosahedral structure with a diameter of 20 to 26 nm and a small DNA genome of approximatly 4.7 kb (7,8). The virion has a density of 1.41 g/cm³ (9), which is conveniently higher than the density of adenovirus (1.35 g/cm³), allowing separation of the two by CsCl density centrifugation. AAV virion particles are also heat resistant (56°C for 1 hour) and stable at low pH and relatively resistant to detergent and proteases (10).

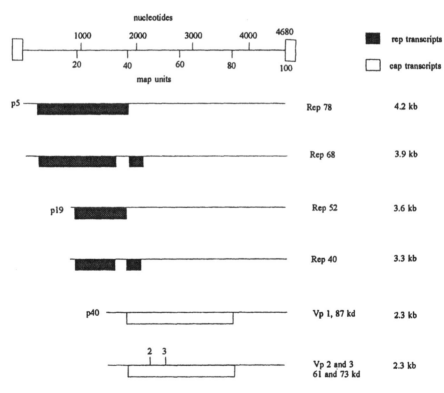

Figure 1 AAV2 genetic map. The AAV2 genome is shown along two scales, by nucleotide and by map units. The open boxes represent the inverted terminal repeats (*itr*). A promoter at map unit 5 (p5 promoter) initiates the Rep78 mRNA transcript, and the alternatively spliced Rep68 transcript. A promoter at map unit 19 (p19 promoter) initiates the Rep52 transcript and the alternatively spliced Rep40 transcript. The promoter at map unit 40 (p40 promoter) initiates all three cap transcripts. VP2 is produced by the same alternative splicing mechanism as the Rep68 and Rep40 transcripts. VP2 utilizes an alternative ACG start codon.

Both polarities of the single-stranded DNA are encapsidated with equal frequency and both are equally infectious (11–14).

The most common and extensively characterized human AAV is AAV2. The complete sequence is 4680 bases (15,16), and a diagram is found in Figure 1. There is an inverted terminal repeat (*itr*) of 145 bases at each end (15,17,18). The terminal 125 bases are palindromic (19). Within these areas are two shorter palindromes such that when the terminal symmetrical sequence is folded onto itself to maximize base-pairing, a T-shaped structure is formed (Fig. 2) (17,20,21). The terminal sequences can exist in one of two configurations called

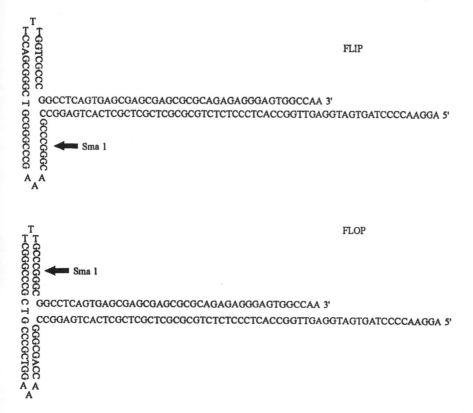

Figure 2 AAV *itr*. The first 125 nucleotides of the 145 base *itr* are palindromic and contain two smaller 21 nucleotide palindromes immediately flanking the axis of symmetry of the larger palindrome. The overall palindrome folds onto itself to maximize base pairing and produces a T-shaped hairpin configuration. There are two isomeric forms commonly referred to as flip and flop with respect to the restriction enzyme Sma1, which cuts only one of the two smaller internal palindromes.

flip or flop refering to the palindrome (note the Sma 1 site in Fig. 2 is closer to the end) and the two occur with equal probability (21).

Between the two *itr* are two nonoverlapping open reading frames (*orf*). As can be seen in Figure 1, the first *orf* codes for regulatory proteins (22–24), and the second *orf* codes for structural proteins. The regulatory *orf* is called the rep gene(s), and the structural *orf* is labeled the cap gene. There are two promoters at map positions 5 and 19, an intron near the 3′ end of the rep genes, and a polyadenylation site at map position 96. This organization produces four different rep genes by splicing mechanisms (reps 78, 68, 52, and 40) (25–27). The structural *orf* uses a promoter, p40, found at map position 40, and produces VP1-3 by splicing in the same frame (16,18,28,29). This basic organization is the same for the autonomous virus genomes that are related to AAV parvoviruses. There is some homology between human parvovirus B19 and AAV in the rep regions, but there is no detectable hybridization between parvoviruses and AAV (30).

The Rep proteins were studied by making mutations in the left-hand open reading frame (22–24,31–38) that resulted in aberrations in viral DNA replication, transactivation of AAV promoters, repression of heterologous gene expression, and AAV coded gene expression. Rep may also be the protein that is responsible for the anti-oncogenic effects associated with AAV infections in vitro (39–45) and in vivo (46–48). It is not clear why the spliced and unspliced versions of the rep proteins exist or what the smaller reps are responsible for (49,50). Rep78 has been found covalently attached to viral DNA on the outside of a preformed virion (51). This is presumably an intermediate since the final infectious particles contain only capsid proteins.

There are three structural coat proteins (VP1, VP2, and VP3) that overlap in their coding sequence and differ only by extension at the amino terminus (16,29,52–54). The smallest one, VP3, is most abundant (90%), but the largest one, VP1, is thought to interact with the DNA via the aminoterminal sequences (55). VP2 and 3 are synthesized from a single spliced transcript using two different initiation sites (56). Sequestration of single stranded genomes from the pool of replicating molecules cannot occur in the absence of either VP2 or 3 (56). In a conceptually different experiment, investigators expressed each of the three capsid proteins alone or in combination by recombinant baculoviruses in Sf9 cells (57). The end point was the production of empty capsids during infection of HeLa cells with AAV and adenovirus and supported a requirement for VP2 in the formation of empty capsids.

The *itr* sequences are required in cis orientation for AAV DNA replication and for rescue or excision from prokaryotic plasmids (36,58,59). They are the minimum sequences required for AAV proviral integration and for packaging into virions (15). Recent evidence is accumulating that the *itr* are sufficient for promoter activity in recombinant AAV virions (60).

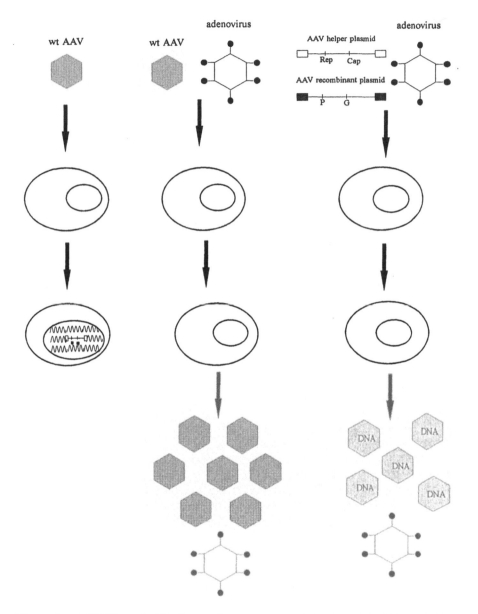

Figure 3 AAV life cycle. The first pathway represents infection with wildtype AAV alone. AAV integrates preferably at the AAVS1 site on human chromosome 19. The second pathway depicts coinfection with wild-type AAV and adenovirus. AAV uses the helper virus to produce a productive infection for AAV and inhibition of adenovirus production. The third pathway is used in the laboratory to construct recombinant AAV vectors. Wildtype adenovirus, a plasmid with rep and cap and the AAV recombinant plasmid are introduced in cell culture.

III. AAV Replication In Vivo

AAV will undergo a productive infection in the presence of a helper virus, either adenovirus or herpes simplex virus (61–63). Helper functions have been more thoroughly studied for adenovirus (64–68). The adenovirus early functions coded for by E1A, E1B, E2A, and E4 all contribute to this helper function in vitro (67–72). It is possible, however, that certain conditions applied to cells can activate a latent infection without helper virus—e.g., UV irradiation or chemical carcinogens (18,73,74).

There are few reports concerning the nature of the cell surface interaction with AAV (75). Thus the receptor if it exists is unknown. Once the virion penetrates to the nucleus, the genome is uncoated and one of two pathways is followed (see Fig. 3). If a helper virus is present, full gene expression occurs. If the cell is nonpermissive, gene expression is repressed, and the viral DNA integrates in the host genome. Typically, a cell that is latent with respect to AAV produces little to no AAV gene mRNA or protein (55,76–78). Upon coinfection with adenovirus, the adenovirus E1A gene product is produced and induces transcription from the AAV p5 and p19 promoters. A small amount of Rep products is produced (23,79). The p5 Rep products induce the synthesis of mRNA from all three AAV promoters to yield higher levels of all AAV products (23,26). Adenovirus production is drastically reduced under these conditions (18,55,80).

The linear single strand genome is replicated by a single strand displacement mechanism diagrammed in Figure 4 (14,81–83). The palindromic *itr* functions as a DNA primer. A duplex intermediate covalently crossed at one end is formed. A nick occurs on the original strand and the 3′ OH at the nick serves as the primer and the displaced hairpin as the template (84,85). Duplex dimers are formed if the original hairpin crosslink is not resolved (82). Capsids are probably formed by self-assembly and likely interact with replicative intermediates to sequester single strands for the final virion (51,57,86–88). The d sequence within the *itr*, a 20-nucleotide stretch exclusive of the hairpin structure, is crucial for high-efficiency rescue and replication of the AAV genome (89).

Figure 4 AAV replication. The working model for AAV replication begins with a single strand form which is always found in nature folded into a partial duplex in the region of the *itr*s. In the partial duplex hairpin conformation, the left hairpin can provide a 3′OH group at the arrowhead for self-primed DNA replication. A duplex replicative intermediate is formed which is covalently crosslinked at one end. The hairpin is nicked at the terminal resolution site and the 3′ end of the parental strand is restored by elongation. A new round can begin with strand separation at the hairpins.

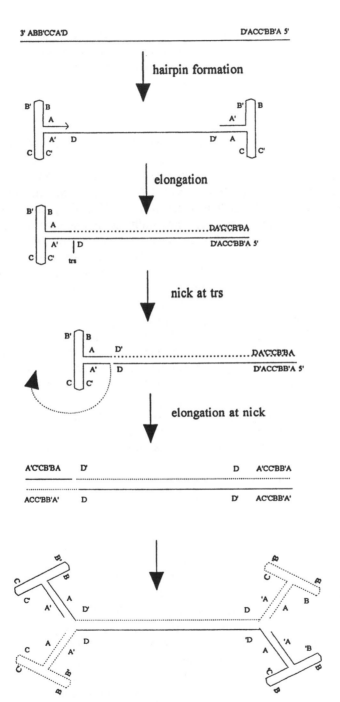

The resolution step can be initiated by either Rep 68 or 78 (31,90–94). Rep 68 can bind to the terminal hairpin DNA, cut the hairpin at the *trs* site, covalently attach to the 5' end of the nick during terminal resolution, and perform DNA helicase activity (90,91).

A. Integration

Evidence of latent AAV infection was first announced by Hoggan et al. (95), who noted that 20% of primary African green monkey kidney cell lots and 1% to 2% of primary human embryonic kidney lines produced AAV when infected with adenovirus. Recently, Grossman et al. (5) used PCR and detected AAV2 genomes in peripheral blood leukocytes from 2 of 55 healthy blood donors and in 2 of 16 hemophilia patients. Experimentally, latency in vitro has been easily established at multiplicities of infection 100 or greater and DNA restriction analysis followed by hybridization with AAV probes demonstrated recombination with the host DNA (96,97). The ability to rescue wild-type AAV from latently infected cell clones is apparently the consequence of tandem copies of the AAV genome (98) since rearrangements of both the host flanking sequences and the AAV *itr* were present (99–101). Most (70% or greater) of the time, AAV integrates at the AAVS1 site on chromosome 19q13.3-qter (99,101–105). This had led gene therapists to speculate that recombinant vectors would behave similarly (see below). Any vector DNA, whatever the source, has the capability of recombining in the host genome, usually randomly and related to activity of gene transcription. The AAVS1 site flanking sequences share homology only for one to five bases, and considerable rearrangements of both viral and flanking cellular DNA are observed (106). At least one intact copy of the AAV genome among the tandem inserts is required for successful rescue.

The preintegration site has been cloned and inserted into an Epstein-Barr virus–based shuttle vector (105) for delivery to the cell. AAV infection led to integration in this episomal shuttle vector when it contained certain critical elements mapping to a 510-bp sequence upstream from the actual integration site. Three sequences present both in the preintegration site and in the AAV have been identified—a $(GCTC)_3$ that is a binding site for Rep 68 and 78; a terminal resolution site that is nicked by rep; and a sequence homologous with a yeast recombination enhancer called M26. Others also have data suggesting that a 4-nucleotide repeat in the host genome is associated with integration in the presence of reps 78 and 68 (105,107).

No pathogenic consequences have been implicated for latent AAV infection. However, a number of observations implicating effects from Rep (as opposed to insertional mutagenesis) are accumulating. CHO cells exhibit increased sensitivity to UV irradiation (108), and this could be attributable to effects of Rep on gene expression. A variety of cell types with latent AAV develop decreased plating efficiency, reduced cell growth, and altered expression of differentiation-associated antigens or oncogenes (39,47,109–113). On the other hand, there is also evidence that latency is associated with a reduction in oncogenic potential (114,115). AAV infection prevents cellular transformation after transfection with oncogenes or infection with DNA tumor viruses (39,40,42,43,45,48,116–119). Coinfection of newborn animals with AAV and adenovirus blocks subsequent tumor formation (116,118).

B. Rescue

One *itr* cis to the rep gene is required for rescue and replication, and no helper viral genes are absolutely required. Instead, the evolving concept of helper function is that the additional virus alters the cell milieu facilitating reactivation. It is not clear whether rescue occurs first, followed by replication, or replication can begin while still in the integrated state.

Rescue has been most extensively studied using a model system in which a linear duplex form of an AAV genome has been inserted into a bacterial plasmid pBR322 and the vector has been transfected into cells infected with adenovirus (120). The model for rescue is presented in Figure 5. The Rep gene was required for rescue and DNA replication and one intact *itr* in cis was also necessary. Confirmation of the prediction that a parental genome with a single flop orientation at both ends would yield progeny with both flip and flop at either end was obtained and supports the inversion at both termini that occur during rescue and replication. Figure 5 presents the working model for the rescue process. Presumably in the plasmid or in the cellular genome, the *itr*s are capable of forming the cruciform structure allowing an intermediate that could be cleaved and resealed to form the excised AAV product. In the in vitro plasmid experimental modeling of rescue, a HeLa cell extract and SV40 T antigen supplied the activity that cut across this structure at the base and resealed the break leading to covalent linkage of previously unlinked sequences (see Fig. 5).

A key issue for gene therapists has been whether subsequent infection in vivo with adenovirus and AAV (both community acquired) would lead to rescue, replication, and loss of transgene. One group has directly addressed this issue in a primate model (see below).

IV. Recombinant AAV Vectors

Figure 6 contains the simplified scheme for production of an AAV vector. The basic requirements are adenovirus, the rep and cap genes, a promoter and polyA signal for a transgene, and the *itr*s. If in the production of a recombinant

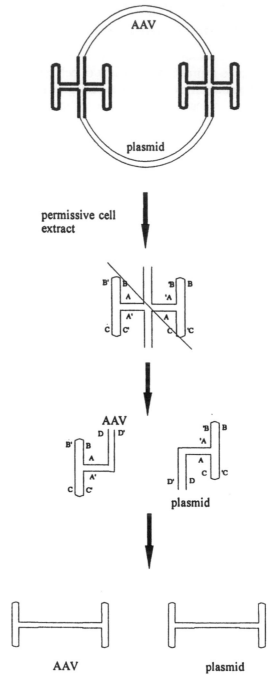

Figure 5 In vitro model for AAV rescue. AAV has been inserted in a plasmid such as pBR322. Factors present in a permissive cell extract (HeLa) catalyze cleavage into the AAV plasmid and pBR322 during a DNA replication assay. The cruciform formation at the *itrs* would be predicted to occur because of negative supercoiling.

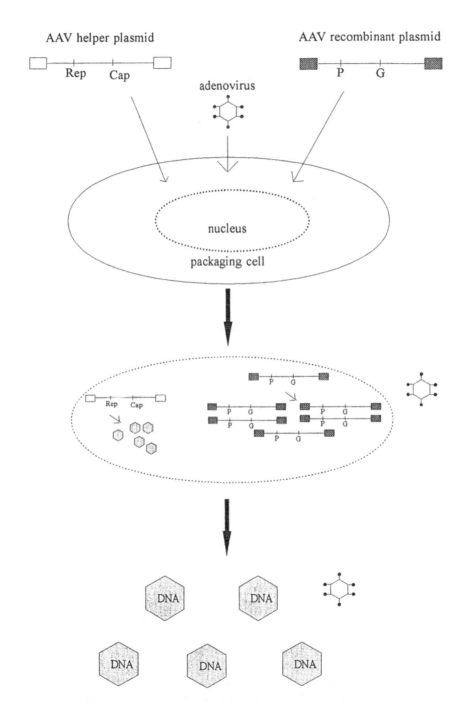

Figure 6 Production of a recombinant AAV vector. A packaging cell line such as 293 is cotransfected with an AAV helper plasmid and an AAV recombinant plasmid, and infected with adenovirus. The recombinant AAV vector is packaged into AAV capsids. Wild-type adenovirus is removed by later CsCl banding and heat inactivation.

AAV stock, a wild-type AAV plasmid (or virus) is used to supply the rep and cap gene products in trans, unacceptably high levels of wild-type AAV virus (10:1 wild-type to recombinant) are produced (121,122). This led to the development of AAV plasmids modified in some way to interfere with wild-type AAV production. Hermonat and Muzycka (123) inserted a 2.5-kb fragment of lambda bacteriophage DNA into the wild-type plasmid, causing it to be too large to package. Others have removed the *itrs* from the wild-type AAV plasmid (121,122). But both approaches still left significant wild-type contamination thought to appear because of recombination between vector and AAV plasmids and deletion of inserted sequences. Samulski (124) constructed the pAAV/Ad complementing plasmid devoid of homology between vector and AAV sequences. It consisted of the AAV coding sequences flanked by the adenovirus 5 terminal repeats. Chejanovsky and Carter (125) isolated an amber mutant pNTC3 in the AAV rep gene which when grown on a monkey cell line did reasonably well with respect to titers.

One of the limiting factors in the consideration of the utility of AAV for gene therapy has been the small upper limit of DNA insert size that is acceptable for packaging into an infectious virion. Where this has been examined, the maximum length is about 4.7 kb. If the transducing gene is close to this size, there is no room for promoter DNA. A CFTR vector was constructed that did not include either an AAV promoter or an exogenous promoter, and detectable expression in vitro was apparently driven by promoter activity in the *itr* (60).

One additional packaging limitation is the absence of stable packaging cell lines. Most laboratories had constructed vectors by using adenovirus-infected cell lines cotransfected with two plasmids—one expressing rep and cap without the *itrs*, and one consisting of the promoter, gene cDNA, and flanking *itrs*. Packaging in this system is limited by the efficiency of the transfection of each of the plasmid constructs, the lack of stable cell lines expressing rep, and the level of expression of rep from the packaging plasmids. Suboptimal rep expression levels are probably explained by Rep 78 and Rep 68 auto downregulation from the p5 promoter in the rep/cap plasmid (pAAV/Ad) (126). Rep itself seems to inhibit neo-resistant colony formation whenever neo selective markers are employed to clone out a stable rep cell line. A Rep packaging plasmid has recently been described that expresses the AAV Rep 68 and 78 proteins from the human immunodeficiency virus type 1 (HIV 1) long terminal repeat (LTR) promoter (127). Clark et al. (128) have been able to establish stable neo-resistant HeLa and 293 cell lines containing AAV Rep and Cap and the recombinant AAV. The packaged virions are produced when the cell line is infected with adenovirus. Yang et al. (129) have created cell lines that express Rep under the control of an inducible mouse metallothionein promoter. Cells recover from the cytostatic effects of Rep when the inducing agent is removed. Another group has doubled their recombinant AAV titer by using adenovirus-polylysine-DNA complexes together with vector DNA and complementation plasmid (130). Clearly, advances in the technology of recombinant AAV virion production would promote the application of AAV vectors to human diseases by increasing yields for adequate dosing.

A. Recombinant AAV Vectors In Vitro

AAV vectors have been reported to have high transduction frequencies in cells of diverse lineages (131–136). Latent wild-type AAV infections have been relatively stable for over 100 serial passages in the absence of selection. Yet, while this would imply stable integration, there is evidence that latency is a dynamic process between integration and episomal species (96). Podsakoff et al. (137) directly measured β-galactosidase expression driven by the RSV long-terminal repeat promoter in dividing and nondividing 293 cells or human diploid fibroblast cell lines. Using chemical or growth-arrest methodologies, they were able to reduce ^{3}H-dNTP incorporation by 88% to 97%. When gene expression was measured histochemically 48 hours after infection at a transducing MOI of 1, the percentage of positive cells (as high as 40% in some examples) from the growth-arrested cultures was at least as high as from the proliferating cells. Also, 293 cells were much more permissive than human diploid fibroblast cultures, and if the mitotic block was lifted from 293 cells, β-galactosidase expression was still detected 9 months later. Regardless of the cell proliferation status at the time of transduction, the ultimate intracellular fate appeared to include integrated copies as assayed by restriction enzyme digestion of genomic DNA and hybridization for transgene sequences. Whether or not the integrated copies are a result of at least one round of cell division at some point later on after lift of the mitotic block is still a subject of debate in the field.

Flotte et al. (135) performed similar studies with two AAV vectors in the cystic fibrosis bronchial epithelial cell line IB3-1. Cultures in which <0.1% of cells were dividing could still be transduced with AAVp5-lacZ or AAVp5-neo up to a maximum of 91% expression at a multiplicity of infection of 3000 vector particles per cell. The fate of vector DNA from the non-dividing cells, as measured by Southern blotting of both low-molecular-weight episomal DNA and high-molecular-weight genomic DNA showed only the nonintegrated forms. A higher efficiency integration pathway may be operative in dividing cells, whereas the less efficient nonintegrating pathway may supervene in growth-arrested cells.

When Russell et al. (138) infected stationary and dividing primary human fibroblast cultures with AAV vectors encoding alkaline phosphatase and neomycin resistance genes, they found that the transduction frequency of S-phase cells is about 200 times greater that non-S-phase cells. Neo^r transduction was equivalent in dividing and nondividing cells, whereas alkaline phosphatase expression was over 20-fold higher in dividing cultures. The authors argued that G418 resistance requires that cells go through division after transduction, wherease the alkaline phosphatase expression does not. Southern blotting of genomic and Hirt extract DNA supported the hypothesis that AAV genomes remain as single-stranded episomes long enough in nondividing cells to then be recruited and expressed once the cell is allowed to enter S-phase. They were able to confirm subsequent random integration in G418-resistant colonies transduced with the alkaline phosphatase vector at 700 particles per cell.

It is not known whether integration is required for gene expression. Both integrated vector sequences and double stranded episomal forms have been detected in the same clones (122,124,139,140). Whether the inefficiency of transduction of nondividing cells reflects these differences is not known. Alexander et al (141) recently demonstrated that transduction efficiency in stationary cultures could be increased up to 750-fold by DNA damaging agents. The mechanism is proposed to involve the repair of DNA during which whatever constitutes the naturally permissive environment such as DNA polymerases or factors modulating cell cycle progression are enhanced or facilitated so that AAV vector transduction and possibly integration is improved.

The potential for integration of recombinant vectors may be underestimated since, in some cells in vitro, integration appears to be a frequent occurrence. Wei et al. (142) used AAV carrying an SV40 promoter and either the human glucocerebrosidase or alrylsulfatase A cDNAs and the Neo^r gene driven by the TK promoter to infect murine and patient fibroblasts. Southern blotting of the G418-selected clones showed that the vector integrated one or two copies per cell genome. Shelling et al. (143) showed site-specific integration of the Neo^r gene in 75% of AAVNeo transfected cell lines and 82% of AAVNeo-infected cell lines. However, Kearns et al. (144,145) noted random integration in immortalized bronchial epithelial cells as detected by FISH.

A number of laboratories have been disappointed with the transduction efficiencies obtained using recombinant AAV vectors (146). Certainly, when compared with adenovirus, AAV is less efficient (147). Some have combined adenovirus proteins in a cationic liposome complex to enhance transduction efficiency (148).

B. Recombinant AAV Vectors In Vivo

Hemoglobinopathies

The human hemoglobinopathies and thalassemias are among the most prevalent single-gene disorders worldwide. The most serious forms of these diseases require frequent erythrocyte transfusions which lead to life-threatening iron overload, antierythrocyte antibodies, and multiple systemic complications.

Pharmacologic manipulation of fetal hemoglobin levels in both the hemoglobinopathies and thalassemias appears to decrease hemolysis and lead to more effective erythropoiesis. Gene transfer of the β- and γ-globin genes to the stem cell population for subsequent specific expression in the erythroid specific lineage would theoretically be highly effective. In the homozygous condition, β-thalassemia is characterized by excessive free α-globin chains which results in premature death of erythroid progenitor cells before they reach the reticulocyte stage. In this situation an antisense strategy is needed to suppress α-globin mRNA transcription.

Hemoglobin is a tetrameric protein formed from two dimeric polypeptides. The α globin gene cluster is located on chromosome 16 within which are located α_1, α_2, and the embryonic ζ. The β-globin family localizes to chromosome 11 and includes the adult δ and β, and the fetal γ^A and γ^B. During normal erythropoiesis, six different hemoglobin proteins are expressed. The process of coordinated expression is the process of hemoglobin switching.

Successful gene transfer with integration and regulated gene expression in hematopoietic lines has been reported. Samulski and co-workers (149) used a recombinant AAV vector containing a marked human γ globin gene and DNAse I hypersensitivity sites from the locus control region (LCR) of the β-globin gene cluster to transduce K562 erythroleukemia cells. High-level, hemin-inducible expression of the marked γ-globin gene was detectable in clonal populations. This same gene was regulatable in BFU-E grown from CD34+ human cells (150). About 20% to 40% of BFU-E carried levels at 4% to 71% of endogenous γ-globin mRNA. When pooled BFU-E from the peripheral blood of a patient with sickle cell anemia was used, the hemoglobin F as measured with HPLC increased from 26% to 40%. AAV has been used to transfer the Neo^R gene to T and B lymphocyte clones (151). Transduction with antisense for the HIV LTR in an HIV-infected cell line inhibited HIV production and replication (152). Suppression of human α globin gene mRNA by 91% in K562 which express high levels in culture was achieved using the α globin promoter driving the antisense sequence (153).

One group has tested an AAV vector carrying parvovirus B19 coding sequences for the ability to infect and suppress erythroid hematopoietic colony formation. This parvovirus replicates only in erythroid progenitor cells in human bone marrow. The recombinant AAV-B19 hybrids expressed B19 genes and inhibited colony formation (154). AAV vectors have been demonstrated in several laboratories to transfect hematopoietic stem cells(122,133,136,155–158).

Although recombinant AAV vector integration in hematopoietic cell clones has been documented by PCR (133,158), integration is rarely site-specific in chromosome 19. Experiments were performed with human or rhesus CD34+ cells infected with an AAV vector carrying the lacZ gene with an efficiency of 60% to 70%. Analysis of 27 individual colonies by PCR techniques found only four with integrated copies in chromosome 19 (158).

V. AAV Vector Expression in Brain

Postmitotic cells of the central nervous system are most amenable to transduction with the DNA viruses, since retroviral vectors require dividing cells. Recombinant herpes simplex virus type I and adenovirus have been studied in mammalian brain; however, both vectors retain many viral genes which can be inflammatory and cytotoxic (159–161). During and colleagues have recently reported the long-term (4-month) transduction of human tyrosine hydroxylase in the denervated striatum of unilateral 6-hydroxydopamine-lesioned rats (162). Both TH immunoreactivity and behavioral recovery were documented, raising hopes for humans with Parkinson's disease. The AAVth vector was created by subcloning the human CMV immediate/early promoter, the human TH cDNA (form II), and the SV40 poly A sequence between the *itrs* in psub201 (124). The virus was stereotactically injected at three sites in 2-μl volumes from a stock with a titer of 5×10^6 infectious units per milliliter. The efficiency of expression was estimated to be 10% by immunocytochemical techniques. It is unknown whether this expression was driven from integrated or episomal copies.

VI. AAV Vectors in Skeletal Muscle

Stable gene transduction in skeletal muscle would be desirable both as a treatment of primary muscle disorders and as an in vivo reservoir for soluble circulating proteins such as insulin in diabetes. One group has developed AAV gene transfer to cultured human myoblasts with the goal of later injection into skeletal muscle (163). AAV was the basis of the plasmid pCKM-gfp, a construct using the human CKM_{muscle} promoter and the green fluorescent protein cDNA as the reporter gene. The cell line was transfected using the gene gun (BioRad) and gold microspheres coated with plasmid DNA. This gene gun accomplished between 1% and 5% transfection efficiency after 4 days in culture. Injection of the naked plasmid DNA directly into murine quadriceps skeletal muscle resulted in much higher efficiency at 4 days, leading the authors to speculate that maximal activity may require differentiated or multinucleated cells. However, it has been observed with other vectors that naked plasmid DNA is very effective in skeletal and cardiac muscle in general (164–172).

VII. Cystic Fibrosis

Cystic fibrosis (CF) is among the most common of the lethal genetic disorders in Caucasians. The mutant gene is located on chromosome 7, and the gene product, the Cystic Fibrosis Transmembrane Conductance Regulator or CFTR, is a large glycoprotein cAMP-regulated chloride channel. The CFTR is expressed at low levels in epithelial and nonepithelial tissues, and the disease, which begins in childhood, is best characterized as a multisystem disorder of exocrine glands and secretory epithelia which become blocked by a viscous mucous. The pulmonary component of the disorder is the most severe, and chronic pulmonary infections with *Pseudomonas aeruginosa* tend to dominate the clinical profile as the patient ages. The median life expectancy for individuals with cystic fibrosis has reached 28.9 years in the United States (173). About one-third of patients in the US Cystic Fibrosis Foundation data registry are over 18 years, and this group has been the focus of in vivo gene transfer trials to the airways for adenoviral, liposomal, and AAV vectors (174).

The airway epithelium, while easily accessible to the gene therapist via aerosol or bronchoscopic lavage, is at a practical disadvantage with respect to its cell biology. Not only is the target epithelial layer composed of multiple cell types, but the submucosal glands, which are particularly rich in normal and mutant CFTR epithelial expression, are buried within the submucosa and theoretically less accessible. The airway epithelial cells in the mature adult human are extremely long-lived with proliferation rates as low as 1% to 5% and $T_{1/2}$ estimated on the order of months (175). The wild-type human CFTR cDNA sequence is roughly 4.5 kb, leaving little room for promoter sequence in the AAV genome. Fortunately the 145 bp *itr* has been demonstrated to provide transcription promoter activity probably related to a consensus initiator sequence (60). This AAVCFTR vector functionally complements CF cell lines in vitro in both immortalized and primary CF nasal polyp cells (134,176).

AAV vectors have theoretical advantages over adenoviral vectors in that adverse immunologic reactions are less likely given that all the AAV viral genes are deleted. However, latent AAV vector in the genome could theoretically be mobilized or rescued by coincident infection by wild-type AAV and adenovirus outbreaks in the community. Afione et al. (177) demonstrated this possibility to be real in an in vivo model where rhesus macaques were transduced in the nose or lung with AAVCFTR. When cells were sampled up to 3 months later, vector DNA was present in episomal form and was able to be rescued in subsequent cell culture by coinfection with wild type AAV2 and adenovirus. Furthermore, rescue was also possible in vivo by similar coinfection. Several interesting findings resulted from this study. First, a 9.4-kb AAVCFTR form consistent with a double-stranded dimer was present in the episomal DNA Hirt extract more than 3 months after primary transduction to the right lower lobe. Since vector DNA was not unequivocably recognizable in the genomic fraction, fluorescent in situ hybridization (FISH) was performed on primary epithelial cells from this animal. Signals were present on interphase nuclei but not metaphase chromosomes, again consistent with episomal rather than integrated vector. It is speculated that the absence of the Rep protein is critical in the selection of the episomal over integration pathway.

Since the adult human airway is lined by a very stable, low-proliferating epithelium, it is likely to require high particle multiplicities of infection on the order of 10^3/cell in order to achieve a therapeutic transduction efficiency for CF. Zeitlin et al. (178) examined AAV vector transduction in an in vivo model of proliferating airway by delivering AAVp5-lacZ to the newborn rabbit lung. Cell proliferation was marked by immunoreactivity in BRDU-labeled lung sections and β-galactosidase expression was colocalized by X-Gal staining. Regions of cell proliferation coincided with regions of lacZ expression in the distal lung. Tracheal epithelium, the site of injection of the recombinant virions, also expressed β-galactosidase in a pattern favoring direct contact with a high particle inoculum in a region that was inactive with respect to BRDU labeling. Differences in regional expression may have reflected differences in viral penetration or cellular uptake, differences in particle multiplicity of infection as the fluid spread distally, and/or differences in permissiveness for vector transduction.

VIII. AAV Vectors in Cancer

During the development of genetically modified tumor vaccines, one group has delivered AAV-Il-2 via liposomes to primary human prostatic tumor cells obtained from radical prostatectomy of cancer patients (179). High-level Il-2 expression was observed compared with retroviral methods sufficient for induction of antitumor immunity in a rat model. Further studies with Il-2 in primary breast, ovarian, and lung tumor cells resulted in 10% to 50% transfection efficiency in vitro (180). In both studies, lethal irradiation to prevent cell proliferation did not inhibit transfection efficiency.

In an attempt to reduce cell proliferation in cervical carcinoma cells in vitro and in vivo, Sun et al. (132) delivered sense and antisense folate receptor cDNA via AAV virions. Sense folate receptor cDNA resulted in an increase in folate receptors, smaller colonies in vitro, smaller tumor volumes in vivo in nude mice with dramatic prolongation of tumor doubling times, and a decrease in cell proliferation.

IX. Summary and Future Directions

AAV vectors have been demonstrated to transduce cells of widely different lineages for prolonged periods of time. Stable integration may not be restricted to the AAVS1 site in human chromosome 19; however long-term expression can also occur from episomal vector. Gene expression is less substantial when compared with adenoviral vectors, but this is compensated for by using higher particle multiplicities of infection. The risk of inflammation appears to be less with AAV vectors, and insertional mutagenesis has yet to be demonstrated in an animal model. The primary technical limitations for AAV in gene therapy are primarily gene size constraints, vector production, and vector purification issues which, when overcome, will greatly improve the potential for human gene therapy in inherited and acquired disease.

References

1. Siegl G, Bates RC, Berns KI, et al. Characteristics and taxonomy of Parvoviridae. Intervirology 1985; 23:61–73.
2. Blacklow NR, Hoggan MD, Sereno MS, et al. A seroepidemiologic study of adenovirus-associated virus infection in infants and children. Am J Epidemiol 1971; 94:359–366.
3. Blacklow NR, Hoggan MD, Kapikian AZ, Austin JB, Rowe WP. Epidemiology of adenovirus-associated virus infection in a nursery population. Am J Epidemiol 1968; 88:368–378.
4. Blacklow NR, Hoggan MD, Rowe WP. Serologic evidence for human infection with adenovirus-associated viruses. J Natl Cancer Inst 1968; 40:319–327.
5. Grossman Z, Mendelson E, Brok-Simoni F, et al. Detection of adeno-associated virus type 2 in human peripheral blood cells. J Gen Virol 1992; 73:961–966.
6. Fisher RE, Mayor HD. The evolution of defective and autonomous parvoviruses. J Theor Biol 1991; 149:429–439.
7. Hoggan M, Blacklow N, Rowe W. Studies of small DNA viruses found in various adenovirus preparations: physical, biological, and immunological characteristics. Proc Natl Acad Sci USA 1966; 55:1457–1471.
8. Hoggan MD, Blacklow NR, Rowe WP. Studies of small DNA viruses found in various adenovirus preparations: physical, biological, and immunological characteristics. Proc Natl Acad Sci USA 1966; 55:1467–1474.
9. de la Maza LM, Carter BJ. Heavy and light particles of adeno-associated virus. J Virol 1980; 33:1129–1137.
10. Bachmann PA, Hoggan MD, Kurstak E, et al. Parvoviridae: second report. Intervirology 1979; 11:248–254.
11. Samulski RJ, Chang LS, Shenk T. A recombinant plasmid from which an infectious adeno-associated virus genome can be excised in vitro and its use to study viral replication. J Virol 1987; 61:3096–3101.
12. Mayor HD, Torikai K, Melnick JL, Mandel M. Plus and minus single-stranded DNA separately encapsidated in adeno-associated satellite virions. Science 1969; 166: 1280–1282.
13. Berns KI, Adler S. Separation of two types of adeno-associated virus particles containing complementary polynucleotide chains. J Virol 1972; 9:394–396.
14. Berns KI, Rose JA. Evidence for a single-stranded adenovirus-associated virus genome: isolation and separation of complementary single strands. J Virol 1970; 5:693–699.

15. Srivastava A, Lusby EW, Berns KI. Nucleotide sequence and organization of the adeno-associated virus 2 genome. J Virol 1983; 45:555–564.
16. Trempe JP, Carter BJ. Alternate mRNA splicing is required for synthesis of adeno-associated virus VP1 capsid protein. J Virol 1988; 62:3356–3363.
17. Lusby E, Fife KH, Berns KI. Nucleotide sequence of the inverted terminal repetition in adeno-associated virus DNA. J Virol 1980; 34:402–409.
18. Berns KI, Linden RM. The cryptic life style of adeno-associated virus. Bioessays 1995; 17:237–245.
19. Bohenzky RA, LeFebvre RB, Berns KI. Sequence and symmetry requirements within the internal palindromic sequences of the adeno-associated virus terminal repeat. Virology 1988; 166:316–327.
20. Spear IS, Fife KH, Hauswirth WW, Jones CJ, Berns KI. Evidence for two nucleotide sequence orientations within the terminal repetition of adeno-associated virus DNA. J Virol 1977; 24:627–634.
21. Lusby E, Bohenzky R, Berns KI. Inverted terminal repetition in adeno-associated virus DNA: independence of the orientation at either end of the genome. J Virol 1981; 37:1083–1086.
22. Hermonat PL, Labow MA, Wright R, Berns KI, Muzyczka N. Genetics of adeno-associated virus: isolation and preliminary characterization of adeno-associated virus type 2 mutants. J Virol 1984; 51:329–339.
23. Tratschin JD, Miller IL, Carter BJ. Genetic analysis of adeno-associated virus: properties of deletion mutants constructed in vitro and evidence for an adeno-associated virus replication function. J Virol 1984; 51:611–619.
24. Yang Q, Kadam A, Trempe JP. Mutational analysis of the adeno-associated virus rep gene. J Virol 1992; 66:6058–6069.
25. Beaton A, Palumbo P, Berns KI. Expression from the adeno-associated virus p5 and p19 promoters is negatively regulated in trans by the rep protein. J Virol 1989; 63:4450–4454.
26. Labow MA, Hermonat PL, Berns KI. Positive and negative autoregulation of the adeno-associated virus type 2 genome. J Virol 1986; 60:251–258.
27. Labow MA, Berns KI. Inhibition by the adeno-associated virus rep gene of replication of an AAV/SV40 hybrid genome in Cos-7 cell. Cancer Cells 1988; 6:101–4 1988.
28. Rose JA, Maizel JV Jr, Inman JK, Shatkin AJ. Structural proteins of adenovirus-associated viruses. J Virol 1971; 8:766–770.
29. Lubeck MD, Lee HM, Hoggan MD, Johnson FB. Adenovirus-associated virus structural protein sequence homology. J Gen Virol 1979; 45:209–216.
30. Berns KI, Labow MA. Parvovirus gene regulation. J Gen Virol 1987; 68(pt 3): 601–614.
31. Owens RA, Carter BJ. In vitro resolution of adeno-associated virus DNA hairpin termini by wild-type Rep protein is inhibited by a dominant-negative mutant of rep. J Virol 1992; 66:1236–1240.
32. Leonard CJ, Berns KI. Cloning, expression, and partial purification of Rep78: an adeno-associated virus replication protein. Virology 1994; 200:566–573.
33. Tratschin JD, Tal J, Carter BJ. Negative and positive regulation in trans of gene expression from adeno-associated virus vectors in mammalian cells by a viral rep gene product. Mol Cell Biol 1986; 6:2884–2894.

34. Owens RA, Trempe JP, Chejanovsky N, Carter BJ. Adeno-associated virus rep proteins produced in insect and mammalian expression systems: wild-type and dominant-negative mutant proteins bind to the viral replication origin. Virology 1991; 184:14–22.
35. Chejanovsky N, Carter BJ. Mutagenesis of an AUG codon in the adeno-associated virus rep gene: effects on viral DNA replication. Virology 1989; 173:120–128.
36. Senapathy P, Tratschin JD, Carter BJ. Replication of adeno-associated virus DNA. Complementation of naturally occurring rep-mutants by a wild-type genome or an ori-mutant and correction of terminal palindrome deletions. J Mol Biol 1984; 179: 1–20.
37. Chejanovsky N, Carter BJ. Mutation of a consensus purine nucleotide binding site in the adeno-associated virus rep gene generates a dominant negative phenotype for DNA replication. J Virol 1990; 64:1764–1770.
38. McCarty DM, Ni TH, Muzyczka N. Analysis of mutations in adeno-associated virus Rep protein in vivo and in vitro. J Virol 1992; 66:4050–4057.
39. Katz E, Carter BJ. Effect of adeno-associated virus on transformation of NIH 3T3 cells by ras gene and on tumorigenicity of an NIH 3T3 transformed cell line. Cancer Res 1986; 46:3023–3026.
40. Ostrove JM, Duckworth DH, Berns KI. Inhibition of adenovirus-transformed cell oncogenicity by adeno-associated virus. Virology 1981; 113:521–533.
41. Mendelson E, Smith MG, Miller IL, Carter BJ. Effect of a viral rep gene on transformation of cells by an adeno-associated virus vector. Virology 1988; 166:612–615.
42. Hermonat PL. The adeno-associated virus Rep78 gene inhibits cellular transformation induced by bovine papillomavirus. Virology 1989; 172:253–261.
43. Hermonat PL. Inhibition of H-ras expression by the adeno-associated virus Rep78 transformation suppressor gene product. Cancer Res 1991; 51:3373–3377.
44. Hermonat PL. Down-regulation of the human c-fos and c-myc proto-oncogene promoters by adeno-associated virus Rep78. Cancer Lett 1994; 81:129–136.
45. Khleif SN, Myers T, Carter BJ, Trempe JP. Inhibition of cellular transformation by the adeno-associated virus rep gene. Virology 1991; 181:738–741.
46. Schlehofer JR. The tumor suppressive properties of adeno-associated viruses. Mutat Res 1994; 305:303–313.
47. Bantel-Schaal U. Adeno-associated parvoviruses inhibit growth of cells derived from malignant human tumors. Int J Cancer 1990; 45:190–194.
48. de la Maza LM, Carter BJ. Inhibition of adenovirus oncogenicity in hamsters by adeno-associated virus DNA. J Natl Cancer Inst 1981; 67:1323–1326.
49. Holscher C, Horer M, Kleinschmidt JA, Zentgraf H, Burkle A, Heilbronn R. Cell lines inducibly expressing the adeno-associated virus (AAV) rep gene: requirements for productive replication of rep-negative AAV mutants. J Virol 1994; 68: 7169–7177.
50. Holscher C, Kleinschmidt JA, Burkle A. High-level expression of adeno-associated virus (AAV) Rep78 or Rep68 protein is sufficient for infectious-particle formation by a rep-negative AAV mutant. J Virol 1995; 69:6880–6885.
51. Prasad KM, Trempe JP. The adeno-associated virus Rep78 protein is covalently linked to viral DNA in a preformed virion. Virology 1995; 214:360–370.

52. Janik JE, Huston MM, Rose JA. Adeno-associated virus proteins: origin of the capsid components. J Virol 1984; 52:591–597.
53. Johnson FB, Whitaker CW, Hoggan MD. Structural polypeptides of adenovirus-associated virus top component. Virology 1975; 65:196–203.
54. Cassinotti P, Weitz M, Tratschin JD. Organization of the adeno-associated virus (AAV) capsid gene: mapping of a minor spliced mRNA coding for virus capsid protein 1. Virology 1988; 167:176–184.
55. Leonard CJ, Berns KI. Adeno-associated virus type 2: a latent life cycle. Prog Nucleic Acid Res Mol Biol 1994; 48:29–52.
56. Muralidhar S, Becerra SP, Rose JA. Site-directed mutagenesis of adeno-associated virus type 2 structural protein initiation codons: effects on regulation of synthesis and biological activity. J Virol 1994; 68:170–176.
57. Ruffing M, Zentgraf H, Kleinschmidt JA. Assembly of viruslike particles by recombinant structural proteins of adeno-associated virus type 2 in insect cells. J Virol 1992; 66:6922–6930.
58. Samulski RJ, Srivastava A, Berns KI, Muzyczka N. Rescue of adeno-associated virus from recombinant plasmids: gene correction within the terminal repeats of AAV. Cell 1983; 33:135–143.
59. Gottlieb J, Muzyczka N. In vitro excision of adeno-associated virus DNA from recombinant plasmids: isolation of an enzyme fraction from HeLa cells that cleaves DNA at poly(G) sequences. Mol Cell Biol 1988; 8:2513–2522.
60. Flotte TR, Afione SA, Solow R, et al. Expression of the cystic fibrosis transmembrane conductance regulator from a novel adeno-associated virus promoter. J Biol Chem 1993; 268:3781–3790.
61. Blacklow NR, Hoggan MD, McClanahan MS. Adenovirus-associated viruses: enhancement by human herpesviruses. Proc Soc Exp Biol Med 1970; 134:952–954.
62. Blacklow NR, Dolin R, Hoggan MD. Studies of the enhancement of an adenovirus-associated virus by herpes simplex virus. J Gen Virol 1971; 10:29–36.
63. Buller RM, Janik JE, Sebring ED, Rose JA. Herpes simplex virus types 1 and 2 completely help adenovirus-associated virus replication. J Virol 1981; 40:241–247.
64. Lipps BV, Mayor HD. Properties of adeno-associated virus (type 1) replicated in rodent cells by murine adenovirus. J Gen Virol 1980; 51:223–227.
65. Carter BJ, Laughlin CA, de la Maza LM, Myers M. Adeno-associated virus autointerference. Virology 1979; 92:449–462.
66. Berns KI, Bohenzky RA. Adeno-associated viruses: an update. Adv Virus Res 1987; 32:243–306.
67. West MH, Trempe JP, Tratschin JD, Carter BJ. Gene expression in adeno-associated virus vectors: the effects of chimeric mRNA structure, helper virus, and adenovirus VA1 RNA. Virology 1987; 160:38–47.
68. Carter BJ, Antoni BA, Klessig DF. Adenovirus containing a deletion of the early region 2A gene allows growth of adeno-associated virus with decreased efficiency. Virology 1992; 191:473–476.
69. Myers MW, Laughlin CA, Jay FT, Carter BJ. Adenovirus helper function for growth of adeno-associated virus: effect of temperature-sensitive mutations in adenovirus early gene region 2. J Virol 1980; 35:65–75.

70. Jay FT, Laughlin CA, Carter BJ. Eukaryotic translational control: adeno-associated virus protein synthesis is affected by a mutation in the adenovirus DNA-binding protein. Proc Natl Acad Sci USA 1981; 78:2927–2931.
71. Ostrove JM, Berns KI. Adenovirus early region 1b gene function required for rescue of latent adeno-associated virus. Virology 1980; 104:502–505.
72. Janik JE, Huston MM, Rose JA. Locations of adenovirus genes required for the replication of adenovirus-associated virus. Proc Natl Acad Sci USA 1981; 78:1925–1929.
73. Bantel-Schaal U. Carcinogen-induced accumulation of adeno-associated parvovirus DNA is transient as a result of two antagonistic activities that both require de novo protein synthesis. Int J Cancer 1993; 53:334–339.
74. Yalkinoglu AO, Zentgraf H, Hubscher U. Origin of adeno-associated virus DNA replication is a target of carcinogen-inducible DNA amplification. J Virol 1991; 65:3175–3184.
75. Ruffing M, Heid H, Kleinschmidt JA. Mutations in the carboxy terminus of adeno-associated virus 2 capsid proteins affect viral infectivity: lack of an RGD integrin-binding motif. J Gen Virol 1994; 75(pt 12):3385–3392.
76. Laughlin CA, Jones N, Carter BJ. Effect of deletions in adenovirus early region 1 genes upon replication of adeno-associated virus. J Virol 1982; 41:868–876.
77. Redemann BE, Mendelson E, Carter BJ. Adeno-associated virus rep protein synthesis during productive infection. J Virol 1989; 63:873–882.
78. Antoni BA, Rabson AB, Miller IL, Trempe JP, Chejanovsky N, Carter BJ. Adeno-associated virus Rep protein inhibits human immunodeficiency virus type 1 production in human cells. J Virol 1991; 65:396–404.
79. Chang LS, Shi Y, Shenk T. Adeno-associated virus P5 promoter contains an adenovirus E1A-inducible element and a binding site for the major late transcription factor. J Virol 1989; 63:3479–3488.
80. Lipps BV, Mayor HD. Defective parvoviruses acquired via the transplacental route protect mice against lethal adenovirus infection. Infect Immun 1982; 37:200–204.
81. Hauswirth WW, Berns KI. Adeno-associated virus DNA replication: nonunit-length molecules. Virology 1979; 93:57–68.
82. Ward P, Berns KI. Minimum origin requirements for linear duplex AAV DNA replication in vitro. Virology 1995; 209:692–695.
83. Hong G, Ward P, Berns KI. Intermediates of adeno-associated virus DNA replication in vitro. J Virol 1994; 68:2011–2015.
84. Straus SE, Sebring ED, Rose JA. Concatemers of alternating plus and minus strands are intermediates in adenovirus-associated virus DNA synthesis. Proc Natl Acad Sci USA 1976; 73:742–746.
85. Berns KI, Hauswirth WW. Adeno-associated viruses. Adv Virus Res 1979; 25:407–449.
86. Myers MW, Carter BJ. Assembly of adeno-associated virus. Virology 1980; 102:71–82.
87. Smuda JW, Carter BJ. Adeno-associated viruses having nonsense mutations in the capsid genes: growth in mammalian cells containing an inducible amber suppressor. Virology 1991; 184:310–318.

88. Myers MW, Carter BJ. Adeno-associated virus replication. The effect of L-canavanine or a helper virus mutation on accumulation of viral capsids and progeny single-stranded DNA. J Biol Chem 1981; 256:567–570.
89. Wang XS, Ponnazhagan S, Srivastava A. Rescue and replication signals of the adeno-associated virus 2 genome. J Mol Biol 1995; 250:573–580.
90. Im DS, Muzyczka N. The AAV origin binding protein Rep68 is an ATP-dependent site-specific endonuclease with DNA helicase activity. Cell 1990; 61:447–457.
91. Wonderling RS, Kyostio SR, Owens RA. A maltose-binding protein/adeno-associated virus Rep68 fusion protein has DNA-RNA helicase and ATPase activities. J Virol 1995; 69:3542–3548.
92. Chiorini JA, Wiener SM, Owens RA, Kyostio SR, Kotin RM, Safer B. Sequence requirements for stable binding and function of Rep68 on the adeno-associated virus type 2 inverted terminal repeats. J Virol 1994; 68:7448–7457.
93. Snyder RO, Im DS, Muzyczka N. Evidence for covalent attachment of the adeno-associated virus (AAV) rep protein to the ends of the AAV genome. J Virol 1990; 64:6204–6213.
94. Snyder RO, Samulski RJ, Muzyczka N. In vitro resolution of covalently joined AAV chromosome ends. Cell 1990; 60:105-113.
95. Hoggan MD, Thomas GF, Johnson FB. Continuous carriage of adeno-associated virus genome in cell culture in the absence of adenovirus. In: Proceedings of the fourth Lepetit Colloquium. Amsterdam: North Holland, 1972:243–249.
96. Cheung AK, Hoggan MD, Hauswirth WW, Berns KI. Integration of the adeno-associated virus genome into cellular DNA in latently infected human Detroit 6 cells. J Virol 1980; 33:739–748.
97. Berns KI, Pinkerton TC, Thomas GF, Hoggan MD. Detection of adeno-associated virus (AAV)-specific nucleotide sequences in DNA isolated from latently infected Detroit 6 cells. Virology 1975; 68:556–560.
98. Kotin RM, Berns KI. Organization of adeno-associated virus DNA in latently infected Detroit 6 cells. Virology 1989; 170:460–467.
99. Kotin RM, Linden RM, Berns KI. Characterization of a preferred site on human chromosome 19q for integration of adeno-associated virus DNA by non-homologous recombination. EMBO J 1992; 11:5071–5078.
100. Kotin RM, Siniscalco M, Samulski RJ, et al. Site-specific integration by adeno-associated virus. Proc Natl Acad Sci USA 1990; 87:2211–2215.
101. Samulski RJ, Zhu X, Xiao X, et al. Targeted integration of adeno-associated virus (AAV) into human chromosome 19. EMBO J 1991; 10:3941–3950.
102. Kotin RM, Menninger JC, Ward DC, Berns KI. Mapping and direct visualization of a region-specific viral DNA integration site on chromosome 19q13-qter. Genomics 1991; 10:831–834.
103. Samulski RJ. Adeno-associated virus: integration at a specific chromosomal locus. Curr Opin Genet Dev 1993; 3:74–80.
104. Samulski RJ, Zhu X, Xiao X, et al. Targeted integration of adeno-associated virus (AAV) into human chromosome 19. EMBO J 1991; 10:3941–3950. (Published erratum appears in EMBO J 1992;11(3):1228.)

105. Giraud C, Winocour E, Berns KI. Site-specific integration by adeno-associated virus is directed by a cellular DNA sequence. Proc Natl Acad Sci USA 1994; 91:10039–10043.
106. Giraud C, Winocour E, Berns KI. Recombinant junctions formed by site-specific integration of adeno-associated virus into an episome. J Virol 1995; 69:6917–6924.
107. Weitzman MD, Kyostio SR, Kotin RM, Owens RA. Adeno-associated virus (AAV) Rep proteins mediate complex formation between AAV DNA and its integration site in human DNA. Proc Natl Acad Sci USA 1994; 91:5808–5812.
108. Winocour E, Puzis L, Etkin S, et al. Modulation of the cellular phenotype by integrated adeno-associated virus. Virology 1992; 190:316–329.
109. Bantel-Schaal U. Infection with adeno-associated parvovirus leads to increased sensitivity of mammalian cells to stress. Virology 1991; 182:260–268.
110. Winocour E, Callaham MF, Huberman E. Perturbation of the cell cycle by adeno-associated virus. Virology 1988; 167:393–399.
111. Bantel-Schaal U. Growth properties of a human melanoma cell line are altered by adeno-associated parvovirus type 2. Int J Cancer 1995; 60:269–274.
112. Bantel-Schaal U, Stohr M. Influence of adeno-associated virus on adherence and growth properties of normal cells. J Virol 1992; 66:773–779.
113. Klein-Bauernschmitt P, zur Hausen H, Schlehofer JR. Induction of differentiation-associated changes in established human cells by infection with adeno-associated virus type 2. J Virol 1992; 66:4191–4200.
114. Bantel-Schaal U, zur Hausen H. Dissociation of carcinogen-induced SV40 DNA-amplification and amplification of AAV DNA in a Chinese hamster cell line. Virology 1988; 166:113–122.
115. Bantel-Schaal U, zur Hausen H. Adeno-associated viruses inhibit SV40 DNA amplification and replication of herpes simplex virus in SV40-transformed hamster cells. Virology 1988; 164:64–74.
116. Kirschstein RL, Smith KO, Peters EA. Inhibition of adenovirus 12 oncogenicity by adeno-associated virus. Proc Soc Exp Biol Med 1968; 128:670–673.
117. Casto BC, Goodheart CR. Inhibition of adenovirus transformation in vitro by AAV-1. Proc Soc Exp Biol Med 1972; 140:72–78.
118. Mayor HD, Houlditch GS, Mumford DM. Influence of adeno-associated satellite virus on adenovirus-induced tumours in hamsters. Nature New Biol 1973; 241:44–46.
119. Cukor G, Blacklow NR, Kibrick S, Swan IC. Effect of adeno-associated virus on cancer expression by herpesvirus-transformed hamster cells. J Natl Cancer Inst 1975; 55:957–959.
120. Samulski RJ, Berns KI, Tan M, Muzyczka N. Cloning of adeno-associated virus into pBR322: rescue of intact virus from the recombinant plasmid in human cells. Proc Natl Acad Sci USA 1982; 79:2077–2081.
121. Tratschin JD, Miller IL, Smith MG, Carter BJ. Adeno-associated virus vector for high-frequency integration, expression, and rescue of genes in mammalian cells. Mol Cell Biol 1985; 5:3251–3260.
122. Lebkowski JS, McNally MM, Okarma TB, Lerch LB. Adeno-associated virus: a vector system for efficient introduction and integration of DNA into a variety of mammalian cell types. Mol Cell Biol 1988; 8:3988–3996.

123. Hermonat PL, Muzyczka N. Use of adeno-associated virus as a mammalian DNA cloning vector: transduction of neomycin resistance into mammalian tissue culture cells. Proc Natl Acad Sci USA 1984; 81:6466–6470.

124. Samulski RJ, Chang LS, Shenk T. Helper-free stocks of recombinant adeno-associated viruses: normal integration does not require viral gene expression. J Virol 1989; 63:3822–3828.

125. Chejanovsky N, Carter BJ. Replication of a human parvovirus nonsense mutant in mammalian cells containing an inducible amber suppressor. Virology 1989; 171:239–247.

126. Kyostio SR, Owens RA, Weitzman MD, Antoni BA, Chejanovsky N, Carter BJ. Analysis of adeno-associated virus (AAV) wild-type and mutant Rep proteins for their abilities to negatively regulate AAV p5 and p19 mRNA levels. J Virol 1994; 68:2947–2957.

127. Flotte TR, Barraza-Ortiz X, Solow R, Afione SA, Carter BJ, Guggino WB. An improved system for packaging recombinant adeno-associated virus vectors capable of in vivo transduction. Gene Ther 1995; 2:29–37.

128. Clark KR, Voulgaropoulou V, Fraley DM, Johnson PR. Cell lines for the production of recombinant adeno-associated virus. Hum Gene Ther 1995; 6:1329–1341.

129. Yang Q, Chen F, Trempe JP. Characterization of cell lines that inducibly express the adeno-associated virus Rep proteins. J Virol 1994; 68:4847–4856.

130. Mamounas M, Leavitt M, Yu M, Wong-Staal F. Increased titer of recombinant AAV vectors by gene transfer with adenovirus coupled to DNA-polylysine complexes. Gene Ther 1995; 2:429–432.

131. Rivadeneira ED, Thierry AR, Popescu NC, et al. Adeno-Associated Virus mediated delivery of anti-HIV-1 genes by transduction or transfection: integration and efficiency studies. Natl Conf Hum Retroviruses Relat Infect (2nd) 1995; 142.

132. Sun XL, Murphy BR, Li QJ, et al. Transduction of folate receptor cDNA into cervical carcinoma cells using recombinant adeno-associated virions delays cell proliferation in vitro and in vivo. J Clin Invest 1995; 96:1535–1547.

133. Zhou SZ, Cooper S, Kang LY, et al. Adeno-associated virus 2-mediated high efficiency gene transfer into immature and mature subsets of hematopoietic progenitor cells in human umbilical cord blood. J Exp Med 1994; 179:1867–1875.

134. Flotte TR, Solow R, Owens RA, Afione S, Zeitlin PL, Carter BJ. Gene expression from adeno-associated virus vectors in airway epithelial cells. Am J Respir Cell Mol Biol 1992; 7:349–356.

135. Flotte TR, Afione SA, Zeitlin PL. Adeno-associated virus vector gene expression occurs in nondividing cells in the absence of vector DNA integration. Am J Respir Cell Mol Biol 1994; 11:517–521.

136. Luo F, Zhou SZ, Cooper S, et al. Adeno-associated virus 2-mediated gene transfer and functional expression of the human granulocyte-macrophage colony-stimulating factor. Exp Hematol 1995; 23:1261–1267.

137. Podsakoff G, Wong KK Jr, Chatterjee S. Efficient gene transfer into nondividing cells by adeno-associated virus-based vectors. J Virol 1994; 68:5656–5666.

138. Russell DW, Miller AD, Alexander IE. Adeno-associated virus vectors preferentially transduce cells in S phase. Proc Natl Acad Sci USA 1994; 91:8915–8919.

139. McLaughlin SK, Collis P, Hermonat PL, Muzyczka N. Adeno-associated virus general transduction vectors: analysis of proviral structures. J Virol 1988; 62: 1963–1973.

140. Walsh CE, Liu JM, Xiao X, Young NS, Nienhuis AW, Samulski RJ. Regulated high level expression of a human gamma-globin gene introduced into erythroid cells by an adeno-associated virus vector. Proc Natl Acad Sci USA 1992; 89:7257– 7261.

141. Alexander IE, Russell DW, Miller AD. DNA-damaging agents greatly increase the transduction of nondividing cells by adeno-associated virus vectors. J Virol 1994; 68:8282–8287.

142. Wei JF, Wei FS, Samulski RJ, Barranger JA. Expression of the human glucocere-brosidase and arylsulfatase A genes in murine and patient primary fibroblasts transduced by an adeno-associated virus vector. Gene Ther 1994; 1:261–268.

143. Shelling AN, Smith MG. Targeted integration of transfected and infected adeno-associated virus vectors containing the neomycin resistance gene. Gene Ther 1994; 1:165–169.

144. Kearns WG, Afione S, Fulmer SB, et al. Recombinant AAV vectors demonstrate episomal expression in immortalized human bronchial epithelial cells. Am Soc Hum Genet 1994; 55:A225.

145. Kearns WG, Afione SA, Fulmer SB, et al. Recombinant adeno-associated virus (AAV-CFTR) vectors do not integrate in a site-specific fashion in an immortal-ized epithelial cell line. Gene Ther 1996; 3:748–755.

146. Fisher KJ, Gao GP, Weitzman MD, DeMatteo R, Burda JF, Wilson JM. Trans-duction with recombinant adeno-associated virus for gene therapy is limited by leading-strand synthesis. J Virol 1996; 70:520–532.

147. DeMatteo RP, Raper SE, Ahn M, et al. Gene transfer to the thymus. A means of abrogating the immune response to recombinant adenovirus. Ann Surg 1995; 222:229–239.

148. Zhou H, Zeng G, Zhu X, et al. Enhanced adeno-associated virus vector expres-sion by adenovirus protein-cationic liposome complex. A novel and high efficient way to introduce foreign DNA into endothelial cells. Chin Med J (Engl) 1995; 108:332–337.

149. Walsh CE, Liu JM, Miller JL, Nienhuis AW, Samulski RJ. Gene therapy for human hemoglobinopathies. Proc Soc Exp Biol Med 1993; 204:289–300.

150. Miller JL, Donahue RE, Sellers SE, Samulski RJ, Young NS, Nienhuis AW. Re-combinant adeno-associated virus (rAAV)-mediated expression of a human gamma-globin gene in human progenitor-derived erythroid cells. Proc Natl Acad Sci USA 1994; 91:10183–10187. (Published erratum appears in Proc Natl Acad Sci USA 1995; 92(2):646.)

151. Muro-Cacho CA, Samulski RJ, Kaplan D. Gene transfer in human lymphocytes using a vector based on adeno-associated virus. J Immunother 1992; 11:231– 237.

152. Chatterjee S, Johnson PR, Wong KK Jr. Dual-target inhibition of HIV-1 in vitro by means of an adeno-associated virus antisense vector. Science 1992; 258:1485– 1488.

153. Ponnazhagan S, Nallari ML, Srivastava A. Suppression of human alpha-globin gene expression mediated by the recombinant adeno-associated virus 2-based antisense vectors. J Exp Med 1994; 179:733–738.
154. Srivastava A. Parvovirus-based vectors for human gene therapy. Blood Cells 1994; 20:531–538.
155. Walsh CE, Nienhuis AW, Samulski RJ, et al. Phenotypic correction of Fanconi anemia in human hematopoietic cells with a recombinant adeno-associated virus vector. (See comments.) J Clin Invest 1994; 94:1440–1448.
156. Wong KK, Lu D, Podsakoff G, Brar D, Chatterjee S. Efficient gene transfer to primary human CD4+ T lymphocytes, monocyte-macrophages, and CD34+ hematopoietic stem cells mediated by adeno-associated virus-based vectors: prospects for the gene therapy of AIDS. Natl Conf Hum Retroviruses Relat Infect (2nd) 1995; 142
157. Dunbar CE, Emmons RV. Gene transfer into hematopoietic progenitor and stem cells: progress and problems. Stem Cells (Dayt) 1994; 12:563–576.
158. Goodman S, Xiao X, Donahue RE, et al. Recombinant adeno-associated virus-mediated gene transfer into hematopoietic progenitor cells. Blood 1994; 84:1492–1500. (Published erratum appears in Blood 1995; 85(3):862.)
159. Simon RH, Engelhardt JF, Yang Y, et al. Adenovirus-mediated transfer of the CFTR gene to lung of nonhuman primates: toxicity study. Hum Gene Ther 1993; 4:771–780.
160. Yang Y, Nunes FA, Berencsi K, Furth EE, Gonczol E, Wilson JM. Cellular immunity to viral antigens limits E1-deleted adenoviruses for gene therapy. Proc Natl Acad Sci USA 1994; 91:4407–4411.
161. Boucher RC, Knowles MR, Johnson LG, et al. Gene therapy for cystic fibrosis using E1-deleted adenovirus: a phase I trial in the nasal cavity. Hum Gene Ther 1994; 5:615–639.
162. Kaplitt MG, Leone P, Samulski RJ, et al. Long-term gene expression and phenotypic correction using adeno-associated virus vectors in the mammalian brain. Nat Genet 1994; 8:148–154.
163. Bartlett RJ, Singer JT, Shean MK, Weindler FW, Secore SL. Expression of a fluorescent reporter gene in skeletal muscle cells from a virus-based vector: implications for diabetes treatment. Transplant Proc 1995; 27:3344
164. Lin H, Parmacek MS, Morle G, Bolling S, Leiden JM. Expression of recombinant genes in myocardium in vivo after direct injection of DNA. Circulation 1990; 82:2217–2221.
165. Buttrick PM, Kass A, Kitsis RN, Kaplan ML, Leinwand LA. Behavior of genes directly injected into the rat heart in vivo. Circ Res 1992; 70:193–198.
166. Acsadi G, Dickson G, Love DR, et al. Human dystrophin expression in mdx mice after intramuscular injection of DNA constructs. (See comments.) Nature 1991; 352:815–818.
167. Wolff JA, Malone RW, Williams P, et al. Direct gene transfer into mouse muscle in vivo. Science 1990; 247:1465–1468.
168. Wolff JA, Williams P, Acsadi G, Jiao S, Jani A, Chong W. Conditions affecting direct gene transfer into rodent muscle in vivo. Biotechniques 1991; 11:474–485.

169. Jiao S, Williams P, Berg RK, et al. Direct gene transfer into nonhuman primate myofibers in vivo. Hum Gene Ther 1992; 3:21–33.
170. Wolff JA, Dowty ME, Jiao S, et al. Expression of naked plasmids by cultured myotubes and entry of plasmids into T tubules and caveolae of mammalian skeletal muscle. J Cell Sci 1992; 103(pt 4):1249–1259.
171. Acsadi G, Jiao SS, Jani A, Duke D, Williams P, Chong W, Wolff JA. Direct gene transfer and expression into rat heart in vivo. New Biol 1991; 3:71–81.
172. Kitsis RN, Buttrick PM, McNally EM, Kaplan ML, Leinwand LA. Hormonal modulation of a gene injected into rat heart in vivo. Proc Natl Acad Sci USA 1991; 88:4138–4142.
173. Rosenstein BJ, Zeitlin PL. Prognosis in cystic fibrosis. Curr Opin Pulmonary Med 1995; 1:444–449.
174. Rosenfeld MA, Collins FS. Gene therapy for cystic fibrosis. Chest 1996; 109: 241–252.
175. Ayers MM, Jeffery PK. Proliferation and differentiation in mammalian airway epithelium. Eur Respir J 1988; 1:58–80.
176. Egan M, Flotte T, Afione S, et al. Defective regulation of outwardly rectifying Cl-channels by protein kinase A corrected by insertion of CFTR. (See comments.) Nature 1992; 358:581–584.
177. Afione SA, Conrad CK, Kearns WG, et al. In vivo model of adeno-associated virus vector persistence and rescue. J Virol 1996; 70(5):3235–3241.
178. Zeitlin PL, Chu S, Conrad C, et al. Alveolar stem cell transduction by an adeno-associated viral vector. Gene Ther 1995; 2:623–631.
179. Vieweg J, Boczkowski D, Roberson KM, et al. Efficient gene transfer with adeno-associated virus-based plasmids complexed to cationic liposomes for gene therapy of human prostate cancer. Cancer Res 1995; 55:2366–2372.
180. Philip R, Brunette E, Kilinski L, et al. Efficient and sustained gene expression in primary T lymphocytes and primary and cultured tumor cells mediated by adeno-associated virus plasmid DNA complexed to cationic liposomes. Mol Cell Biol 1994; 14:2411–2418.

5

Cellular Responses to Adenovirus Entry

SUSANNA CHIOCCA and MATTHEW COTTEN

Institute for Molecular Pathology
Vienna, Austria

I. Introduction

Adenovirus has shown great potential for gene delivery, exciting a large number of investigators to prepare recombinant vectors bearing therapeutic genes. These vectors are used in two types of applications—gene correction applications, and vaccine applications. In gene correction, the vector is used to deliver a gene whose function is absent in the patient. The hope is that adenovirus vectors will behave appropriately: enter all of the desired target tissue, express the therapeutic gene for weeks and months, and perform these functions with no side effects. Unfortunately, the initial clinical experiences with adenovirus vectors have shown limited functional gene transfer and an alarming degree of inflammatory side effects. A better understanding of the virus entry process will allow investigators to modify these reagents for better applications. On the other hand, the inflammatory side effects of adenovirus vectors can be quite a positive feature when these vectors are used to provoke immune responses in vaccine applications. Successful clinical applications of adenovirus vectors either for gene correction or as vaccines will require a better understanding of why adenovirus is such an antagonistic vector.

Adenovirus-based gene delivery systems are the most efficient vectors for gene delivery available to date. No other virus system can survive the bloodstream or can function efficiently in intratumoral applications. The process and the consequences of adenovirus entry into eukaryotic cells is not fully understood at the molecular level. There is very little information available on the mechanisms used by any virus to enter its eukaryotic host. The subgroup C adenoviruses can initiate infection from a single particle, demonstrating that the entry functions of this virus are very effective. In addition to the deposition of novel nucleic acid in the nucleus of a host cell, the entry of adenovirus can trigger a series of important host responses including the activation of transcription factors involved in inflammatory responses (e.g. NF-κB) and the initiation of apoptotic responses. A clear analysis of the events following adenovirus entry will provide information about both apoptosis and inflammation.

In this article we will review the adenovirus entry literature and ask if there is a way that adenovirus can efficiently target an entire cell monolayer without activating these inflammatory responses. We will present the idea that in the wild, adenovirus actually relies upon local inflammation to facilitate cell infection and virus replication

II. The Virus Entry Mechanism

The barriers that an adenovirus must surmount to successfully transduce a cell are summarized in Figure 1. The virus must reach the target cell and remain in the vicinity long enough for appropriate receptor/virus interactions to occur. The virus must enter the cell by some form of endocytosis, pass across the limiting membrane through the cytoplasm, and enter the nucleus. Appropriate dismantling of the incoming virion must occur. Once inside the target cell, gene expression must occur, and, most importantly, this transduced cell must survive in the host long enough to perform the desired clinical function. These topics are discussed in detail below.

A. Receptor Interactions

All viruses have capsid components that bind to receptors on the surface of the cells that the virus must enter for replication. These structures can be either proteins or carbohydrates. Subgroup C adenovirus (adenovirus types 2 and 5) have a major capsid ligand, the fiber, that binds to an as yet unidentified cellular receptor, and the penton base, which contains an exposed arginine/glycine/aspartic acid (RGD) motif and thus binds to certain integrin subgroups.

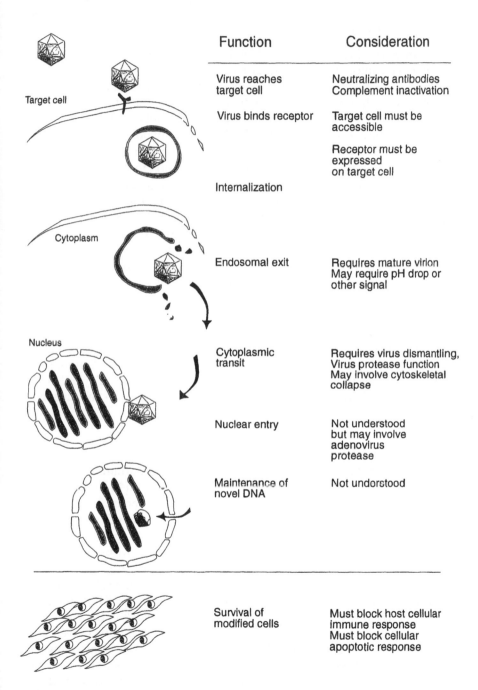

	Function	Consideration
Target cell	Virus reaches target cell	Neutralizing antibodies Complement inactivation
	Virus binds receptor	Target cell must be accessible
		Receptor must be expressed on target cell
	Internalization	
Cytoplasm	Endosomal exit	Requires mature virion May require pH drop or other signal
Nucleus	Cytoplasmic transit	Requires virus dismantling, Virus protease function May involve cytoskeletal collapse
	Nuclear entry	Not understood but may involve adenovirus protease
	Maintenance of novel DNA	Not understood
	Survival of modified cells	Must block host cellular immune response Must block cellular apoptotic response

Figure 1 Summary of requirements for successful transduction by an adenovirus vector. See text for details.

Fiber Interactions

The adenovirus fiber has long been known as a major determinant of host cell binding (1). Although the Ad5 fiber receptor has not yet been identified, the portions of the fiber essential for receptor binding have been identified (2,3). Furthermore, strategies to alter fiber/receptor interactions for targeting vectors or altering the binding specificity are being developed. These include ablating the natural receptor binding using blocking antibodies (4), switching the fiber molecule on a vector to a fiber from a serotype with altered cell binding (5), and attempting to genetically alter the fiber knob to display a novel targeting motif (6). It may also be possible to circumvent immune responses generated against an earlier adenovirus or adenovirus vector exposure by using subsequent applications of vectors bearing fiber components from serotypes not recognized by the host. However, in initial experiments with Ad5/Ad7 fiber swaps, it appears that neutralizing antibody responses were not directed toward fiber epitopes and altering the fiber molecule did not bypass the blocking immune response (5). Because neutralizing antibodies are generally directed toward hexon and penton base epitopes (5,7), more successful serotype switching strategies will likely involve the use of alternate serotype hexons, penton bases, or complete virions (8).

An additional method of altering the cell binding properties of adenovirus is to affix appropriate ligands to adenovirus complexes. Antibodies to CD3 have proven useful for targeting adenovirus complexes to T-lymphocytes (9). Lectins, such as wheat germ agglutinin, have been used to enhance myoblast delivery (10) or tumor cell delivery (11).

Integrin Interactions

Integrin/virus interactions have been proposed to be important for Ad5 entry (12–16). An integrin-binding arginine-glycine-aspartic acid (RGD) motif is present in the Ad5 penton base and is essential for the penton/integrin interactions. Competition experiments with antibodies and peptides suggest that the penton base RGD may interact with a subclass of integrins during viral infection (13,14) and have led to a two-step model of adenovirus entry (15,16). However, the inhibition in virus infectivity observed in these experiments is typically 80% and never more than 1 log, suggesting that the role of integrin-penton base interaction is minor. Experiments in which the RGD motif was directly mutated demonstrate a slowdown, but not a block in the virus entry (12). Compelling evidence that the integrin/penton base interactions are involved in virus entry in certain cell types come from studies demonstrating that upregulating integrins by treatment of the target cells with various cytokines can enhance adenovirus infectivity into these cells (17). There is also evidence that blocking the integrin/penton base interactions is important in triggering alterations in the incoming virion essential for virus uncoating. Blocking the integrin/penton base interactions with competing peptide interferes with virus disassembly (18). Alterations in the integrin binding motif of the adenovirus penton base may facilitate targeting of vectors to specific cell types (19) however, these alterations will most likely only be useful when combined with alterations in the more potent fiber/receptor interactions. Integrin engagement or crosslinking can also activate cellular signals such as changes in calcium stores or alterations in pH which may also influence the infection process (20, 21). However, the integrin binding motif is not a strictly conserved penton base motif when other adenovirus serotypes are examined, suggesting that integrin signalling may not be generally required for adenovirus entry.

B. Other Factors That Influence Adenovirus Transduction

It has been demonstrated that in intact, noninflamed tissue, the transduction efficiency of adenovirus vectors is low (22). Mechanical damage of the tracheal epithelium was necessary to expose the basal cells susceptible to adenovirus transduction (22). Future strategies will require either methods to increase the transduction efficiency of the epithelial cell types that are naturally accessible to instilled virus, or to find ways to target the adenovirus to the basal cells that are more readily transduced. Perhaps one consideration for the gene therapy applications of adenovirus is that we are asking adenovirus to behave in a manner in which it does not behave in the wild. The virus probably infects a small focus of cells, perhaps relying on the inflammatory upregulation of receptors provided by a bacterial infection. Local replication of virus increases the inflammatory response in neighboring cells, upregulating transcription factors and viral receptors, thus providing enhanced possibilities for virus spread. It is perhaps not surprising that in intact, noninflamed tissue the transduction rate is sluggish.

Researchers have probably selected for cell lines and virus strains that behave well in culture so that it is generally found that adenovirus binding and cellular entry is efficient. However, adenovirus infection rates in vivo may not be as rapid. The inclusion of an inert agent that increase the local virus concentration and cell contact time can produce large increases in transduction efficiencies both in vitro and in vivo (23). This suggests relatively simple methods of enhancing adenovirus transduction rates by physically maintaining the virus at high concentrations in the neighbourhood of the target cell.

Rather than try to modify adenovirus vectors to perform transduction of a particular tissue a perhaps more productive strategy is to ask what targeting can be obtained with the current Ad5 vectors, and then select applications where this targeting can be exploited. A catalog of Ad5 targeting has been presented (24) and should be a useful basis for designing therapies.

III. The Virus Is Internalized and Endosomal Escape Must Occur

Adenoviruses have selected receptors that provide an entry site into the cell. Once fiber and penton base binding to their receptors occurs, endocytosis of the receptor/virus complex internalizes the virus (reviewed in 15,16). However, the virus is topologically still outside of the cell, with a membrane separating it from the cytoplasm. A number of years ago it was discovered that during many types of viral infection, including adenovirus, a transient permeabilization of the endosome occurs, with other material coendocytosed with the viral particles gaining entry to the cytoplasm (25,26; reviewed in 27,28). Because substantial improvements in a number of gene delivery systems have been obtained with this phenomenon (29–34), understanding the mechanism of adenovirus-enhanced endosomal release is of some interest.

How could this occur? (Here is some pure speculation.) Unfortunately, a clear mechanistic description of the endosomal lysis reaction is not yet available. Some speculations on possible mechanisms include the following:

1. The lower pH encountered by the virus in the endosome might increase the hydrophobicity of regions of the capsid by protonating negatively charged amino acid residues. However, as discussed below, the low pH is not essential. An alternative signal might involve changes in the reducing environment as the virus is internalized. Note that the activity of the adenovirus protease is influenced by the reducing environment (35–39; reviewed in 40,41) and the activation of the proteinase upon cellular entry maybe important in the entry process.

2. Increased hydrophobicity of capsid components might induce a curvature of the endosomal membrane beyond the angles that are energetically allowed by a lipid bilayer. This distortion triggers a local membrane rearrangement that allows virion passage into the cytoplasm.

3. Adenovirus capsid proteins interact with ion pumps in the endosomal membrane, with the resulting osmotic shock disrupting the endosome. The weak base chloroquine probably facilitates transfection into K562 cells by such an osmotic lysis mechanism (42).

4. The adenovirus virion contains an enzymatic activity such as a phospholipase that directly modifies and disrupts the endosomal membrane.

Immature adenovirus particles, which have not undergone proteolytic maturation, are defective at membrane lysis at low pH (43) and accumulate in the endosome (44,45). Any mechanism explaining the endosomal lysis should thus account for the failure of immature particles to exit the endosome. Finally, it should be noted that the endosomal disruption delivery trick works with other viruses including rhinovirus (46,47), poliovirus, vaccinia, vesicular stomatitis virus, and others (reviewed 27,28).

IV. The Role of a Low pH Step in Adenovirus Entry Is Not Yet Clear

There is a body of literature supporting a low pH step in the adenovirus infection process (40,43,48–50). However, Pérez and Carrasco (51) demonstrate that adenovirus infection is not inhibited by bafilomycin A1 (BFA), an inhibitor of endosomal acidification by the vacuolar H^+-ATPase. These BFA results support the conclusion that a low pH step is not mandatory for adenovirus infection. Experiments supporting a low pH step usually demonstrate modest (less than 10-fold) declines in virus production or virus infectivity whereas one would expect a several log decline in virus production if an inhibitor is truly effective. Probably the safest conclusion is that adenovirus relies on several signals to activate endosome disruption and virus entry. A low pH will certainly facilitate capsid/membrane interactions as negatively charge residues on the surface of hexon are neutralized and the protein becomes more hydrophobic. However, in the absence of this low pH exposure, sufficient membrane perturbations can still occur to allow virus passage into the cytoplasm (51).

V. Cytoplasmic Transit

A substantial alteration in the infected cell cytoskeleton was observed during productive adenovirus entry (45). These alterations were not seen with subgroub B adenovirus, which may enter cells by an alternate mechanism, nor were the changes observed with immature particles, which are unable to leave the endosome. It is important to determine these cytoskeletal alterations required for the efficient nuclear entry of the infecting adenovirus particle and if these changes are involved in any of the inflammatory responses observed upon virus entry. In addition, if the virus is altering the cell to increase nuclear entry of itself, it may also inadvertently trigger the nuclear uptake of other cellular component. Since at least two classes of transcription factors (NF-κB and the heat shock factors) are controlled by sequestration in the cytoplasm, a stimulation of non-specific nuclear uptake might alter the transcriptional environment of the cell.

VI. Virus Uncoating and Nuclear Entry: The Role of the Adenovirus Protease

It is becoming clear that an adenovirus-encoded protease must function during entry of adenovirus (18,43). Although initial studies of this process have identified virion proteins that are cleaved during entry (50,52), the viral protease has been demonstrated to cleave at least one host cytoskeletal protein late in

infection (53,54) and it remains possible that the collapse of the vimentin filament network observed during productive entry of Ad2 (45) is a consequence
of viral protease activity. Cleavage of additional host proteins might occur and
might be required for the nuclear entry of the adenovirus particle.

VII. How These Changes Can Influence the Health of the Cell

A. Interferon Responses

During lytic infection, adenovirus requires the infected cell to survive for only
24 to 48 hours, long enough to complete the synthesis of progeny virions, resulting in a 1000-fold amplification of the virus. Adenovirus has evolved strategies to alter the cellular antiviral responses within this time frame. For example, a translational inhibition due to activation of the interferon pathway
occurs during adenovirus infection (55). This is thought to be due to activation
of the interferon-inducible kinase p68 by double-stranded RNA molecules
generated during the adenovirus life cycle. Kinase activity results in the inactivation of translation initiation factor eIF2α, and can potently downregulate
translation. One adenovirus method of blocking this translational block involves the viral-encoded VA RNA molecule, which prevents double-stranded
RNA activation of p68 (56–58; reviewed in 59). The adenovirus induction of
interferon-responsive genes is also blocked at the transcriptional level by products of the E1A region (60–64). The E1A functions are not included in viral
vectors. Furthermore, in the absence of E1A and viral DNA replication, full
VA gene expression is not achieved.

B. Stress Responses

Adenovirus upregulates heat shock promoters through the action of the E1A
gene products (65–68). Although there is no evidence that adenovirus relies
on activating a stress response for productive infection, it is clear from other
studies that modest activation of stress responsive genes can provide beneficial
protection against various environmental challenges such as oxidative stress
or TNF (reviewed in 69). Unfortunately, these protective responses are not
functional in the E1A negative vectors used for therapy.

That adenovirus carries genes that block interferon responses and upregulate heat shock genes suggests that these responses are encountered during a wild-type infection. It is not clear, however, what component of the adenovirus infection is responsible for the activation of the interferon response.
The timing of the interferon response suggests that the activation coincides
with the onset of late gene expression and perhaps significant quantities of
double-stranded RNA are then produced (64). This is an argument that

further genetic inactivation of adenoviral vectors to remove the potential to generate ds RNA molecules is a worthwhile effort. Indeed, substantial progress in this direction has been made with viruses combined defective E1,E2, E1, E4 and E1,E2, E4 are being developed. Carrying this strategy to the extreme will generate a "gutless" or generation X vector bearing only the viral sequences required for encapsidation and initial efforts in this direction have been made, although titers are still below practical levels (70). This type of recombinant adenovirus vector approaches the viruses available with AAV and retroviral systems.

C. Capsid Induction of an Inflammatory Response

There is also evidence that the virus capsid itself, or changes produced in the cell upon virus capsid entry, can activate an inflammatory response. The application of empty adenovirus particles or adenovirus particles with the DNA inactivated by psoralen treatment can generate lung inflammatory responses comparable to the application of noninactivated virus particles (71). This suggests that the capsid itself in the absence of virus gene expression could be responsible for some of the inflammatory responses and efforts to limit inappropriate virus gene expression will not completely eliminate the response to virus entry.

The viral anti-interferon defences have been removed (E1A, VA) the virus's ability to upregulate the stress response has been removed (E1A), the viral E3-encoded anti-TNF and anti-CTL functions have been removed (72–74); and the viral antiapoptotic functions are gone (E1B 55K, E1B 19K, E3 14.7K, 10.5K, and 14.5K gene products; reviewed in 75). Clearly some of the inflammatory and apoptotic functions have also been removed from these vectors, such as the synthesis of dsRNA and the E1A effects on RB, cell proliferation and apoptosis, and the E1A effects on the AP-1 pathway and the E3 11.6K death protein (76). However, evidence is accumulating that the virus capsid components possess the ability to activate some of the inflammatory and apoptotic response. A major concern then is that by altering the balance of gene expression in the recombinant adenovirus vectors we have created viral particles that have enhanced rather than decreased inflammatory and apoptotic responses.

VIII. Efforts to Modulate Immune Responses to Adenovirus Vectors, Survival of Modified Cells, Avoiding the Immune System

It has become clear that a major limit to adenovirus-mediated gene correction is the host cellular and humoral immune response to both the viral vector and to the modified cells (77–83; reviewed in 84). These same immune responses may plague other types of gene delivery systems. It appears that many of the problems facing gene therapy are very similar to those facing transplantation biology. Fortunately, advances in our understanding of the immune system are providing some powerful tools to manipulate these immune responses. These include the use of the cytokines IL-12 and interferon gamma to prevent humoral responses (85), and the use of immunosuppressants such as cyclosporin, cyclophosphamide, or dexamethasone to prevent a cellular immune responses (78,80). The coadministration of CTLA4 Ig, which can block signaling between T-cells and antigen-presenting cells, has been shown to be effective in preventing the cellular immune response to transduced cells (86). The most exciting part of these discoveries is the observation that the immunosuppressive agents need only be applied for a short period to obtain long term benefits. Thus in a clinical application, the treated patient is not permanently immunosuppressed but only modulated for a short time. The success of these initial immunosuppressive strategies bodes well for this approach.

IX. Conclusions, Future Directions

After listing the variety of problems associated with adenovirus vectors, one must weigh the benefits associated with adenovirus use with this substantial list of drawbacks. It is clear that there is a substantial list of benefits associated with adenovirus vectors, benefits that far outweigh any of the drawbacks described in this article. Largely because of its use as a tool for molecular biology, the subgroup C adenoviruses are among the most thoroughly characterized human viruses. There is a substantial clinical experience with adenovirus vaccines. The vectors are genetically stable and can be easily constructed, grown, and purified to clinical standards. In fact, if adenovirus did not have such a vast potential as a gene therapy vector, we would not bother to examine in such detail the problems associated with the virus. We can also argue that many of the problems listed here are not unique to adenovirus vectors and will certainly be encountered with other vectors, both viral and nonviral. Solving these problems once for adenovirus will provide a number of tools applicable to other systems. Finally, we should mention that many of these problems of immune system activation are actually benefits when adenovirus is used as a vaccine.

References

1. Philipson L, Longberg-Holm K, Petterson U. Virus-receptor interaction in an adenovirus system. J Virol 1968; 2:1064–1075.

2. Stevenson SC, Rollence M, White B, Weaver L, McClelland A. Human adenovirus serotypes 3 and 5 bind to two different cellular receptors via the fiber head domain. J Virol 1995; 69:2850–2857.
3. Henry LJ, Xia D, Wilke ME, Deisenhofer J, Gerard RD. Characterization of the knob domain of the adenovirus type 5 fiber protein expressed in *Escherichia coli*. J Virol 1994; 68:5239–5246.
4. Michael SI, Huang CH, Romer MU, Wagner E, Hu PC, Curiel DT. Binding-incompetent adenovirus facilitates molecular conjugate-mediated gene transfer by the receptor-mediated endocytosis pathway. J Biol Chem 1993; 268:6866–6869.
5. Gall J, Kass-Eisler A, Leinwand L, Falck-Pedersen E. Adenovirus type 5 and 7 capsid chimera: Fiber replacement alters receptor tropism without affecting primary immune neutralization epitopes. J Virol 1996; 70:2116–2123.
6. Michael SI, Hong JS, Curiel DT, Engler JA. Addition of a short peptide ligand to the adenovirus fiber protein. Gene Ther 1995; 2:660–668.
7. Wohlfart C. Neutralization of adenovirus: kinetics, stoichiometry, and mechanisms. J Virol 1988; 62:2321–2328.
8. Mastrangeli A, Harvey B-G, Yao J, et al. "Sero-Switch" adenovirus-mediated in vivo gene transfer: circumvention of anti-adenovirus humoral immune defenses against repeat adenovirus vector administration by changing the adenovirus serotype. Human Gene Ther 1996; 7:79–87.
9. Buschle M, Cotten M, Kirlappos H, et al. Receptor-mediated gene transfer into human T-lymphocytes via binding of DNA/CD3 antibody particles to the CD3 T cell receptor complex. Human Gene Ther 1995; 6:753–761.
10. Cotten M, Wagner E, Zatloukal K, Birnstiel ML. Chicken adenovirus (CELO virus) particles augment receptor-mediated DNA delivery to mammalian cells and yield exceptional levels of stable transformants, JVirol 1993; 67:3777–3785.
11. Batra RK, Wang-Johanning F, Wagner E, Garver RI, Curiel DT.Receptor-mediated gene delivery employing lectin-binding specificity. Gene Ther 1994; 1:255–260.
12. Bai M, Harfe B, Freimuth P. Mutations that alter an Arg-Gly-Asp (RGD) sequence in the adenovirus type 2 penton base protein abolish its cell rounding activity and delay virus reproduction in flat cells. J Virol 1993; 67:5198–5205.
13. Wickham TJ, Mathias P, Cheresh DA, Nemerow GR. Integrins $a_v\beta_3$ or $a_v\beta_5$ promote adenovirus internalization but not virus attachment. Cell 1993; 73:309–319.
14. Belin M-T, Boulanger P. Involvement of cellular adhesion sequences in the attachment of adenovirus to the HeLa cell surface. J Gen Virol 1993; 74:1485–1497.
15. White JM. Integrins as virus receptors. Curr Biol 1993; 3:596–599.
16. Nemerow GR, Cheresh DA, Wickham TJ. Adenovirus entry into host cells: a role for av integrins. Trends Cell Biol 1994; 4:52–55.
17. Huang S, Endo RI, Nemerow GR. Upregulation of integrins avß3 and avß5 on human monocytes and T-lymphocytes facilitates adenovirus-mediated gene delivery. J Virol 1995; 69:2257–2263.
18. Greber UF, Webster P, Weber J, Helenius A. The role of the adenovirus protease on virus entry into cells. EMBO J 1996; 15:1766–1777.
19. Wickham TJ, Carrion ME, Kovesdi I. Targeting of adenovirus penton base to new receptors through replacement of its RGD motif with other receptor-specific peptide motifs. Gene Ther 1995; 2:750–756.

20. Schwartz MA, Ingber D, Lawrence M, Springer TA, Lechene C. Multiple integrins share the ability to induce elevation of intracellular pH. Exp Cell Res 1991; 195: 533–535.
21. Schwartz MA. Spreading of human endothelial cells on fibronectin or vitronectin triggers elevation of intracellular free calcium. J Cell Biol 1993; 120:1003–1010.
22. Grubb BR, Pickles RJ, Ye H, et al. Inefficient gene transfer by adenovirus vectors. Nature 1994; 271:802–806.
23. March KL, Madison JE, Trapnell BC. Pharmokinetics of adenoviral vector-mediated gene delivery to vascular smooth muscle cells: modulation by poloxamer 407 and implications for cardiovascular gene therapy. Human Gene Ther 1995; 6:41–53.
24. Huard J, Lochmüller H, Acsadi G, Jani A, Massie B, Karpati G. The route of administration is a major determinant of the transduction efficiency of rat tissues by adenoviral recombinants. Gene Ther 1995; 2:107–115.
25. Fernández-Puentes C, Carrasco L. Viral infection permeabilizes mammalian cells to protein toxins. Cell 1980; 20:769–775.
26. Fitzgerald D, Padmanabhan R, Pastan I, Willingham M. Adenovirus-induced release of epidermal growth factor and *Pseudomonas* toxin into the cytosol of KB cells during receptor-mediated endocytosis. Cell 1983; 32:607–617.
27. Carrasco L. Entry of animal viruses and macromolecules into cells. FEBS Lett 1994; 350:151–154.
28. Carrasco L. Modification of membrane permeability by animal viruses. Adv Virus Res 1995; 45:61–112.
29. Curiel DT, Agarwal S, Wagner E, Cotten M. Adenovirus enhancement of transferrin-polylysine mediated gene delivery. Proc Natl Acad Sci USA 1991; 88:8850–8854.
30. Wagner E, Zatloukal K, Cotten M, et al. Coupling of adenovirus to transferrin-polylysine/DNA complexes greatly enhances receptor-mediated gene delivery and expression of transfected cells. Proc Natl Acad Sci USA 1992; 89:6099–6103.
31. Cristiano RJ, Smith LC, Woo SL. Hepatic gene therapy: adenovirus enhancement of receptor-mediated gene delivery and expression in primary hepatocytes. Proc Natl Acad Sci USA 1993; 90:2122–2126.
32. Cristiano RJ, Smith LC, Kay MA, Brinkley BR, Woo SL. Hepatic gene therapy: efficient gene delivery and expression in primary hepatocytes utilizing a conjugated adenovirus-DNA complex. Proc Natl Acad Sci USA 1993; 90:11548–11552.
33. Yoshimura K, Rosenfeld M, Seth P, Crystal R. Adenovirus-mediated augmentation of cell transfection with unmodified plasmid vectors. J Biol Chem 1993; 268: 2300–2303.
34. Fisher KJ, Wilson JM. Biochemical and functional analysis of an adenovirus-based ligand complex for gene transfer. Biochem J 1994; 299:49–58.
35. Mange WF, Toledo DL, Brown MT, Martin JH, McGrath WJ. Characterization of three components of human adenovirus proteinase activity in vitro. J Biol Chem 1996; 271:536–543.
36. McGrath WJ, Abola AP, Toledo DL, Brown MT, Mangel WF. Characterization of human adenovirus proteinase activity in disrupted virus particles. Virology 1996; 217:131–138.

37. Rancourt C, Tihanyi K, Bourbonniere M, Weber JM. Identification of active site residues of the adenovirus endopeptidase. Proc Natl Acad Sci USA 1994; 91:844–847.
38. Webster A, Hay R, Kemp G. The adenovirus protease is activated by a virus-coded disulphide-linked peptide. Cell 1993; 72:97–104.
39. Webster A, Leith IR, Hay RT. Activation of adenovirus-coded protease and processing of preterminal protein. J Virol 1994; 68:7292–7300.
40. Weber JM, Tihanyi K. Adenovirus endopeptidases. Methods Enzymol 1994; 244: 595–604.
41. Weber JM. Adenovirus endopeptidase and its role in virus infection. In: Doerfler W, Böhm P, eds. *Current Topics in Microbiology and Immunology*. Berlin: Springer, 1995:227–235.
42. Cotten M, Längle-Rouault F, Kirlappos H, et al. Transferrin-polycation-mediated introduction of DNA into human leukemic cells: stimulation by agents that affect the survival of transfected DNA or modulate transferrin receptor levels. Proc Natl Acad Sci USA 1990; 87:4033–4037.
43. Cotten M, Weber JM. The adenovirus protease is required for virus entry into host cells. Virology 1995; 213:494–502.
44. Miles BD, Luftig RB, Weatherbee JA, Weihing RR, Weber J. Quantitation of the interaction between adenovirus types 2 and 5 and microtubules inside infected cells. Virology 1980; 105:265–269.
45. Defer C, Belin M, Caillet-Boudin M, Boulanger P. Human adenovirus-host cell interactions: comparative study with members of subgroups B and C. J Virol 1990; 64:3661–3673.
46. Zauner W, Blaas D, Küchler E, Wagner E. Rhinovirus mediated endosomal release of transfection complexes. J Virol 1995; 69:1085–1092.
47. Prchla E, Plank C, Wagner E, Blaas D, Fuchs R. Virus-mediated release of endosomal content in vitro: different behavior of adenovirus and rhinovirus serotype 2. J Cell Biol 1995; 131:111–123.
48. Seth P, Willingham MC, Pastan I. Adenovirus-dependent release of 51Cr from KB cells at an acidic pH. J Biol Chem 1984; 259:14350–14353.
49. Seth P, Pastan I, Willingham MC. Adenovirus-dependent increase in cell membrane permeability. J Biol Chem 1985; 260:9598–9602.
50. Greber UF, Willetts M, Webster P, Helenius A. Stepwise dismantling of adenovirus 2 during entry into cells. Cell 1993; 75:477–486.
51. Pérez L, Carrasco L. Involvement of the vacuolar H^+-ATPase in animal virus entry. J Gen Virol 1994; 75:2595–2606.
52. Greber UF, Singh I, Helenius A. Mechanisms of virus uncoating. Trends Microbiol 1994; 2:52–56.
53. Chen P, Ornelles D, Shenk T. The adenovirus L3 23-kilodalton proteinase cleaves the amino-terminal head domain from cytokeratin 18 and disrupts the cytokeratin network of HeLa cells. J Virol 1993; 67:3507–3514.
54. Zhang Y, Schneider RJ. Adenovirus inhibition of cell translation facilitates release of virus particles and enhances degradation of the cytokeratin network. J Virol 1994; 68:2544–2555.

55. Huang J, Schneider RJ. Adenovirus inhibition of cellular protein synthesis is prevented by the drug 2-aminopurine. Proc Natl Acad Sci USA 1990; 87:7115–7119.
56. Thimmappaya B, Weinberger C, Schneider RJ, Shenk T. Adenovirus VAI RNA is required for efficient translation of viral mRNAs at late times after infection. Cell 1982; 31:543–551.
57. Svensson C, Akusjärvi G. Adenovirus VA RNAI: a positive regulator of mRNA translation. Mol Cell Biol 1984; 4:736–742.
58. Manche L, Green SR, Schmedt C, Mathews MB. Interactions between double-stranded RNA regulators and the protein kinase DAI. Mol Cell Biol 1992; 12: 5238–5248.
59. Mathews MB Shenk T. Adenovirus virus-associated RNA and translation control. J Virol 1991; 65:5657–5662.
60. Reich NC, Pine R, Levy D, Darnell JE Jr. Transcription of interferon-stimulated genes is induced by adenovirus particles but is suppressed by E1A gene products. J Virol 1988; 62:114–119
61. Akrill AM, Foster GR, Laxton CD, Flavell DM, Stark GR, Kerr IM. Inhibition of the cellular response to interferons by products of the adenovirus type 5 E1A oncogene Nucleic Acids Res 1991; 19:4387–4393.
62. Gutch MJ, Reich NC. Repression of the interferon signal transduction pathway by the adenovirus E1A oncogene. Proc Natl Acad Sci USA 1991; 88:7913–7917.
63. Kalvakolanu DVR, Bandyopadhyay SK, Harter ML, Sen GC. Inhibition of interferon-inducible gene expression by adenovirus E1A proteins: block in transcriptional complex formation. Proc Natl Acad Sci USA 1991; 88:7459–7463.
64. Daly C, Reich NC. Double-stranded RNA activates novel factors that bind to the interferon-stimulated response element. Mol Cell Biol 1993; 13:3756–3764.
65. Wu B, Hurst H, Jones N, Morimoto R. The 13S product of adenovirus 5 activates transcription of the cellular human HSP70 gene. Mol Cell Biol 1986; 6:2994–2999.
66. Lum LSY, Hsu S, Vaewhongs M, Wu B. The hsp70 gene CCAAT-binding factor mediates transcriptional activation by the adenovirus E1A protein. Mol Cell Biol 1992; 12:2599–2605.
67. Kraus VB, Moran E, Nevins JR. Promoter-specific trans-activation by the adenovirus E1A 12S product involves separate E1A domains. Mol Cell Biol 1992; 12:4391–4399.
68. Agoff SN, Wu B. CBF mediates adenovirus E1a trans-activation by interacting at the C-terminal promoter targeting domain of conserved region 3. Oncogene 1994; 9:3707–3711.
69. Parsell DA, Lindquist S. Heat shock proteins and stress tolerance. In: *The Biology of Heat Shock Proteins and Molecular Chaperones*. Cold Spring Harbor, NY: Cold Spring Harbor Laboratory Press, 1994:457–494.
70. Fisher KJ, Choi H, Burda J, Chen S-J, Wilson JM. Recombinant adenovirus deleted of all viral genes for gene therapy of cystic fibrosis. Virology 1996; 217:11–22.
71. McCoy RD, Davidson BL, Roessler BJ, et al. Pulmonary inflammation induced by incomplete or inactivated adenoviral particles. Human Gene Ther 1995; 6:1553–1560.
72. Wold WSM, Gooding LR. Region of E3 of adenovirus: a cassette of genes involved in host immunosurveillance and virus-cell interactions. Virology 1991; 184:1–8.

73. Gooding LR. Virus proteins that counteract host immune defenses. Cell 1992; 71:5–7.
74. Wold WSM. Adenovirus gene that modulate the sensitivity of virus-infected cells to lysis by TNF. J Cell Biochem 1993; 53:329–335.
75. White E, Gooding LR. Regulation of apoptosis by human adenoviruses. In: Apoptosis II: The Molecular Basis of Apoptosis in Disease. Cold Spring Harbor, NY: Cold Spring Harbor Laboratory Press. 1994.
76. Tollefson AE, Scaria A, Hermiston TW, Ryerse JS, Wold LJ, Wold WS. The adenovirus death protein (E3-11.6K) is required at very late stages of infection for efficient cell lysis and release of adenovirus from infected cells. J Virol 1996; 70: 2296–2306.
77. Barr D, Tubb J, Ferguson D, et al. Strain related variations in adenovirally mediated transgene expression from mouse hepatocytes in vivo: comparisons between immunocompetent and immunodeficient inbred strains. Gene Ther 1995; 2:151–155.
78. Dai Y, Schwarz EM, Gu D, Zhang WW, Sarvetnick N, Verma IM. Cellular and humoral immune responses to adenoviral vectors containing factor IX gene: tolerization of factor IX and vector antigens allows for long-term expression. Proc Natl Acad Sci USA. 1995; 92:1401–1405.
79. Yang Y, Li Q, Ertl HCJ, Wilson JM. Cellular and humoral immune responses to viral antigens create barriers to lung-directed gene therapy with recombinant adenoviruses. J Virol 1995; 69:2004–2015.
80. Zsengeller ZK, Wert SE, Hull WM, et al. Persistence of replication-deficient adenovirus-mediated gene transfer in lungs of immune-deficient (nu/nu) mice. Human Gene Ther 1995; 6:457–467.
81. DeMatteo RP, Markmann JF, Kozarsky KF, Barker CF, Raper SE. Prolongation of adenoviral transgene expression in mouse liver by T lymphocyte subset depletion. Gene Ther 1996; 3:4–12.
82. Sawchuk SJ, Boivin GP, Duwel LE, et al. Anti-T cell receptor monoclonal antibody prolongs transgene expression following adenovirus-mediated in vivo gene transfer to mouse synovium. Human Gene Ther 1996; 7: 499–506.
83. Dong J-Y, Wang D, Van Ginkel FW, Pascual DW, Frizzell RA. Systemic analysis of repeated gene delivery into animal lungs with a recombinant adenovirus vector. Human Gene Ther 1996; 7:319–331.
84. Wilson C, Kay MA. Immunomodulation to enhance gene therapy. Nature Med 1995; 1:887–889.
85. Yang Y, Trinchieri G, Wilson JM. Recombinant IL-12 prevents formation of blocking antibodies to recombinant adenovirus and allows repeated gene therapy to mouse lung. Nature Med 1995; 1:890–893.
86. Kay MA, Holterman AX, Meuse L, et al. Long-term hepatic adenovirus-mediated gene expression in mice following CTLA4Ig administration. Nature Genet 1995; 11:191–197.

6

Cationic Lipid-Based Gene Delivery: An Update

Xiang Gao

Vanderbilt University School of Medicine
Nashville, Tennessee

I. Introduction

With the advent of our understanding of human inherited and acquired diseases at a molecular level, the use of therapeutic genes to restore a missing function or to treat certain disease stages is rational. Currently, however, a major hurdle for successful gene therapy is lack of a safe and efficient gene delivery technology for human uses. Replication-defective viral vectors and synthetic nonviral systems are two major DNA delivery systems for gene therapy. The viral vectors, such a retrovirus, adenovirus, or adeno-associated virus (AAV), which are capable of introducing foreign DNA into cells to achieve transient or stable expression with high efficiency, are currently being tested in human clinical trials. Safety issues and immunological responses are major concerns for repeated administration of the viral vectors. The synthetic vectors, mainly represented by cationic lipid/liposomes and cationic polymers, are an alternative to the viral vectors. They have low toxicity and low immunogenicity, they are capable of delivering large DNA to cells, and they are easy to produce in large quantities—all major advantages over viral systems. The use of cationic liposomes has become popular because of their high transfection efficiency in vitro (1–6), reported their successful transfection results in vivo (7–12), their simplicity of use, and their ready availability from various commercial sources. Cationic liposome-mediated gene transfer, better known as "lipofection" (1) or "cytofection" (13), was developed by Dr. Philip Felgner a decade ago. Since then, many cationic lipids have been reported as in vitro and in vivo gene transfer reagents (2–6). Based on initial preclinical studies, several human gene therapy clinical trials using cationic liposome-DNA complexes targeted to human diseases, including cancer and cystic fibrosis, have been conducted or are in progress (14,15). Over the past few years, several review papers have been published which discuss structure-activity aspects of cationic lipids and results of clinical trials (13,16–18). This discussion will focus on the most recent advances in this field, including new lipids and formulations, new applications, and recent discoveries related to mechanisms. Readers are encouraged to obtain comprehensive information from the earlier reviews.

II. New Cationic Lipids and Formulations

Figure 1a illustrates the structure of some well-known cationic lipids that have been widely used over the years. A functional cationic lipid contains a positively charged head group, a spacer, a linkage bond, and a hydrophobic domain. Each individual structural domain seems to be important for the transfection activity of a cationic lipid. For example, the number of protonable nitrogen atoms and the pKa of each nitrogen atom determine the charge multiplicity of the head group; a spacer provides accessibility of the head group to the negative charge of phosphate groups on the nucleic acid backbone; the linkage bond determines the stability and potential biodegradability of the lipids; and the hydrophobic portion of the molecule, which can be either two long hydrocarbon chains or a cholesterol ring structure, determines the packing order of lipids and membrane flexibility [for review see (17)].

Some new cationic lipids are shown in Figure 2. Some of these lipids have structures similar to those in Figure 1, but some have rather unusual structures. Several new monovalent cationic lipids are reported to be more efficient at

---

Abbreviations: DC-chol, 3β[(N',N'-dimethylaminoethane)carbamoyl]-cholesterol; DMRIE, 1,2-dimyristyloxypropyl-3-dimethyl-hydroxyl ammonium chloride; DOGS, dioctadecyldimethylammonium chloride; DOPE, dioleoylphosphatidylethanolamine; DOPC, dioleoylphosphatidylethanolcholine; DOSPER, 1,3-dioleoyoxy-2-(6-carboxy-spermyl)-propylamide; DOSPA, 2,3-dioleoyloxy-N-[2(sperminecarboxamido)ethyl]-N,N-dimethyl-1-propanamanium trifluoroacetate; DOTAP, 1,2-dioleoyloxy-3-(trimethylammonio)propane; DOTIM, 1-[2-(oleoyoxy)-ethyl]-2-oleyl-3-(2-hydroxyethyl)imidazolinium chloride; DOTMA, N-[2,3-(dioleoyloxy)propyl]-N,N,N-trimethylammonium.

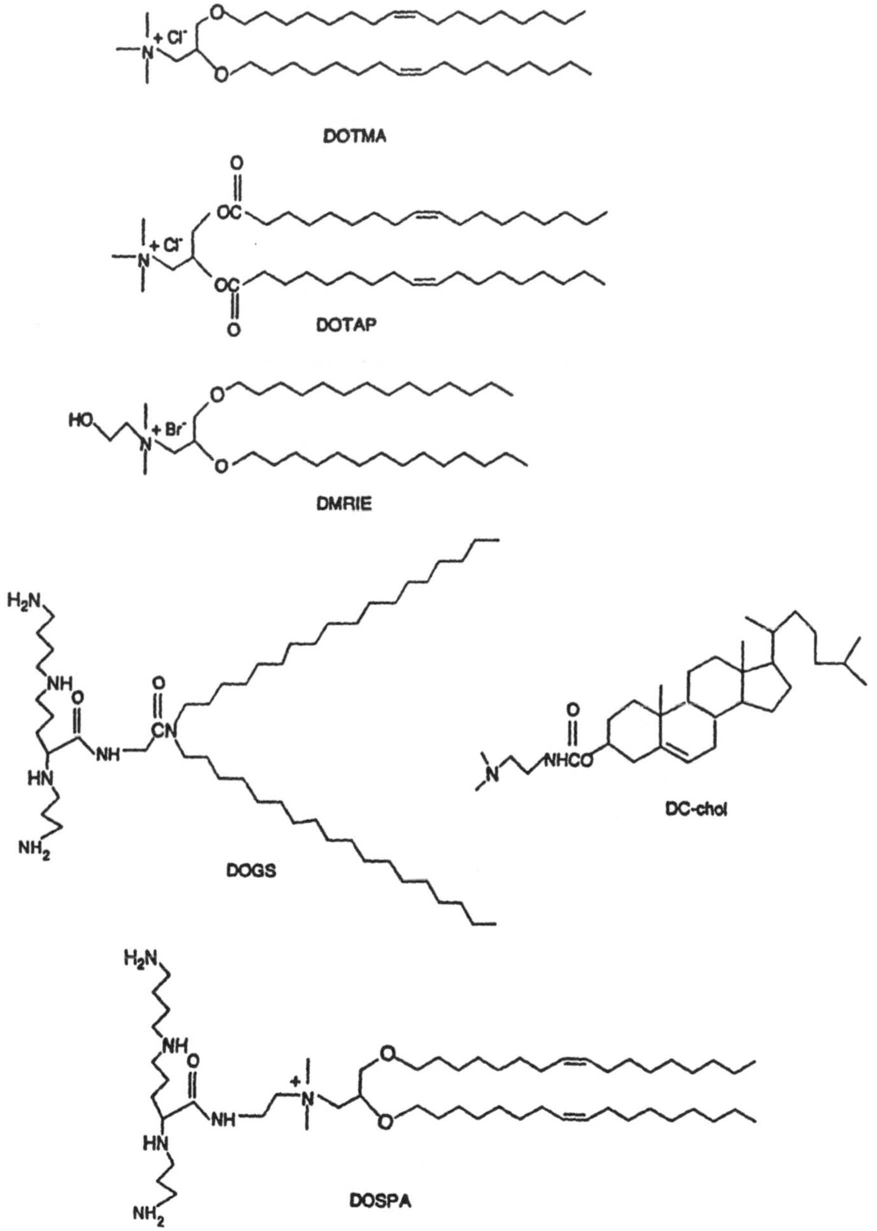

Figure 1 Chemical structure of commonly used cationic lipids.

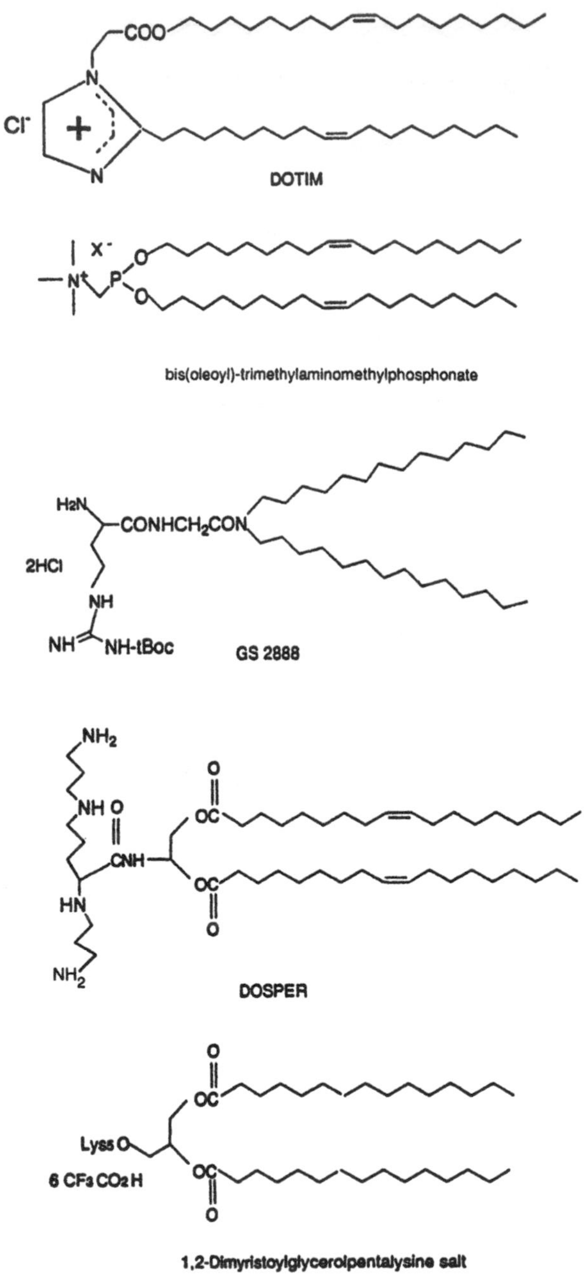

Figure 2 Chemical structure of newly reported cationic lipids.

Spermine-cholesterol

Cholesteryl-Spermindine

Lipid 67

Y= spermine, pentamine or hexamine
R= α-glucoside

polycationic facial amphiphiles

transfection in vitro under serum-free conditions (19) or in the presence of serum (20) than current formulations. An interesting series of 1,3-dialkylated imidazolinium derivatives was reported to be active in vivo in mouse organs after DNA/lipid complexes were administered intravenously (21).

Among these new lipids, a multivalent polyamine head group such as spermine or spermidine seems to be the most popular choice. The spermine or the spermidine group was either linked to the hydrophobic domain through the terminal amine group in a linear polyamine configuration (12,22) or coupled through the central secondary amino group in a T shape (12,23,24). These polyamine head groups have been used previously in lipids known to have excellent transfection activity in vitro, such as DOGS* (2) and DOSPA (5). Not surprisingly, new cationic lipids with either a linear or a T-shaped polyamine head group can mediate efficient gene (12,23,24) and oligonucleotide delivery in vitro (22). Similarly, other multivalent cationic head groups, such as a pentamine or hexamine (25) and a pentalysine head group (26) were also found active in vitro. It is known that multivalent spermine groups can condense DNA, resulting in smaller and more compact DNA/lipid complexes, which may partially explain high transfection efficiency.

Recently, an elegant structure-activity study on a series of multivalent cationic lipids synthesized by a group from Genezyme Inc. revealed that while many multivalent cationic lipids could transfect cultured cells in vitro, only a few of these lipids showed significant in vivo transfection activity when delivered through airways to mouse lungs (12). High in vivo transfection activity was determined by several structural characteristics of the cationic lipids. It was shown that cationic lipids with multivalent head groups (lipid 67) were 100-fold more active than monovalent cationic lipids with a similar overall structure (DC-chol), although the differences with in vitro transfection were much smaller. In contrast to in vitro transfection, only cationic lipids with T-shaped multivalent head groups and not the ones with linear polyamine head groups showed significant activity in vivo. The chemical linker appeared to be critical as well. For instance, the most active lipid of this series (lipid 67), which has a T-shaped spermine head group lined to a cholesterol motif through a carbamate linkage, was far more active than similar lipids with different chemical linkages, such as urea, amide, or amine bonds. In vivo studies on mouse lungs also demonstrated that a cholesterol hydrophobic domain appears to be a better choice than double hydrocarbon chains in this series of lipids.

Besides these important structural features of cationic lipids, other factors also contribute to high in vivo transfection activities. These include the degree of protonation of the head group amines, the ratio of the cationic lipid to a neutral lipid, DOPE, and the ratio of DNA-to-lipids. The highest in vivo transfection activity of lipid 67 was found when it was used in its free base form together with DOPE at a molar ratio of 1:2 and a DNA to lipid molar ratio equal to 8:1 (nucleotide to lipid). These results indicate that in vivo transfection activity of cationic liposomes is closely related not only to the chemical structure of the cationic component but also to the way the lipid/DNA complex was formulated. The exact mechanism behind the phenomenon of rather narrow structural and formulation requirements for high levels of in vivo transfection activity of this series of cationic lipids is not clear.

III. New DNA/Lipid Formulations

Although significant effort has been made to synthesize and screen new cationic lipids, several groups have attempted to improve transfection efficiency by improving DNA/lipid formulations using existing cationic lipids (11,27,28). A standard lipid/DNA formulation is usually prepared by mixing dilute solutions of small unilamellar liposomes, prepared from lipid by sonication or microfluidization, and DNA, followed by a short incubation. It has been reported that complexes formed between mixtures of DNA and multilamellar liposomes are as efficient at transfection as complexes formed between DNA and small unilamellar liposomes (6). Plasmid DNA has also been entrapped inside large unilamellar cationic liposome vesicles or in the space between bilayer membranes of multilamellar vesicles. The resulting particles showed a slight increase in transfection activity and in some cases reduced toxicity (27). Taking a similar approach, Thierry and colleagues recently developed a DOPE-lipospermine system (DLS) based on DOGS/DOPE multilamellar liposomes (11). The complexes were prepared by hydrating a thin lipid film of DOGS and DOPE at 1:1 molar ratio with concentrated DNA solution followed by a vortex. The resulting DNA-lipid complexes of 300–1000 nm in size were reported to be active in vivo when delivered intravenously (11,29).

Hofland et al. took a different approach to making stable cationic lipid/ DNA complexes (28). By using initial complexation of DNA with DOSPA/ DOPE/octylglucoside mixed micelles followed by removal of octylglucoside detergent from the complex by dialysis, stable and biologically active DNA/ lipid complexes were obtained. This new formulation was more active and more resistant to inhibitory effects of serum than freshly prepared DNA/lipid complexes using traditional methods.

IV. DNA/Lipid/Polymer Formulations

One common problem associated with cationic liposome/DNA delivery systems is that cationic liposomes have a tendency to form large unstable particles when complexed with DNA, especially when high concentrations of DNA and liposome are used. The relatively large size of liposomes (about 100–300 nm in diameter), insufficient DNA condensation for some cationic liposome formulations, and aggregation of unstable DNA/lipid intermediates could contribute to this problem. Ultrastructural studies have provided direct evidence for less condensed DNA/lipid structures, such as elongated DNA coated with a lipid tube attached to semi-intact liposomes (31) or DNA coated with lipid and aligned into hexagonal bundles (32,33). Morphological changes of DNA-lipid complexes over time also imply formation of the DNA/lipid intermediates (31).

On the contrary, it is known that some cationic polymers can condense DNA into small particles of 15–100 nm in size (34,35). These DNA/polymer complexes, however, are relatively inefficient at transfection, unless they are accompanied by agents, such as chloroquine, that perturb endosome maturation, or treatments like osmotic shock or viral fusion proteins which destroy endosomal membranes (35–38). Since it is generally believed that DOPE-containing cationic liposomes have intrinsic cell membrane fusion activity (1), an interesting question was asked: Would a combination of polycations, which have superior DNA condensation capability, and cationic liposomes, which can rupture endosomes, result in a new generation of DNA delivery systems? To explore such a possibility, Gao and Huang recently tested several high-molecular-weight cationic polymers for their ability to modulate DNA/lipid complex formulation and to improve gene transfection activity (39). Several polymers such as polylysine or protamine significantly potentiated transfection activity by twofold to 28-fold in a number of cell types with several cationic liposomes in vitro, including DC-chol/DOPE and Lipofectin, particularly at lower than the usual optimal ratios of lipid to DNA used for transfection. Polylysine potentiated transfection was less for LipofectAMINE, a liposome formulation that condenses DNA well. Mechanistic studies revealed that polycations could drastically reduce the size of DNA-lipid complexes and that DNA, in the form of a polycation/liposome complex, became more resistant to nuclease activity present in fetal bovine serum. Interestingly, ultracentrifugation analysis of the DNA/polylysine/DC-chol liposome complexes on a sucrose density gradient revealed that purified liposome/polylysine/DNA (LPD) complexes contain small, condensed structures of 30–70 nm in size. These purified particles are three- to ninefold more potent in transfection compared to unpurified complexes based on the same amount of DNA. Electron microscopic images of these purified particles clearly showed that lipidic structures were associated with a proportion of the particles.

Two other groups, Worton and his colleagues (40) and Wolff and his colleagues (41) conducted similar studies independently with the same conclusion. In studies by Wolff's group, calf thymus histone H1 and a recombinant human histone H1 carrying an additional nuclear localization signal appeared to be the best polycations (enhancing transfection by Lipofectin as much as 45-fold), followed by histone 2A/B, polylysine, and protamine in that rank order. They also concluded that much less Lipofectin is needed when polycations are included. The different potentiation effect of these polycations could be due to different biocompatability or differences in DNA condensation activity (41). The results from these studies clearly demonstrate that incorporating polycations into cationic liposomes could dramatically alter the structure and transfection activity of DNA/lipid complexes.

Earlier, several studies demonstrated that ligand conjugated cationic polymers may provide an opportunity for ligand-specific cell uptake (34–38, 42). For example, a polylysine conjugate using a monoclonal antibody against a mouse endothelial antigen has been shown to facilitate the targeting of plasmid DNA administered intravenously to endothelial cells (42). In an in vitro study of cultured mouse lung cells using the same antibody-polylysine conjugate, addition of DC-chol liposome was also shown to enhance transfection efficiency 10- to 20-fold (43). Taking into account that current effort to synthesize new and better cationic lipids is slow and costly, and largely relies on trial and error, using polycations to improve the transfection efficiency of existing cationic liposome formulations is an interesting approach.

V. Recent Progress on Mechanisms of Lipid-Mediated Transfection

After nearly 10 years of research, development, and numerous applications based on the technology of cationic lipid-mediated gene transfer, many fundamental questions related to the mechanisms of transfection are still not well addressed. Critical steps that lead to a successful gene transfection include initial interaction of the complex with the cell surface and entry into the cell, cell membrane penetration, DNA/lipid dissociation, and nuclear translocation. Over the past few years, a series of studies using different approaches has provided evidence that endocytosis is a major mechanism of DNA/lipid complex uptake (30,33,44). These include morphological studies on cells at different stages of transfection using electron microscopy and effects of various pharmacological interventions on cellular endocytosis activity. Zhou and Huang showed that the majority of DNA/lipid complexes entered the cell first through coated pits, then traveled through early endosomes and late endosomes, and eventually reached lysosomes (44). A small fraction of the complexes managed to escape from endosomes before being degraded in lysosomes. Two recent electron microscopic studies by other groups support the observations of Zhou and Huang (30,33). The results of morphological studies by Zhou and Huang were also supported by the fact that chloroquine treatment, which slows down maturation of endosomes to lysosomes, enhanced transfection.

Very little is known about how lipid-DNA complexes escape the endosomal compartment. Some monovalent cationic lipid-based formulations require a neutral helper lipid, DOPE, for efficient transfection activity. Substitution of DOPE with DOPC, a phospholipid with a choline instead of an ethanolamine head group, drastically reduced transfection activity in some formulations (1,46). DOPE appears to play an important role in the endosomal release process due to its fusogenic activity. DOPE has an unusual non-bilayer-forming property at physiological conditions, while DOPC is a bilayer-forming lipid. At pH 7.0, pure DOPE exists in a hexagonal configuration, one of the lipid polymorphisms thought to be important in various cell membrane fusion events (46). An early study using fluorescence markers by Felgner and his colleagues indicated that extensive fusion occurred between cell membranes and free DOTMA/DOPE liposomes, but not with DOTMA/DOPC liposomes (1). Recently, Wrobel and Collins demonstrated that DC-chol/DOPE liposomes fused with cell membranes primarily within endosomes, not at the cell surface, as fusion was inhibited by treatments that reduced cellular endocytosis (47). The different results from these two studies could be a result of using different formulations. DC-chol is a weak base with a titratable tertiary amine head group. Inside the endosome, DC-chol becomes more protonated, which may trigger the fusion of DC-chol/DOPE liposomes with endosomal membranes more extensively than on the cell surface. DOTMA has a quaternary ammonium head group, making it a strong base or an abligated cation. DOTMA/DOPE liposomes could directly interact with plasma membranes. Although these cell-liposome fusion experiments were conducted with free liposomes, these results may be relevant to DNA/lipid complex transfection in cells as well. Farhood et al. showed that if one round of complex binding and endocytosis was permitted, fourfold more liposomes were needed to reach maximal transfection than the amount required to mediate maximal binding of DNA (48). What these data imply is that when there are excess free liposomes, free liposomes may enter the endosomal compartment with DNA/lipid complexes, thus facilitating release of complexes from the endosome, possibly via a liposome-endosome membrane fusion event like that observed by Wrobel and Collins (47). The mechanism of DOPE-independent, cationic liposome-mediated transfection is even less definitive. The less basic secondary amine group in the lipospermine molecule (pKa~5.4) has been thought to buffer acidification and slow down the process of endosome maturation (18).

Elegant work by Xu and Szoka uncovered a very important potential cellular mechanism that may facilitate the DNA-lipid dissociation process. In their model system, a panel of ionic molecules was tested for the ability to facilitate dissociation of DNA/cationic lipid complexes (49). Surprisingly, given the hypothesis that DNA molecules may be displaced by endogenous high concentration polycations such as histones in the nucleus, the most potent DNA dissolution agents turned out to be anionic lipid molecules abundantly located on the inner surface of the plasma membrane. To a lesser extent, other acid polymers, such as polysaccharides, heparin, and dextran sulfate, but not RNA or single-strand DNA, could also release DNA from cationic lipids. Xu and Szoka proposed that within the endosomal compartment, anionic cellular lipids could mix with cationic lipids and form a charge-neutral ion pair, thus competing with DNA for cationic lipids; as a result, DNA would be released in its free form. Their hypothesis is consistent with several earlier observations, including the observation that biologically functional plasmid DNA, presumably in its free form, is found in the cytoplasm after transfection, and that DNA-lipid complexes delivered to nuclei by direct microinjection were weaker in gene expression than the same complexes delivered into the cytoplasm. This suggests that intereaction in the cytoplasm leads to DNA-lipid dissociation and is important for restoration of DNA activity from DNA/lipid.

VI. Problems and Future Development

In summary, three major potential problems seem to limit the overall transfection efficiency of cationic liposomes. First, all of the reported electron microscopy studies have revealed a low frequency of DNA-lipid complex escaping from the endosomal compartment, and have shown that most of the particles are transported to lysosomes and degraded. Further development of DNA delivery systems aimed at increasing endosomal release of DNA-lipid complexes should be an interesting direction for research related to cationic liposome-mediated gene transfer. Another important issue is how to minimize the inhibitory effect of serum or other biological fluids. There are some initial observations indicating that certain chemical properties of cationic lipids (20) and certain DNA-lipid formulations (28) could render the system resistant to serum inhibition. Hopefully, more basic studies will help to elucidate the mechanisms of serum resistance and will provide a rationale for synthesis of new lipids or improved formulations. The third issue is a safety issue. Although cationic liposomes have been tested and proven to be safe for humans in relatively small doses, toxicity of doses likely to be required for clinical uses has not been thoroughly tested. Recently reported animal studies indicate that toxicity ranging from mild inflammation to death can occur with certain cationic liposome formulations (11,12) at high doses (11). Careful pharmacology and toxicology studies are needed to find the precise causes of toxicity so that cationic liposome-mediated gene delivery technology suitable for use in humans can be developed.

Acknowledgments

This work is supported by grants HL 45151, HL 19153, and AI 31900 from the National Institutes of Health. The author thanks Dr. Kenneth L. Brigham for critical review of the manuscript, and Tamara Lasakow for editorial assistance.

References

1. Felgner PL, Gadek TR, Holm M, et al. Lipofection: a highly efficient, lipid-mediated DNA transfection procedure. Proc Natl Acad Sci USA 1987; 84:7413–7417.
2. Behr J-P, Demeneix B, Loeffler J-P, Perez-Mutul J. Efficient gene transfer in mammalian primary endocrine cells with lipopolyamine-coated DNA. Proc Natl Acad Sci USA 1989; 86:6982–6986.
3. Gao X, Huang L. A novel cationic liposome reagent for efficient transfection in mammalial cells. Biochem Biophys Res Commun 1991; 179:280–285.
4. Leventis R, Silvius JR. Interactions of mammalian cells with lipid dispersions containing novel metabolizable cationic amphiphiles. Biochim Biophys Acta 1990; 1023:124–132.
5. Hawley-Nelson P, Ciccarone V, Gebeyehu G, Jessee J. LipofecAMINE reagent: a new, higher efficiency polycationic liposome transfection reagent. Focus 1993; 15:73–79.
6. Felgner JH, Kumar R, Sridhar CN, et al. Enhanced gene delivery and mechanism studies with a novel series of cationic lipid formulations. J Biol Chem 1994; 269: 2550–2561.
7. Brigham KL, Meyrick B, Christman B, Magnuson M, King G, Berry LC. Rapid communication: in vivo transfection of expression of murine lungs with a functioning prokaryotic gene using a liposome vehicle. Am J Med Sci 1989; 298:278–281.
8. Zhu N, Liggitt D, Liu Y, Debs R. Systemic gene expression after intravenous DNA delivery into adult mice. Science 1993; 261:209–211.
9. Plautz GE, Yang Zy, Wu BY, Gao X, Huang L, Nabel GL. Immunotherapy of malignancy by in vivo gene transfer into tumors. Proc Natl Acad Sci USA 1993; 90:4645–4649.
10. Canonico A, Conary J, Meyrick B, Brigham KL. Aerosol and intravenous transfection of human α1 antitrtypsin gene to lungs of rabbits. Am J Respir Cell Mol Biol 1994; 10:24–29.
11. Thierry AR, Lunardi-Iskandar Y, Bryant JL, Rabinovich P, Gallo RC, Mahan LC. Systemic gene therapy: Biodistribution and long-term expression of a transgene in mice. Proc Natl Acad Sci USA 1995; 92:9742–9746.
12. Lee ER, Marshall J, Siegel CS, et al. Detailed analysis of structure and formulations of cationic lipids for efficient gene transfer to the lung. Human Gene Ther 1996; 7:1701–1717.
13. Felgner PL. Cationic lipid/nucleotide condensates for in vitro and in vivo polynucleotide delivery- the cytofectins. J Liposome Res 1993; 3:3–16.
14. Nabel GL, Nabel EG, Yang Z, Fox BA, Plautz GE, Gao X, Huang L, Shu S, Gordon D, Chang AE. Direct gene transfer with DNA liposome complexes in melanoma: Expression, biological activity, lack of toxicity in humans. Proc Natl Acad Sci USA 1993; 90:11307–11311.
15. Caplen NJ, Alton EFWW, Middleton PG, et al. Liposome-mediated CFTR gene transfer to the nasal epithelium of patients with cystic fibrosis. Nature Med 1995; 1:39–46.
16. Ledley FD. Nonviral gene therapy: the promise of genes as pharmaceutical products. Human Gene Ther 1995; 6:1129–1144.

17. Gao X, Huang L. Cationic liposome-mediated gene transfer. Gene Ther 1995; 2:710–722.
18. Behr J-P. Gene transfer with synthetic cationic amphiphiles: prospects for gene therapy. Bioconjugate Chem 1994; 5:383–389.
19. Le Bolch G, Le Bris N, Yaouane J-J, Clement J-C, des Abbayes H. Cationic phosphonolipids as non viral vectors for DNA transfection. Tetrahedron Lett 1995; 36:6681–6684.
20. Lewis JG, Lin KY, Kothavale A, et al. A serum resistant cytofectin for cellular delivery of antisense oligodeoxynucleotides and plasmid DNA. Proc Natl Acad Sci USA 1996; 93:3176–3181.
21. Solodin I, Brown CS, Bruno MS, et al. A novel series of amphiphilic imidazolinium compounds for in vitro and in vivo gene delivery. Biochemistry 1995; 34:13537–13544.
22. Guy-Caffey JK, Bodepudi V, Bishop JS, Jayaraman K, Chaudhary N. Novel polyaminolipids enhance the cellular uptake of oligonucleotides. J Biol Chem 1995; 270:31391–31396.
23. Moradpour D, Schauer JI, Zurawski VR Jr, Wands JR, Boutin RH. Efficient gene transfer into mammalian cells with cholesteryl-spermidine. Biochem Biophys Res Commun 1996; 221:82–88.
24. Buchberger B, Fernholz E, Bantle E, et al. DOSPER liposomeal transfection reagent: a reagent with unique transfection properties. Biochemica 1996; 2:7–9.
25. Walker S, Sofia MJ, Kakasrla R, et al. Cationic facial amphiphiles: a promising class of transfection agents. Proc Natl Acad Sci USA 1996; 93:1585–1590.
26. Weibel J-M, Kichler A, Remy J-S, et al. Synthesis and evaluation as a gene transfer agent of a 1,2-dimyristoyl-sn-glycerol-3-pentalysine salt. Chem Lett 1995; 6:473–474.
27. Koshizaka T, Hayashi Y, Yagi K. Novel liposomes for efficient transfection of ß-galactosidase gene into Cos–1 cells. J Clin Biochem Nutr 1989; 7:185–192.
28. Hofland HE, Shepherd L, Sullivan SM. Formation of stable cationic lipid/DNA complexes for gene transfer. Proc Natl Acad Sci USA 1996; 93:7305–7309.
29. Aksentijevich I, Pastan I, Lunardi-Iskandar Y, Gall RC, Gottesman MM, Thierry AR. In vitro and in vivo liposome-mediated gene transfer leads to human MDR1 expression in mouse bone marrow progenitor cells. Human Gene Ther 1995; 7: 1111–1122.
30. Friend DS, Papahadjopoulos D, Debs RJ. Endocytosis and intracellular processing accompanying transfection mediated by cationic liposomes. Biochim Biophys Acta 1995; 1278:41–50.
31. Sternberg B, Sorgi FL, Huang L. New structures in complex formation between DNA and cationic liposomes visualized by freeze-fracture electron microscopy. FEBS Lett 1994;356:361–366.
32. Felgner PL, Tsai YJ, Sukhu L, et al. Improved cationic lipid formulations for in vivo gene therapy. Ann NY Acad Sci 1995; 772:126–139.
33. Zabner J, Fasbender AJ, Morninger T, Poellinger KA, Welsh MJ. Cellular and molecular barriers to gene transfer by a cationic lipid. J Biol Chem 1995; 270: 18997–19007.

34. Wagner E, Cotten M, Foisner R, Birnstiel ML. Transferrin-polycation conjugates as carriers for DNA and uptake into cells. Proc Natl Acad Sci USA 1991; 88:4255–4259.
35. Perales JC, Ferkol T, Beegen H, Ratnoff OD, Hanson RW. Gene transfer in vivo: sustained expression and regulation of genes introduced into the liver by receptor-targeted uptake. Proc Natl Acad Sci USA 1994; 91:4086–4090.
36. Wagner E, Zenke M, Cotten M, Beug H, Birnstiel MT. Transferrin-polycation conjugates as carriers for DNA uptake into cells. Proc Natl Acad Sci USA 1990; 87:3410–3414.
37. Wu GY, Wu CH. Receptor-mediated in vitro gene transformation by a soluble DNA carrier system. J Biol Chem 1987; 262:4429–4432.
38. Curiel DT, Agarwal S, Wagner E, Cotten M. Adenovirus enhancement of transferrin-polylysine-mediated gene delivery. Proc Natl Acad Sci USA 1991; 88:8850–8854.
39. Gao X, Huang L. Potentiation of cationic liposome-mediated gene delivery by polycations. Biochemistry 1996; 35:1027–1036.
40. Vitiello L, Chonn A, Wasserman JD, Duff C, Worton RG. Condensation of plasmid DNA with polylysine improved liposome-mediated gene transfer into established and primary muscle cells. Gene Ther 1996; 3:396–404.
41. Fritz JD, Herweijer H, Zhang G, Wolff JA. Gene transfer into mammalial cells using histone-condensed plasmid DNA. Human Gene Ther 1996; 7:1395–1404.
42. Trubeskoy VS, Torchilin VP, Kennel S, Huang L. Cationic liposomes enhance targeted delivery and expression of exogenous DNA mediated by N-terminal modified poly-L-lysine-antibody conjugate in mouse endothelial cells. Biochim Biophys Acta 1992; 1131:311–313.
43. Trubeskoy VS, Torchilin VP, Kennel S, Huang L. Use of N-terminal modified poly-L-lysine-antibody conjugate as a carrier for targeted gene delivery in mouse lung endothelial cells. Bioconjugate Chem 1992; 3:323–327.
44. Zhou X, Huang L. DNA transfection mediated by cationic liposomes containing lipopolylysine: characterization and mechanism of action. Biochim Biophys Acta 1994; 1189:195–203.
45. Farhood H, Bottega R, Epand RM, Huang L. Effect of cationic cholesterol derivatives on gene transfer and protein kinase C activity. Biochim Biophys Acta 1992; 1111:239–246.
46. Litzinger DC, Huang L. Phosphatidylethanolamine liposomes: drug delivery, gene transfer and immunodiagnostic applications. Biochim Biophys Acta 1992; 1113: 201–227.
47. Wrobel I, Collins D. Fusion of cationic liposomes with mammalian cells occurs after endocytosis. Biochim Biophys Acta 1995; 1235:296–304.
48. Farhood H, Serbina NS, Huang L. The role of dioleoyl-phosphatidylethanolamine in cationic liposome mediated gene transfer. Biochim Biophys Acta 1995; 1235: 289–295.
49. Xu Y, Szoka FC Jr. Mechnism of DNA release from cationic liposome/DNA complex used in cell transfection. Biochemistry 1996; 35:5616–5623.

7

Liposome/Viral Hybrid Gene Delivery Systems

ARLENE A. STECENKO

Vanderbilt University School of Medicine
Nashville, Tennessee

I. Introduction

Liposomes are one of the safest gene delivery systems currently available. However, low efficiency of gene transfer and short duration of transgene expression may limit liposomes' therapeutic utility. Certain viral vectors, such as recombinant adenovirus, deliver DNA more efficiently than liposomes while other viral vectors, such as recombinant adeno-associated virus (AAV), produce a longer duration of transgene expression. In contrast, viral vectors are less safe than liposomes. One strategy to improve the capability of liposomes as a gene delivery system is to marry the two technologies. Specific viral functions that enable the virus to perpetuate itself in the targeted host cell could be usurped to design viral/liposome hybrids with improved efficiency and/or duration of expression. For example, designing plasmid/liposome complexes that emulate a virus's ability to target and enter a specific cell type, to rupture the endosome, or to replicate its foreign genome in a mammalian cell could improve the performance characteristics of liposomes as a gene delivery system. By adding only the specific portion of the virus necessary for viral mimicry to the plasmid/liposome construct, the excellent safety profile of liposomes may not necessarily be compromised. In this chapter, we will show that liposome-mediated gene transfer can be improved significantly using this strategy of viral mimicry.

II. Cell Targeting and Entry

As described in Chapter 6, cationic liposomes complex with anionic plasmid DNA via charge/charge interactions. In contrast, anionic liposomes encapsulate plasmid DNA within a lipid vesicle. An advantage of anionic liposomes over cationic ones is that they can be composed of naturally occurring lipids and therefore are biodegradable. Also, anionic liposomes have an even better safety profile than cationic liposomes: they have been given as an aerosol to mice on a chronic basis (1), to sheep at high doses (2), and to normal human volunteers on an acute basis (3) without any alterations in lung or alveolar macrophage function. Finally, specific proteins can be inserted on the outer surface of the anionic liposome-encapsulated DNA in order to improve the delivery capability of the liposome. The disadvantage of anionic liposomes as a DNA delivery system is that it is technically difficult to encapsulate large amounts of the liked-charge DNA within a small anionic liposome.

Nicolau et al. (4) provided one of the earliest reports of successful in vivo transfection using anionic liposomes. Large anionic liposomes encapsulating a plasmid expressing the insulin gene were given intravenously to mice. Transgene expression was found only in the liver and spleen, presumably because of phagocytosis of these large particles by the reticuloendothelial system.

For anionic liposomes to be useful as a gene delivery system to the lung, they should be formulated to target and enter lung cells, the majority of which are non-phagocytic. Viral proteins can be inserted on the outer surface of anionic liposomes to target specific lung cells and also to promote fusion of the lipid layer of the liposome with the cell membrane. Such viral/liposome hybrids are often termed virosomes.

We reasoned that virosomes made using respiratory syncytial virus (RSV) attachment (G) and fusion (F) proteins would target and fuse with bronchial epithelial cells, the host cell for the virus, when given via the airway. Since RSV F protein is thought to mediate fusion directly at the cell surface at neutral pH, the contents of the liposome would be delivered directly to the cytoplasm, bypassing the possibility of being degraded in the endocytic compartment. To test this notion, we assessed the ability of RSV F and G virosomes to deliver the fluorescent marker 5,6-carboxyfluorescein into HEp-2 cells (5), non-phagocytic human laryngeal epidermoid cells, a cell line that permits RSV entry and replication. The virosomes were small (200 nm in diameter), unilamellar, anionic liposomes with affinity chromatography-purified RSV F and G inserted on the surface. As shown in Figure 1, unencapsulated fluorophore

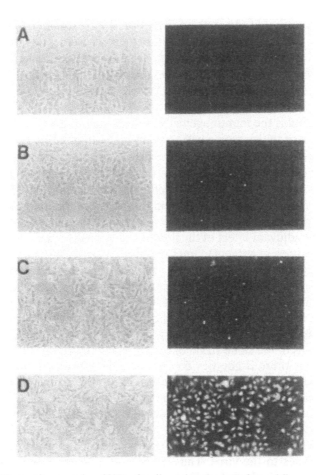

Figure 1   Photomicrographs of HEp-2 cells under phase (left) and fluorescent (right) microscopy after 1 hour incubation with free carboxyfluorescein (A) or carboxyfluorescein-encapsulated anionic liposomes with RSV G protein (B), F protein (C), or both F and G proteins (D). Reprinted with permission from Stecenko et al. (5).

did not enter HEp-2 cells. When the fluorophore was encapsulated in an anionic liposome with only the attachment protein or only the fusion protein, intracellular fluorescence was minimal after a 1-hour incubation. Complete virosomes with F and G viral proteins delivered the fluorophore to virtually every cell in the monolayer.

These data support the notion that specific viral proteins can enhance the delivery capability of liposomes. However, there are several questions raised regarding the utility of this delivery system in the treatment of lung disease. For example, what is the immunogenicity of these virosomes and will mounting an immune response to the foreign viral proteins preclude repetitive administrations? Will virosomes work in vivo? And, can one achieve a sufficient encapsulation efficiency of plasmid DNA to make large-scale production feasible?

First, what is the immunogenicity of these virosomes? The viral protein on the surface of the virosome will certainly elicit an immune response, most likely synthesis of specific serum antibodies. Liposomes act as an adjuvant, enhancing the humoral immune response to foreign protein. For example, the humoral immune response of mice to influenza virus hemagglutinin protein is significantly enhanced when the protein is inserted in the lipid bilayer of the liposome, especially when it is in the same orientation as in the viral envelope (6). Thus, the adjuvant effect of the liposome appears to reside in its ability to appropriately present antigen to immune effector cells. This humoral immune response theoretically would prevent effective gene transfer with repetitive administrations of the virosome. This is similar to what is seen with recombinant adenovirus vector given repetitively (7). Strain-specific antiadenovirus serum antibodies produced following the first administration of the vector can prevent productive infection (and thus transgene expression) with subsequent administrations of the same strain of recombinant virus.

However, there is good evidence that RSV virosomes will be exempt from this effect. RSV is a somewhat unique virus as repetitive infections with the exact same strain of the virus can be readily achieved. Hall et al. (8) were able to reinfect the nasal passages of normal human adult volunteers on multiple occasions with the same strain of RSV, even when the repeat inoculation was separated by as little as 60 days. This was possible in the presence of high serum antibodies to the virus. Thus, RSV is able to enter and replicate in respiratory cells in the face of a complete and competent immune system. Since G and F proteins are key for respiratory cell entry of the virus, it would be entirely reasonable to expect RSV virosomes to enter respiratory epithelial cells in the presence of RSV-specific humoral immunity.

There is a final note regarding the potential for an adverse immune response to virosomes. Since there is no new synthesis of viral proteins following administration of the virosome, it is unlikely that a cytotoxic T-cell response will occur. This cellular immune response is associated with inflammation and termination of transgene expression following in vivo administration of recombinant adenovirus vectors (9,10). Although recombinant adenovirus is rendered replication-deficient (i.e., no infectious virions are produced), there is synthesis of viral proteins which elicits a cytotoxic T lymphocyte response, which in turn triggers an inflammatory response and extinguishes transgene expression.

The bottom line here is that by understanding the basic biology of the replication cycle of a virus, virosomes can be formulated which target and enter specific lung cells and which potentially could evade the destructive powers of a competent immune system.

The next question is will virosomes work in vivo as a gene delivery system? Studies using Sendai virosomes suggest the answer to this question may be yes. Sendai virus (which is also called hemagglutinating virus of Japan; HVJ) is an enveloped virus with two surface proteins required for cell entry (11). An attachment protein mediates initial binding of the viral envelope to sialic acid residues on the cell membrane. A fusion protein mediates interactions between the viral envelope and the plasma membrane at neutral pH and thus introduces the viral genome into the cytoplasm. Sendai virosomes are manufactured in a slightly cruder fashion than RSV virosomes (12). Basically, the virus is propagated in chick eggs, the viral envelope harvested, and then lysed with detergent. Liposomes are prepared separately using various combinations of naturally occurring lipids. Plasmid DNA, the disrupted viral envelopes, and the liposomes are added together and, through physical methods, reconstituted viral envelopes encapsulating DNA are formed. These Sendai virosomes are large (1 μm in diameter) and contain extraneous viral components (but no active virus). In vitro, Sendai virosomes are much more efficient at delivering their payload to the interior of the cell than similarly constructed liposomes without viral envelope components (11).

Sendai virosomes have been used successfully to deliver genes in vivo. Sendai virosomes encapsulating a plasmid encoding neuropeptide Y, an abundant peptide whose most potent action is to stimulate food intake, was injected directly into the brains of rats (13). Immunoreactive neuropeptide Y was found in the rat brains for up to 50 days after transfection. More importantly, there was sufficient expression of the peptide to have a physiologic effect. Food intake and body weight were significantly greater in animals receiving Sendai virosomes encapsulating the neuropeptide Y plasmid compared to virosomes containing empty vector (Fig. 2). This physiologic effect of transgene expression lasted for 3 weeks after injection.

Others have used Sendai virosomes to successfully deliver a reporter gene to the liver of rats when virosomes are injected under the perisplanchnic membrane (14). This latter report underscores an additional advantage of virosomes: molecules can be coencapsulated within the virosome in order to enhance transgene expression. These authors had previously shown that nonhistone nuclear protein (HMG1) enhanced migration of plasmid DNA into the nucleus of cells. They reasoned that Sendai virosomes that encapsulated both HMG1 and plasmid DNA would result in improved transgene expression because the HMG1 would escort the plasmid to the nucleus. To test this notion, they compared the level of transgene expression using Sendai virosomes

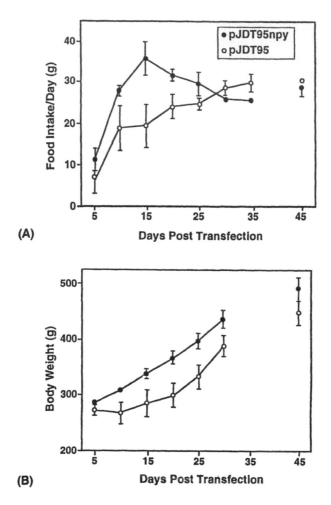

Figure 2 Food intake (A) and body weight (B) after bilateral injection into the para-ventricular nucleus of rat brains using Sendai virosomes encapsulating either a plasmid expressing neuropeptide y (● pJDT95npy) or empty vector (○ pJDT95). Values are mean ± SEM of three or four animals per group. Reprinted with permission from Wu et al. (13).

with and without coencapsulation of HMG1. Expression of ß-galactosidase was three times greater when HMG1 was in the interior of the virosome.

If virosomes can be constructed to safely target and penetrate respiratory cells, an important issue to address is whether the large-scale production of virosomes necessary for clinical trials is technically feasible. Issues of possible contamination with viable virus, purity of viral proteins (and conversely lack of extraneous viral components), and the poor encapsulation efficiency of plasmid DNA need to be addressed. This latter problem of low encapsulation efficiency requires a few additional comments. A low encapsulation efficiency results in a tremendous waste of plasmid as unencapsulated plasmid does not enter most cells. Also, a low encapsulation efficiency might mean that the production of sufficient quantities of viral proteins to construct the virosomes is prohibitively expensive. The reason for this is that the amount of viral protein used depends on the number of liposomes needed to administer a given amount of DNA. For example, improving encapsulation efficiency from the usual 5% to 10% (4,14) to 50% would result in using 5- to 10-fold less viral protein. Unfortunately, even to achieve a 5% to 10% encapsulation efficiency requires a large virosome of 1 μm in diameter, which is probably too large to enter nonphagocytic lung cells readily. With smaller liposomes, encapsulation efficiency plummets.

To circumvent the problem of low encapsulation efficiency, we reasoned the following. Anionic DNA could be simply complexed with sufficient quantities of cationic polylysine to render the complex cationic. This process produces condensed DNA/polylysine complexes (15). Then, small (100 nm) anionic RSV virosomes with the fusion and attachment protein on the surface would complex to the cationic plasmid DNA/polylysine complex via simple charge/charge interactions. Thus, all DNA would be entrapped by virosomes. These RSV virosome/plasmid DNA/polylysine complexes significantly enhanced gene delivery and expression in vitro in transformed human cystic fibrosis (CF) airway epithelial cells (Fig. 3). However, no such enhancement was found in normal human bronchial epithelial cells.

In discussing the utility of virosomes as a delivery system, we have focused on the use of viral proteins to target and promote fusion with the cell. The delivery capabilities of virosomes are also affected by the lipid composition of the virosome itself. Schreier et al. (16) designed HIV virosomes that targeted CD4-positive cells via HIV gp120 protein tagged to the surface of the liposome. The authors compared intracellular trafficking of the virosome using virosomes containing a 7:1 molar ratio of phosphatidylcholine:cholesterol or using virosomes with a lipid composition and a phospholipid:cholesterol molar ratio similar to that found in the HIV envelope (1:1 phospholipid:cholesterol molar ratio). Cell trafficking was assessed by encapsulating a fluorophore within the virosome and examining the cells under confocal fluorescent microscopy. Virosomes constructed with similar lipids to the HIV viral envelope were transported well into the interior of the cell and were found in late endosomes and lysosomes. Virosomes with a 7:1 phospholipid:cholesterol ratio were confined to endosomes that remained close to the outer cell membrane. Since the two types of virosomes had the same targeting ligand responsible for

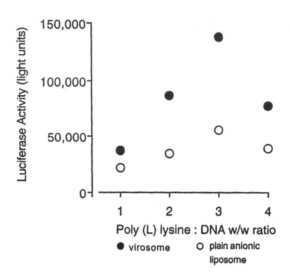

Figure 3 Transgene expression (measured as light units) in transformed human cystic fibrosis airway epitheial cells 2 days after transfection using increasing amounts of poly(L)lysine complexed with plasmid DNA and either plain anionic liposomes (o) or RSV virosomes constructed with viral attachment and fusion proteins (•). From Stecenko A, unpublished data.

binding and internalization, the authors concluded that the lipid composition of the virosome was important for determining subsequent trafficking of the virosome in the endocytic compartment.

In summary, proteins responsible for specific viral functions, such as targeting a given cell or entering a cell via fusion with the cell membrane, can be used to enhance liposome-mediated gene delivery. Furthermore, careful selection of the type of virus mimicked may provide a means of obviating the effect of the immune response to the viral proteins. The greatest technical hurdle to overcome is improving the entrapment efficiency of plasmid DNA by virosomes. Until this hurdle can be overcome, the potential therapeutic utility of virosomes for in vivo gene delivery to the lung cannot be addressed adequately.

III. Endosomolysis

Cationic liposome/plasmid complexes enter many different cell types readily, mainly through endocytosis. However, the low fraction of cells that express the delivered transgene suggest that escape from the endosome and trafficking of functional plasmid to the nucleus may be a limiting factor in achieving efficient gene transfer. The means by which certain viruses escape the endosome before the lysosomal acidic environment inactivates the virus can be mimicked to create a liposome/viral hybrid which similarly escapes lysosomal degradation.

As discussed in Chapter 3, a number of investigators have used adenovirus's ability to rupture the endosome to design a more efficient gene delivery system. Plasmid DNA is coupled to the viral capsid of replication-deficient adenovirus, often using a ligand to target specific cell types. This approach has been used to transfect hepatocytes (17,18) and respiratory epithelial cells in vitro (19). Intratracheal instillation of these complexes in the cotton rat results in transgene expression lasting over 1 week (20).

This system can also be manipulated to potentially eliminate the pulmonary inflammatory response found with administration of recombinant adenovirus. As mentioned previously, upon administration of replication-deficient adenovirus there is synthesis of viral proteins. This elicits a pulmonary inflammatory response and a cytotoxic T-lymphocyte response, which eliminates the transfected cells and terminates transgene expression (9,10). Adenovirus can be inactivated by UV light and psoralen treatment so that viral protein replication is undetectable. Molecular conjugates made with inactivated adenovirus continue to enhance plasmid DNA expression in vitro (21,22). Completely inactivated adenovirus does not cause inflammation in vivo (9).

The advantages of this system are (1) much larger transgenes can be inserted in the plasmid compared to the adenoviral vector; (2) using inactivated adenovirus might eliminate the inflammatory response induced by viral protein synthesis; and (3) specific cells types can be targeted by adding a ligand. A ligand may be necessary to target ciliated respiratory epithelial cells as recombinant adenovirus may not be as tropic for respiratory epithelium as originally thought (23). The disadvantage of this technique is that it is technically complex, although easier than manipulating the large adenoviral genome.

We reasoned that linking adenovirus directly to a plasmid/cationic liposome complex would couple the endosomolytic activity of adenovirus with the ability of liposomes to enter a wide variety of cell types, including respiratory epithelium (24,25). The cationic liposome would serve to link the plasmid DNA with the virus as well as to deliver the complex to respiratory cells. The plasmid would serve as the transgene expression system and the virus as an endosomolytic agent preventing inactivation of the plasmid in the endosome. In an earlier study, it was shown that plasmid DNA/cationic liposome/replication-deficient adenovirus complexes transfected vascular endothelial cells in vitro and in vivo more efficiently than plasmid/liposome alone (26). Therefore, this approach might prove useful for in vivo gene delivery to the lung.

We complexed replication-deficient adenovirus with unmodified plasmid DNA and cationic liposomes via simple charge/charge interactions. Using this tripartite complex of plasmid/liposome/adenovirus, the efficiency of transgene expression in normal and CF human airway epithelial cell lines was enhanced 3- to 3050-fold compared to expression with plasmid/liposome alone (27,28). The degree of enhancement with virus was greater for the normal cells (median fold enhancement = 26) compared to the CF cells (median fold enhancement = 7.5).

The time course of expression during the first week following transfection with these complexes was different depending on whether or not virus was present. Adenovirus enhancement of gene transfer was present at day 1 in CF cells and day 5 in normal cells (Fig. 4). Once peak expression was achieved with the plasmid/liposome/adenovirus complex, there was little decline in the level of transgene expression. (Although there are several possible explanations for this latter observation, results from subsequent experiments suggest that this prolongation of expression might be due to virus-mediated transactivation of the plasmid.) There was the usual decline in expression over the first week of transfection with plasmid/liposome alone.

Cytotoxicity was an issue in these experiments. The tripartite complex enhanced liposome-mediated gene transfer at the price of converting an extremely safe, but inefficient, gene delivery system (liposomes) into a system which was toxic. Beginning on the third day following transfection, cells exposed to adenovirus developed mild abnormalities in cell morphology. Increase release in LDH and cell lysis appeared on days 5 and 7. Because there were no immune effector cells in our experiments, direct toxicity of the virus to bronchial epithelial cells is likely. The timing of the toxicity (day 5 to 7) and the morphology of the cells (rounded up cells followed by detachment and cell lysis) suggest replication of virions.

Although the tripartite complex worked extremely well for in vitro enhancement of transgene expression, in vivo application requires a safer system. Given the work on molecular conjugates utilizing adenovirus where the genome has been inactivated yet endosomolytic function maintained (21,22), using an inactivated adenovirus may decrease the toxicity of this system. Another approach would be to link the viral protein responsible for disruption of the endosomal membrane, the penton base protein (29), directly to the plasmid/liposome complex. In both these situations (use of inactivated virus or penton base protein), viral proteins may elicit a specific serum immunoglobulin response which could significantly reduce the efficacy of gene transfer with subsequent administrations (7).

In summary, different investigators have taken advantage of the endosomolytic property of adenovirus to design gene delivery systems with good results. However, the immune response to adenovirus, whether to the foreign viral protein itself or to the later synthesis of new viral protein, as well as direct

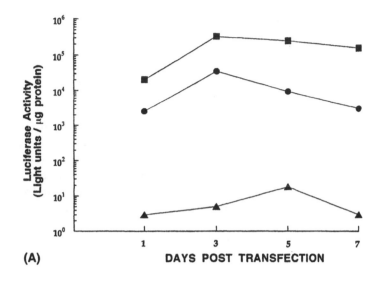

(A)

DAYS POST TRANSFECTION

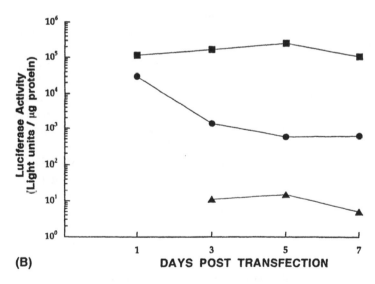

(B)

DAYS POST TRANSFECTION

Figure 4 Time course of transgene expression (measured as light units per μg cellular protein) in transformed cystic fibrosis (A) or normal (B) human airway epithelial cells. Cells were transfected with naked plasmid (▲), plasmid/catinic liposome complexes (●), or plasmid/cationic liposome/replication-deficient adenovirus complexes (■). From Stecenko A, unpublished data.

adenoviral toxicity, may make this system too complex for practical use as gene therapy for lung diseases.

IV. Retention/Replication of Foreign DNA

The duration of liposome-mediated transgene expression is limited to a few weeks both in vitro and in vivo, presumably due to two mechanisms. In non-dividing cells, the foreign DNA is either inactivated or destroyed by the cell. In dividing cells, the plasmid is diluted in each subsequent generation as plasmids cannot replicate in mammalian cells. In other words, transgene expression is terminated by both a failure to retain functional plasmid and a failure of the plasmid to replicate in a controlled fashion and be passed to daughter cells.

Some viruses have developed the ability to remain latent in cells yet replicate in concert with the parent cell cycle so that daughter cells contain the same number of virions as the parent cell. Epstein-Barr virus (EBV) is such a virus. Almost all adults worldwide are asymptomatic carriers of EBV (30). When EBV infects its host cell, the B lymphocyte, the linear, viral DNA becomes circularized and is maintained in the nucleus as an episome (31). Initially, through some unknown mechanism, the genome is overamplified. Soon thereafter, there is a constant copy number per B cell, usually between 5 and 500, which is maintained through several generations (31). Controlled, non-lytic replication of EBV requires three elements—a 1700-bp EBV genomic sequence termed the plasmid maintenance origin (*oriP*); the EBV nuclear antigen protein 1 (EBNA-1); and the cell machinery necessary for DNA replication (32). The *oriP* sequence has two areas of multiple EBNA-1 binding sites separated by approximately 1000 bp of DNA. EBNA-1 binds to these multiple sites on *oriP*, bringing the two areas in close proximity, and viral DNA replication is initiated (32). It appears that cellular mechanisms during the S-phase control viral replication initiated by *oriP* and EBNA-1, ensuring that only once during S-phase one copy of the EBV plasmid is made (32). In contrast, *ori-Lyt* and viral DNA polymerase are responsible for amplification of plasmids and subsequent lysis of the cell (32). Finally, up to 8 or 10 viral genes are responsible for in vitro transformation of B cells (33): these viral sequences potentially are related to the tumorigenesis capabilities of the virus in vivo.

Given this knowledge on the replication cycle of EBV, one could postulate that in the presence of *oriP* and EBNA-1, a transgene would be duplicated in mammalian cells in a controlled fashion when the cell divides and would be passed on to the daughter cells thus maintaining a constant copy number per cell in actively dividing cells. Banerjee et al. (33) constructed a mini-EBV recombinant vector with *oriP* and *ori-Lyt* sequences, transfected human B lymphoblast cell lines which were transformed with EBV and therefore produced EBNA-1 protein, and showed episomal replication of the transgene. Expression was retained at the rate of 98% per generation in actively growing cells.

Clearly, recombinant EBV cannot be used for gene therapy for the lung as EBV does not infect respiratory cells. However, we reasoned that plasmids with *oriP* and EBNA-1 sequences as well as a desired transgene could be delivered to respiratory cells using cationic liposomes. When the cell divides, one copy of the plasmid should be produced maintaining a constant plasmid number per cell. Furthermore, expression of the EBNA-1 protein in nondividing cells might cause retention of the plasmid, as it does for viral plasmid. Production of this foreign viral protein should not produce an immune response and potentially terminate transfection for the following reasons. About one B cell/ml of blood from a healthy person is infected with EBV (34,35). Since almost all adults are carriers of EBV, most adults would have some B cells producing EBNA-1. One of the mechanisms whereby EBV evades immune surveillance is via the Gly-Ala internal repeats in EBNA-1 (36).

Our contention was that plasmids containing EBV *oriP* and EBNA-1 sequences as well as a reporter gene would produce a longer duration of transgene expression, either through replication of the plasmid in dividing eukaryotic cells and/or through retention of the plasmid in nondividing cells. We transfected transformed human airway epithelial cells using cationic liposomes and a plasmid with the EBV sequences for *oriP* and EBNA-1, either a CMV or a rous sarcoma virus (RSV) promoter, and the reporter gene luciferase (pCEP+ or pREP+, respectively) (37). As a control plasmid, we excised almost all the *oriP* and EBNA-1 sequences from pCEP+ or pREP+ to construct pCEP– and pREP–. Transgene expression was significantly prolonged using plasmids containing the full *oriP* and EBNA-1 sequences: expression lasted 7 to 10 days longer with these plasmids compared to those with the EBV sequences deleted (Fig. 5). Given that less than 5% of the cells are initially transfected using standard cationic liposomes, this effect is not trivial.

To eliminate the problem of low transfection efficiency and to determine whether plasmid replication, as opposed to retention, was responsible for the prolongation in transgene expression, cells were transfected and then grown in media containing hygromycin (38), which is toxic to cells. Since all four plasmids contained a hygromycin-resistance gene, only transfected cells would survive hygromycin. If plasmids replicated when cells divided, subsequent daughter cells would survive in hygromycin. With only retention of plasmids (as opposed to replication), plasmids would be diluted in subsequent generations, and daughter cells would begin to die. Finally, a hygromycin-resistance gene in a non-replicating plasmid could induce integration of the plasmid into the host genome and production of a population of dividing, hygromycin-resistant (and luciferase expressing) cells. As a control for the number of cells

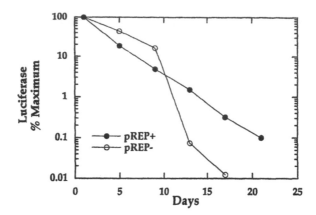

Figure 5 Time course of transgene expression (expressed as percent of maximum luciferase activity) in transformed normal human bronchial epithelial cells transfected with a plasmid containing the full *oriP* and EBNA-1 sequences (●) or a plasmid with the sequences deleted (○). The slope of the two lines are statistically different ($P < .05$). Reprinted with permission from Shih et al. (37).

which might spontaneously develop resistance to hygromycin, cells were also transfected with the RSVluc plasmid, a plasmid expressing the luciferase gene under the control of the RSV promoter but not expressing the hygromycin-resistance gene.

As seen in Figure 6A, cell count per well rapidly dropped in the first week following transfection with all three plasmids. However, cells transfected with plasmids containing the *oriP* and EBNA-1 sequences showed a different growth pattern thereafter. After an initial 10-fold drop in cell count, the cells began to divide, and by 7 weeks posttransfection were 40% above baseline levels. In contrast, there was a 500-fold decrease in cell count in the first 3 weeks following transfection using plasmids with the sequences deleted. Thereafter, cells began to divide and by 7 weeks were returning to the baseline level. When cells were transfected using RSVluc, most cells were killed by 3 weeks. In the subsequent weeks, a few cells spontaneously became hygomycin resistant and began to divide.

Transgene expression, measured as light units per cell, increased by almost 500-fold over a few weeks and stayed fairly constant over the 7 week period when pCEP+ was used (Fig. 6B). Thus, with the EBV sequences, luciferase activity increased while cells were actively dividing until a point was reached when luciferase activity per cell was constant. This is strong evidence that the plasmid was replicating in mammalian cells in a controlled fashion.

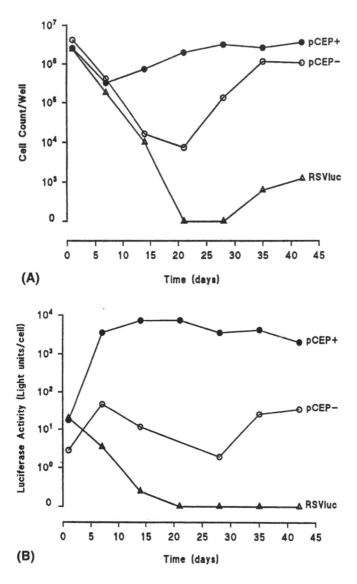

Figure 6 Time course of hygromycin-induced cell death, expressed as cell count/well (A) and of transgene expression, expressed as light units/cell (B) in transformed normal human bronchial epithelial cells transfected with a plasmid containing the full *oriP* and EBNA-1 sequences and a hygromycin-resistance gene (●); a plasmid with the sequences delted but with the hygromycin-resistance gene maintained (○); or a plasmid with the RSV promoter, luciferase as the reporter gene, and no hygromycin-resistance gene (△). Twenty-four hours following transfection, cells were maintained in complete media containing hygromycin. From Brigham K, unpublished data.

Without the full EBV sequences, luciferase activity/cell decreased to low, but detectable levels suggesting hygromycin selection of cells with the plasmid integrated into the host genome. Transgene expression using the plasmid with no EBV sequence and without the hygromycin-resistance gene disappeared by 3 weeks following transfection. Finally, there was no evidence of cellular toxicity using the replicating plasmid.

What is important about this work with the recombinant EBV vector and with the plasmid containing the EBV *oriP* and EBNA-1 sequences is the demonstration that knowledge of viral replication can be used to design an improved gene delivery and expression system while maintaining the safety of plasmid vectors. Duration of transgene expression was prolonged in our studies presumably owing to replication of foreign DNA in a controlled fashion in dividing human cells.

Similarly, plasmids have been constructed using the human papovavirus origin of replication and the transactivator of the origin. Such plasmids replicate episomally in a bladder carcinoma cell line (39) and prolong transgene expression for 3 months when given intravenously with cationic liposomes to mice (40). This latter observation of prolongation of expression of episomally replicating plasmids when given in vivo is a critical step toward the therapeutic goal of prolonged gene expression for chronic pulmonary diseases.

Another virus that has been used to construct a plasmid which might replicate in eukaryotic cells is AAV. As discussed in greater detail in Chapter 4, AAV is a replication-deficient parvovirus which is nonpathogenic to humans. The wild-type AAV genome stably integrates into the host genome at specific chromosome sites. Philip et al. (41) reasoned that plasmids could be constructed using the genomic sequences of AAV, the inverted terminal repeats (ITR), which join the cellular DNA to the viral DNA. Delivery of these plasmids using cationic liposomes might result in site-specific integration of the transgene into the host genome and thus long-term expression. The results were somewhat unexpected (41,42). Plasmids with the AAV ITRs flanking a transgene driven by a CMV promoter produced 3- to 10-fold greater efficiency of transfection compared to plasmids without the ITRs. The duration of expression was significantly prolonged in nondividing, but not dividing, cells. There was no evidence that the plasmid replicated. The authors speculated that there was enhancement in the amount of functional plasmid that reached the nucleus and that this plasmid was retained in the nucleus in an episomal form. Although they did not achieve their original goal (stable integration), their results are remarkable. The efficiency of transgene expression using cationic liposome delivery of the ITR containing plasmid was as high as 80% of cells transfected in cell culture, compared to 5% or less of cells using plasmids without the AAV ITRs. Also, transgene expression was still present, although at a lower level than initially, at 1 month after transfection.

In summary, plasmid design can be improved significantly by borrowing specific viral sequences that confer upon the virus the ability to replicate in a controlled fashion in mammalian cells. A thorough knowledge of the biology of the virus is key in selecting the appropriate sequences. However, as exemplified by the AAV plasmids, it can never be assumed that the specific selected viral function will be retained when the rest of the virus is eliminated.

Acknowledgments

Dr. Stecenko gratefully acknowledges the support from the Cystic Fibrosis Foundation and the National Institutes of Health (HL 50258 and AI 31900).

References

1. Myers MA, Thomas DA, Straub L, et al. Pulmonary effects of chronic exposure to liposome aerosols in mice. Exp Lung Res 1993; 19:1–19.
2. Schreier H, McNicol KJ, Ausborn M, et al. Pulmonary delivery of amikacin liposomes and acute liposome toxicity in the sheep. Int J Pharmaceut 1992; 87:183–193.
3. Thomas DA, Myers MA, Wichert B, Schreier H, Gonzalez-Rothi RJ. Acute effects of liposome aerosol inhalation on pulmonary function in healthy human volunteers. Chest 1991; 99:1268–1270.
4. Nicolau C, Le Pape A, Soriano P, Fargette F, Juhel M-F. In vivo expression of rat insulin after intravenous administration of the liposome-entrapping gene for rat insulin I. Proc Natl Acad Sci USA 1983; 80:1068–1072.
5. Stecenko AA, Walsh EE, Schreier H. Fusion of artificial viral envelopes containing respiratory syncytial virus attachment (G) and fusion (F) glycoproteins with HEp-2 cells. Pharm Pharmacol Lett 1992; 1:127–129.
6. Stahn R, Schafer H, Kunze M, Malur J, Ladhoff A, Lachmann U. Quantitative reconstitution of isolated influenza haemagglutinin into liposomes by the detergent method and the immunogenicity of haemagglutinin liposomes. Acta Virol 1992; 36: 129–144.
7. Kaplan JM, St George JA, Pennington SE, et al. Humoral and cellular immune responses of nonhuman primates to long-term repeated lung exposure to Ad2/CFTR-2. Gene Ther 1996; 3:117–127.
8. Hall CB, Walsh EE, Long CE, Schnabel KC. Immunity to and frequency of reinfection with respiratory syncytial virus. J Infect Dis 1991; 163:693–698.
9. Yang Y, Nunes FA, Berencsi K, Furth EE, Gonczol E, Wilson JM. Cellular immunity to viral antigens limits E1-deleted adenovirus for gene therapy. Proc Natl Acad Sci USA 1994; 91:4407–4411.
10. Yang Y, Jooss KU, Su Q, Ertl HCJ, Wilson JM. Immune response to viral antigens versus transgene product in the elimination of recombinant adenovirus-infected hepatocytes in vivo. Gene Ther 1996; 3:137–144.

11. de Fiebre CM, Bryant SO, Notabartolo D, Wu P, Meyer EM. Fusogenic properties of Sendai virosome envelopes in rat brain preparations. Nerochem Res 1993; 18: 1089–1094.
12. Volsky DJ, Loyter A. An efficient method for reassembly of fusogenic Sendai virus envelopes after solubilization of intact virions with Triton X-100. FEBS Lett 1978; 92:190–194.
13. Wu P, de Fiebre CM, Millard WJ, et al. An AAV promoter-driven neuropeptide Y gene delivery system using Sendai virosomes for neurons and rat brains. Gene Ther 1996; 3:246–253.
14. Kato K, Nakanishi M, Kaneda Y, Uchida T, Okada Y. Expression of hepatitis B virus surface antigen in adult rat liver. J Biol Chem 1991; 266:3361–3364.
15. Wolfert MA, Seymour LW. Atomic force microscopic analysis of the influence of the molecular weight of poly(L)lysine on the size of polyelectrolyte complexes formed with DNA. Gene Ther 1996; 3:269–273.
16. Schreier H, Moran P, Caras IW. Targeting of liposomes to cells expressing CD4 using glycosylphosphatidylinositol-anchored gp120. J Biol Chem 1994; 12:9090–9098.
17. Cristiano RJ, Smith LC, Woo SL. Hepatic gene therapy:adenovirus enhancement of receptor-mediated gene delivery and expression in primary hepatocytes. Proc Natl Acad Sci USA 1993; 90:2122–2126.
18. Curiel DT, Wagner E, Cotten M, et al. High efficiency gene transfer mediated by adenovirus coupled to DNA-polylysine complexes. Hum Gene Ther 1992; 3:147–154.
19. Curiel DT, Agarwal S, Romer MU, et al. Gene transfer to respiratory epithelial cells via the receptor-mediated endocytosis pathway. Am J Respir Cell Mol Biol 1992; 6:247–252.
20. Gao L, Wagner E, Cotten M, et al. Direct in vivo gene transfer to airway epithelium employing adenovirus-polylysine-DNA complexes. Hum Gene Ther 1993; 4:17–24.
21. Cotten M, Wagner E, Zatloukal K, Phillips S, Curiel DT, Birnstiel ML. High-efficiency receptor-mediated delivery of small and large (48kb) gene constructs using the endosome disruption activity of defective or chemically inactivated adenovirus particles. Proc Natl Acad Sci USA 1992; 89:6094–6098.
22. Cotten M, Saltik M, Kursa M, Wagner E, Maass G, Birnstiel ML. Psoralen treatment of adenovirus particles eliminates virus replication and transcription while maintaining the endosomolytic activity of the viral capsid. Virology 1994; 205:254–261.
23. Goldman MJ, Wilson JM. Correlation of adenoviral mediated gene transfer efficiency with various airway epithelial cell phenotypic markers. Am J Respir Crit Care Med 1995; S151:670.
24. Canonico AE, Conary JT, Meryick BO, Brigham KL. Aerosol and intravenous transfection of human alpha 1-antitrypsin gene to lungs of rabbits. Am J Respir Cell Mol Biol 1994; 10: 24–29.
25. Lu L, Zeitlin PL, Guggino WB, Craig RW. Gene transfer by lipofection in rabbit and human secretory epithelial cells. Pflugers Arch 1989; 415:198–203.
26. Raja-Walia R, Weber JC, Chapman GD, Naftilan J, Naftilan AJ. Enhancement of liposome-mediated gene transfer into vascular tissue by replication-deficient adenovirus. Hum Gene Ther 1995; 2:521–530.

27. Raja-Walia R, King G, Peters M, et al. Enhancement of plasmid/liposome gene transfer to lung epithelial cells with replication-deficient adenovirus. Pediatr Pulmonol 1994; 10S:233.
28. Stecenko A, Brigham K, King G, Persmark M, Schreier H, Raja-Walia R. Characterization of adenovirus enhancement of liposome-mediated gene transfer in bronchial epithelial cells. Am J Respir Crit Care Med 1995; 151S:20.
29. Bai M, Harfe B, Freimuth P. Mutations that alter an arg-gly-asp (RGD) sequence in the adenovirus type 2 penton base protein abolish its cell-rounding activity and delay virus reproduction in flat cells. J Virol 1993; 67:5198–5205.
30. Kieff E, Liebowitz D. Epstein Barr virus and its replication. In: Fields BN, Knipe DM, eds. Virology. 1889–1920. New York: Raven, 1990.
31. Middleton T, Sugden B. EBNA1 can link the enhancer element to the initiator element of the Epstein-Barr virus plasmid origin of DNA replication. J Virol 1992; 66:489–495.
32. Yates JL, Guan N. Epstein-Barr virus-derived plasmids replicate only once per cell cycle and are not amplified after entry into cells. J Virol 1991; 65:483–488.
33. Banerjee S, Livanos E, Vos J-MH. Therapeutic gene delivery in human B-lymphoblastoid cells by engineered non-transforming infectious Epstein-Barr virus. Nature Med 1995; 1(12):1303–1308.
34. Miyashita EM, Yang B, Lam KMC, Crawford DH, Thorley-Lawson DA. A novel form of Epstein-Barr virus latency in normal B cells in vivo. Cell 1995; 80:593–601.
35. Chen F, Zou J-Z, di Renzo L, et al. A subpopulation of normal B cells latently infected with Epstein-Barr virus resembles Burkitt lymphoma cells in expressing EBNA-1 but not EBNA-2 or LMP1. J Virol 1995; 69:3752-3758.
36. Levitskaya J, Coram M, Levitsky V, et al. Inhibition of antigen processing by the internal repeat region of the Epstein-Barr virus nuclear antigen-1. Nature 1995; 375:685–688.
37. Shih KC, Conary JT, LeFevre BM, Mitchell JA, Stecenko AA, Brigham KL. Prolonged transgene expression using plasmids with Epstein-Barr virus (EBV) origin of replication. J Invest Med 1995; 43S:329.
38. Shih KC, Massey JM, Stecenko AA, Brigham KL. Plasmids with Epstein-Barr virus (EBV) replication initiation sequences replicate in normal and cystic fibrosis bronchial epithelial cells. Am J Respir Crit Care Med 1996; 153S:114.
39. Cooper MJ, Miron S. Efficient episomal expression vector for human transitional carcinoma cells. Hum Gene Ther 1993; 4:557–566.
40. Thierry AR, Lunardi-Iskandar Y, Bryant JL, Rabinovich P, Gallo RC, Mahan LC. Systemic gene therapy: biodistribution and long-term expression of a transgene in mice. Proc Natl Acad Sci USA 1995; 92:9742–9746.
41. Philip R, Brunette E, Kilinski L, et al. Efficient and sustained gene expression in primary T lymphocytes and primary and cultured tumor cells mediated by adeno-associated virus plasmid DNA complexed to cationic liposome. Mol Cell Biol 1994; 14:2411–2418.
42. Vieweg J, Boczkowski D, Roberson KM, et al. Efficient gene transfer with adeno-associated virus-based plasmids complexed to cationic liposomes for gene therapy of human prostate cancer. Cancer Res 1995; 55:2366–2372.

Part Three

**GENE THERAPY IN THE LUNG:
THE NEXT GENERATION**

8

Manipulating the Intracellular Trafficking of Nucleic Acids

KATHLEEN E. B. MEYER, LISA S. UYECHI, and FRANCIS C. SZOKA, JR.

University of California at San Francisco
San Francisco, California

I. Introduction

The delivery of nucleic acids for therapeutic purposes must surmount multiple cellular barriers that restrict efficient gene delivery. These include transfer across the plasma membrane, movement through the cytoskeletal network as well as high macromolecule concentrations within the cytoplasm, and finally transfer across the nuclear envelope. Manipulation of the biology of the cell and association of various effectors with the nucleic acid or utilization of low-molecular-weight nucleic acids have been used to increase gene expression. In spite of these efforts we believe DNA uptake into the nucleus remains the rate-limiting barrier for efficient gene delivery by nonviral gene delivery systems.

In this review we describe (1) how molecules diffuse through the cytoplasm, (2) the status of oligonucleotide and plasmid DNA trafficking in cells, (3) factors that regulate such macromolecular transport through the cytoplasm, (4) mechanisms of import/export of proteins and RNA through the nuclear membrane, and finally (5) how the nuclear membrane is reorganized during cell division. Knowledge of these processes provides a basis to understand how they are exploited by viruses to efficiently deliver nucleic acids into the nucleus.

135

Furthermore, this information is useful to evaluate strategies for manipulation of the cell or the plasmid to increase transgene import into the nucleus. It is assumed that gene delivery efficiency, evaluated as expression level per delivered gene, depends on the concentration of transgene nucleic acid in the nucleus and that methods to increase the transgene concentration in the nucleus will lead therefore to higher levels of transgene expression.

In the early literature, translocation of a plasmid molecule across the plasma membrane was recognized as a formidable obstacle so microinjection was employed to introduce DNA into the cell (1). Recent advances using various physical techniques (2,3), cationic lipids (4,5), pH-sensitive liposomes (6) and cationic polymers (7–13) permit cytoplasmic delivery of significant amounts of DNA. Strategies to improve cytoplasmic delivery are in place. Thus, a fundamental limitation to gene expression in most cell types using current nonviral systems is the inability of DNA in the cytoplasm to migrate facilely into the nucleus. This limitation was elegantly demonstrated by the early work of Capecchi, who showed plasmids microinjected into the cytoplasm of cultured cells are very poorly expressed whereas plasmids microinjected into the nucleus are highly expressed (1).

II. Review of the Cytoplasm

A. Characteristics of the Cytoplasm

An understanding of intracellular macromolecular transport requires a general familiarity of the structure and properties of the cytoplasm, which consists of a filamentous network embedded in an aqueous, gel-like matrix. The matrix is composed of a concentrated mixture of soluble proteins and RNA and contains membrane-bound organelles that are linked to cytoskeletal fibers (14–18). The cytoplasm has the properties of a viscoelastic, contractile gel, making it difficult to model the diffusion of macromolecules. Measurement of the viscosity of the cytoplasm using low-molecular-weight molecules in several mammalian cell types has indicated viscosities similar to that of bulk water (19–21). However, macromolecular crowding is also an important feature of the cytoplasmic environment (Fig. 1). The protein concentration in the cytoplasm may be as high as 200 to 300 mg/ml, which gives it a colloid density similar to that found in solutions of 12.4% Ficoll (22) and 13% dextran (23). In this crowded milieu the translational diffusion of molecules the size of proteins is predicted to be reduced approximately threefold owing to macromolecular collisions (22,24–26).

B. Macromolecular Diffusion and Transport in the Cytoplasm

Macromolecular transport mechanisms within the cytoplasmic space can be explained by diffusion and active transport associated with the cytoskeleton.

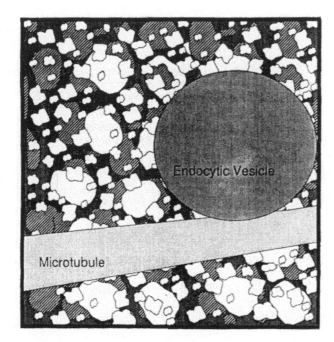

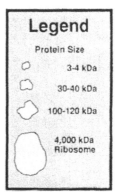

Figure 1 The crowded cytoplasm. This schematic depicts the density of protein macromolecules present in the cytoplasm of a mammalian cell. The sizes range from large, 4200-kDa ribosomes to small, 3-kDa proteins. Here a typical endocytic vesicle, 100 nm in diameter, is transported along a microtubule in a sea of protein molecules. The assumptions necessary to visualize this model are discussed in Ref. 189.

Considerable effort has been directed to studying cytoplasmic diffusion of small inert particles that are microinjected into living cells (27–33). These studies provide insight into the viscosity and dimensions of the sievelike, filamentous spaces within the cytoplasm as well as that of the nuclear pore. In particular, fluorescent analog cytochemistry and fluorescent microphotolysis (for review see Ref. 30) are useful tools to measure the diffusion of inert fluorescent molecules within cells. These methods permit measurement of both diffusional mobility and membrane transport at the resolution of the light microscope (30). Peters (30) used fluorescent microphotolysis to measure the movement of a range of particle sizes and reported the diffusion rate drops rapidly in an exponential manner (Table 1). Small molecules such as sucrose (432 Da) with a radius of 4.4 Å travel rapidly at 200 μm^2/sec; midsize dextran molecules (10 kDa) with a radius of 23.3 Å diffuse at a rate of 25 μm^2/sec; large dextran molecules (157 kDa) with a radius of 90.7 Å travel 1.6 μm^2/sec (see Table 1).

Table 1   Intracellular Diffusion Coefficients

Molecule	Molecular mass (Da)	Radius (Å)	Dw (μm²/sec)	Dc (μm²/sec)	Sample	References
Water	18	—	2400	630	Human RBC	201
Sucrose	324	4.4	520	200	Frog oocyte cytoplasm	202
Dextran	3,600	12.0	180	35	Frog oocyte cytoplasm	204
Inulin	5,500	13.0	150	30	Frog oocyte cytoplasm	203
Dextran	10,000	23.3	92	25	Frog oocyte cytoplasm	33
	20,000	—	101	18	Cytoplasm of 3T3 cell	205
	41,000	46.1	46	4.5	Cytoplasm of HTC cell	29
	62,000	55	39	2	Cytoplasm of HTC cell	29
BSA	67,000	36	69	1	Cytoplasm of human fibroblast	200
IgG	160,000	35	40	0.9	Cytoplasm of human fibroblast	200
Dextran	157,000	90.7	24	1.6	Cytoplasm of HTC	29
				3.8	Nucleus of HTC cell	29

Adapted from Ref. 30.

Luby-Phelps and colleagues (31,32) used inert, size-fractionated fluorescein-conjugated dextrans and Ficoll, and reported that diffusion of inert particles in cytoplasm is hindered in a size-dependent manner and that the apparent hydrodynamic radius of tracer particles was a good indicator of the dimensions of these particles in the cytoplasm. Small fluorescein-labeled dextrans with a hydrodynamic radius of 32 Å are 99.5% mobile while large dextrans with a radius of 148 Å are only 63% mobile. Extrapolation of these data suggests that rigid particles whose smallest radial dimension is larger than 260 Å are practically immobile in the cytoplasmic space of living cells.

III. Transport of Oligodeoxynucleotides

Studies utilizing oligodeoxynucleotides (oligos) and plasmid DNA have provided insight into DNA trafficking mechanisms. Oligodeoxynucleotides are short segments of single-stranded DNA molecules usually 15 to 30 bases in length with corresponding molecular weights of 4950 to 9900 Da. Antisense oligos are designed to be complementary in sequence to a specific region of mRNA or DNA, and hybridization of these oligos with the target sequence can result in inhibition of gene expression. Because of the potential therapeutic nature of these molecules, considerable effort has been directed toward understanding cellular uptake and intracellular fate of oligos (34). Oligos are typically internalized by endocytosis, which is reflected in punctate cytoplasmic staining of intracellular compartments when fluorescent oligos are added to cells (35,36).

Cytoplasmic migration of fluoresence-labeled oligos (9240 Da) was shown to be initially slightly slower than diffusion of coinjected fluorescence-labeled dextrans (10,000 Da) when microinjected into the cytoplasm of living cells (36). The oligos, however, diffused through the cytoplasm and accumulated in the nucleus while dextran reached a state of equilibrium between the cytoplasm and nucleus (36–38). The accumulation resulted in a substantial fraction of the microinjected phosphorothioate oligonucleotides remaining in the nucleus for prolonged periods (c. 48 hours).

The general characteristics of nuclear accumulation of oligos are rapid uptake (noticable within 5 min) and a relative temperature independence (36, 39,40); and uptake does not appear to be sequence-specific. Nuclear accumulation is energy independent (36,37), and inhibitors of protein transport through the nuclear pore such as wheat germ agglutinin failed to diminish oligo accumulation in the nucleus (36). This profile suggests that the oligos are freely diffusing through the nuclear pore complex located on the nuclear membrane.

The interactions between oligos and nuclear structures (36,40,41) can provide insight into why oligos are retained in the nucleus. Leonetti and colleagues (37) demonstrated the presence of four oligo-binding proteins isolated from nuclear fractions. The absence of similar proteins in the cytoplamic fraction suggests the lack of transport proteins in the cytoplasm. Clarenc and coworkers (40) incubated isolated nuclei with a selection of ^{32}P-labeled oligo analogs and demonstrated that phosphorothioate and phosphodiester oligomer analogs compete for the same nuclear binding sites. In addition, these analogs exhibit two categories of binding sites, with similar numbers and affinity. An estimated 0.25×10^6 sites per nucleus ($K_D = 3 \times 10^{-10}$) and 0.35×10^6 sites per nucleus ($K_D = 1 \times 10^{-9}$) were obtained with phosphorodiester oligos. These proteins may correspond to the nuclear binding proteins identified by Leonetti and colleagues (37). Thus oligos readily diffuse into and are retained in the nucleus owing to electrostatic interaction with at least two types of nuclear proteins.

The size limitation for nuclear diffusion of linear oligonucleotides has not yet been defined, but duplexes of up to 28 mers appear in the nucleus (38). Phosphodiester oligos, however, are rapidly degraded in the nucleus ($T_{1/2}$ < 15 min) (36,38), which would limit the use of short oligo sequences for gene expression unless a chemical modification to portions of the oligos could improve their stability while permitting transcription. If binding of the short oligos to the proteins in the nucleus does not interfere with gene expression and they can be stabilized against degradation, this strategy might be extended to provide sufficient genetic information for antisense production or ribozyme synthesis, or to direct the synthesis of short peptides. Thus, many of the transport barriers to nuclear delivery may be circumvented by the use of low-molecular-weight DNA.

Recently Yoon and co-workers (42) reported short oligos are able to correct a gene defect by a homologous recombination mechanism. This group employed a hybrid ribo/deoxyribo-oligonucleotide to direct the correction of a single-base defect in the liver/bone/kidney alkaline phosphatase gene. The frequency of correction was quite high, which may be due to the increased entry of the oligo into the nucleus. The advantage of exploiting recombination to correct a genetic defect would be the maintainence of the gene regulation as well as long-term expression since the corrected gene could be transferred to daughter cells after cell division. A similar approach using a 490-base fragment of the gene coding for the cystic fibrosis transmembrane conductance regulator was demonstrated by Kunzelmann and co-workers (43). This may be a promising avenue to bring about the correction of genetic defects that arise from single-basepair mutations or from short deletions in the gene.

IV. Nucleocytoplasmic Transport of Plasmid DNA

Plasmid DNA is a double-stranded supercoiled DNA molecule, and plasmids used in mammalian cell transfection studies range from 3 to 15 kb (1980 to 9900 kDa). These macromolecules clearly exceed the 40-kDa diffusion limit of nuclear pores (30), yet nuclear entry is evident as plasmid DNA is capable of transfecting cultured cells as well as cells in vivo. The transfer of plasmid DNA from the cytoplasm to the nucleus is an inefficient process in most cells. Capecchi (1) reported that microinjection of 125 plasmid DNA molecules into the nuclei of a mouse cell line resulted in 50% to 100% of the cells expressing the gene product; when the plasmid DNA was microinjected into the cytoplasm, gene expression was observed in less than 0.001% of cells. This low expression level after cytoplasmic microinjection of plasmid DNA has been confirmed by many other investigators (44–47). The nuclear membrane may, however, not be a barrier in all cell types. In particular, the surprising observation that direct injection of DNA into muscle results in long-term expression of transgene in mice (48) and other species suggests that at least in myotubes, DNA is able to reach the nucleus relatively easily. Microinjection studies utilizing rat myotubes have shown up to 70% cell transfection efficiencies after microinjection of plasmid DNA into the cytoplasm (47). This is dependent on the dose of DNA and the proximity of the injection site to the nucleus (47). The 70% transfection level was seen at a dose of 10 mg/ml, which corresponded to about 4×10^7 plasmids per cell as about 2 pL of material was microinjected per cell. Expression was observed in about 5% of the cells when cells were microinjected with approximately 36 plasmids per cell. The authors found no difference in expression when the reporter gene contained an intron in the coding region. This supported their assumption that the nucleus was the site of transcription. Microinjection near the boundary of the nucleus resulted in a 56% rate of transfection, whereas injections made at the same DNA concentration but 60 to 90 nm distal to the nucleus resulted in an 8% rate of transfection (47).

This supports the view of a crowded cytoplasm, which poses a substantial barrier to cytoplasmic movement of nucleic acids. Expression was depressed by coinjection of wheat germ agglutinin (WGA), a known inhibitor of nuclear translocation via the protein mediated nuclear uptake system (see below) as well as by a decrease in the incubation temperature during injection. Dowty and colleagues interpreted this finding as implicating the protein-mediated nuclear localization system in this facile uptake of DNA into the myotube nucleus. This observation raises the question whether transcription factors or other proteins that function in the nucleus might associate with plasmid DNA in the cytoplasm and, in essence, permit the DNA to piggy-back on the nuclear localization transport system to enter the nucleus. In spite of the extensive expression, biotin-labeled DNA was detected within nuclei too infrequently and at such low levels that it could not be quantitatively analyzed. Gold-labeled DNA, however, could be detected at the EM level in about 25% of the microinjected cells (47). Evidence for a transport mechanism was also seen when isolated yeast nuclei were investigated for DNA uptake using radiolabeled DNA (49). This group showed an ATP/GTP-dependent association of linear DNA in isolated nuclei. There was a counterintuitive preference for larger DNA strands (1.5 to 3.8 kbp); however, the authors did not differentiate between surface bound and internalized DNA, so it is difficult to evaluate the significance of this finding for gene delivery.

Loyter and co-workers (50) showed that calcium phosphate precipitation can transfect 1 in 10,000 cells in which 1% to 5% of cells exhibit fluorescence from a DNA binding dye, originally associated with the plasmid, appearing in the nucleus after a few hours in culture. Thus either (1) the fluorescent dye is not a good indicator of the location of the plasmid, (2) only a small fraction of nuclear-localized plasmid DNA molecules are transcribed, or (3) a critical number of plasmid DNA molecules are required to produce detectable levels of reporter protein.

Ethidium bromide-labeled plasmid DNA was introduced into cultured cells by cationic liposomes (51). The ethidium bromide rapidly accumulates in the cell nucleus following cationic lipid-mediated transfection and 5% of cells had ethidium-stained nuclei (51) after 30 min of incubation. This observation correlates with the low levels of transgene expression found in cells exposed to lipid-DNA complexes for 30 min (52,53). The level of labeled nuclei increases up to 6 hours and remains constant up to 24 hours (51). However, these authors were unable to detect plasmid DNA in the nuclei of transfected cells by EM techniques. This raises questions concerning the use of ethidium bromide to reliably identify the location of the introduced plasmid since the dye might transfer from the plasmid into cellular nucleic acids during the incubation period. Our unpublished data show that even high-affinity DNA stains such as YOYO and TOTO can transfer between plasmids during incubation in the test tube.

In a complementation of function study, Thorsness and Fox (54) shows a preferential migration of mitochondrial DNA to the nuclei of yeast cells rather than in the opposite direction. By using microprojectile bombardment, plasmid DNA coding for both nuclear and mitochondrial genes is introduced into yeast cells. Transformants were isolated by their ability to complement one mutant function, and subsequent generations were monitored for the complementation of a second mutant function. The propensity for DNA transfer from mitochondria to the nucleus is 2×10^{-5} events per generation, which is 1 $\times 10^5$ times greater than transfer in the reverse direction. This result suggests that a property exists for DNA escape and preferential nuclear targeting, albeit at a low frequency of nuclear entry.

The intracellular localization of plasmid DNA after delivery by polylysine complexes targeted to hepatocytes in rats has been investigated by Wu and co-workers (55–57). These authors observed that the majority of DNA initially distributed into a light membrane and lysosomal compartment. The plasmid was rapidly degraded and by a few hours postinjection could be detected in the light membrane fraction only by Southern analysis. Partial hepatectomy or pharmacological treatments caused a fraction of the DNA to associate with a vesicle fraction that persisted in the cytoplasm. Although significant expression was detected in the liver, the authors were unable to detect plasmid DNA in the nuclear fraction by Southern analysis but could detect plasmid sequences by PCR techniques. The in vivo system has multiple barriers to plasmid delivery, and the results do not directly support nuclear uptake as the rate-limiting barrier but reinforce the view that nuclear delivery of plasmid DNA is a rare event.

Delivery of plasmid DNA can be facilitated with the coadministeration of viral particles and nuclear DNA binding proteins, as shown by Kaneda and colleagues (58). In these experiments lipid-DNA complexes are fused, via Sendai virus, with red blood cell ghosts previously loaded with protein. By using HMG-1, a nonhistone chromosomal protein, three to five times more DNA was detected in the nuclei of rat liver cells (per portal vein injection) when compared to transfections using nonnuclear BSA. Increased levels of gene expression were demonstrated by immunostaining of tissue sections from rats treated with HMG-1 protein compared to BSA-treated rats. The Sendai viral component is critical; without it the transfections yield negative results.

V. Expression of DNA in the Cytoplasm as a Means to Circumvent Low Nuclear Uptake

To circumvent inefficient nuclear delivery of plasmid DNA, cytoplasmic transcription of plasmids with the appropriate promoters has been driven by polymerase that can be introduced into the cytoplasm. T7 polymerase introduced into the cytoplasm either by a vaccinia virus expression system (59) or as the purified protein (60) could be utilized to drive cytoplasmic transcription of the plasmid DNA (51,61). Gene expression was detected in the cytoplasm of cells containing T7 polymerase that had subsequently received plasmid DNA delivered by cationic lipids. One hundred times more plasmid DNA was required to obtain significant expression in the absence of the T7 polymerase, and only 10% of these cells expressed the transgene (60,61). When T7 polymerase was introduced as the protein, the peak expression level occurred sooner than when the T7 polymerase was introduced on an expression vector, and duration of expression was less than 60 hours while the expression T7 vector showed a duration of greater than 5 days. The conclusion from such experiments is that the increased level of gene expression demonstrates that a considerable number of transcriptionally viable plasmid DNA molecules are present in the cytoplasm, and these molecules are not readily transported into the nucleus. This strategy could be extended to use other polymerases, but, as with any xenogenic protein, the immune response to the polymerase may ultimately lead to loss of transgene expression. Thus, such systems may best be employed in applications that require short-term expression and situations in which the patient will not be repeatedly exposed to the gene transfer system.

VI. The Mechanics of Movement in the Cytoplasm

The question remains how organelles and large molecules (>3300 kDa), including nucleic acids and proteins, once released into the cytoplasm, are physically transported to the nucleus. Diffusion alone cannot account for transport since macromolecules the diameter of plasmids would be nearly immobile in the crowded cytoplasm. The most likely candidiates for transport mechanisms involve participation of the cytoskeletal components—microtubules and actin filaments.

Microtubules and actin filaments are believed to maintain intracellular distribution of organelles and to facillitate trafficking between organelles (for review see Ref. 62). Microtubules can be viewed as tracks for the movement of organelles and their cargo, where movement is driven by protein motors fueled by ATP (63,64). Microtubules radiate out from the microtubule organizing center (MTOC) into the peripheral regions of the cytoplasm, thus forming an extensive network of fibers throughout the cell (Fig. 2). The fast-growing

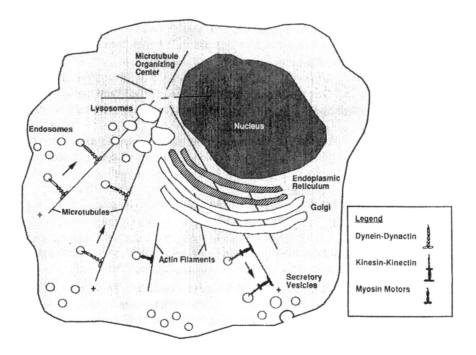

Figure 2   A model of actin filament and microtubule involvement in organelle transport. Vesicle transport is driven by protein motors interacting with either actin filaments or microtubules. Myosin motors are responsible for transport along actin filaments, while kinesin and dynein direct transport along microtubules. The receptor proteins kinectin and dynactin are believed to join the kinesin and dynein, respectively, to the vesicle membrane.

ends (plus ends) of the microtubule are located at the cell periphery while the slow-growing ends (minus ends) are found at the MTOC. Differential distribution of organelles and transport vesicles are observed in the microtubule network with endosomes found near the plus ends at the cell periphery, whereas Golgi, late endosomes, and lysosomes are clustered near the minus ends near the nucleus.

Kinesin and cytoplasmic dynein are the principal ATPase microtubule-based motors involved in organelle transport (65–68). Each type of motor associates with membranous organelles and directs movement along the length of the microtubule. Dynein drives movement toward the minus end of the microtubule, yielding a net inward flow, whereas kinesin drives movement towards the plus end, generating movement toward the cell periphery (69). Dynein is responsible for movement of endosomal and lysosomal vesicles toward the cell nucleus, while kinesin has been implicated in maintaining the extended distribution of the endoplasmic reticulum, the shape of the Golgi complex, the extension of lysosomes, and trafficking of proteins from Golgi to endoplasmic reticulum (70–72).

The motor proteins are believed to attach to receptor proteins anchored in organelle membranes, and two such receptor proteins have been identified and are currently under investigation (73). The kinectin protein is thought to join kinesin to the organelle membrane. Kinectin is predicted by its amino acid sequence to form a coiled-coil motif and is anchored in the membrane by its N-terminal sequence (74). Antibodies to kinectin, blocking potential binding sites for kinesin, inhibit plus-end-directed organelle movement in vitro by 90% (75). The dynactin complex, consisting of 10 different polypeptides, is capable of interacting with cytoplasmin dynein as well as microtubules, kinetochores, and membranes (73), yet the molecular events involved in attachment have not been elucidated. The interaction of microtubules with membranous organelles also appears to involve linker protein, or CLIPS, that bind to membranes via protein receptors (73,76). It remains to be determined whether additional kinectin- or dynactinlike proteins exist, which receptor proteins are found on specific organelles, and how organelle movement is orchestrated.

Actin fibers are shorter than microtubules, and actin-based systems maintain membranous organelles at particular locations within a cell (63,64) and provide movement to local sites along axons. Actin fibers are associated with unconventional myosins that function as motors, and it is the tail domain of the myosin molecule that is predicted to interact with the organelle membrane (77). Conformational changes of myosin that occur during ATP hydrolysis lead to a step of movement along the actin filament. Using optical laser traps, one hydrolysis event was shown to displace the myosin molecule 11 nm (78).

The microtubule and actin motility systems are believed to be closely interrelated with regard to structure and function (62,63). It may be possible to utilize these systems to transport plasmid DNA to the nucleus to facilitate gene delivery. Motor proteins, motor protein receptors, or the relevant peptide sequences may be conjugated to or complexed with plasmid DNA. This may result in association of plasmids with microtubules or actin filaments for more efficient transport through the cytoplasm to regions bordering the nucleus. Release of the plasmid from the microtubules at the destination could be a challenging problem in protein design. The observed accumulation of plasmids in the vicinity of the nuclear pores in myotubes (47) may be due to some type of interaction of the plasmids with the microtubule systems.

Disruption of the microtubule system by partial hepatectomy or colchicine pretreatment of rats has been shown to increase the level and the duration of gene expression from a liver-targeted polylysine gene delivery system (55,79,80). Intact microtubules are required for the translocation of endosomes to lysosomes; thus, microtubule disruption leads to the inhibition of degradation of endosomal contents. The increased duration of intact plasmid DNA in the endosomes of treated rats accounts for the enhancement in expression levels and duration. This group has developed a bioconjugate based on the asialoglycoprotein receptor that can target colchicine to the hepatocytes and may afford a facile approach to altering the intracellular trafficking of plasmid DNA in the liver (81).

VII. Review of the Nucleus

A. Overview of the Nucleus

The nucleus is bounded by the nuclear envelope, which encloses chromatin and the machinery necessary for gene transcription. The nucleus is a dynamic structure, which disassembles at the onset of mitosis and reassembles during telophase. The envelope consists of two membrane bilayers, posing a considerable hydrophobic barrier to macromolecular transport (Fig. 3). The outer nuclear membrane is continuous with the endoplasmic reticulum, and with the inner nuclear membrane forms a perinuclear space that is continuous with the endoplasmic reticulum lumen (82,83). The inner nuclear membrane is supported internally by the nuclear lamina, a network of lamin proteins that lines the inner side of the envelope (82). The lamina also is thought to provide attachment sites for chromatin (84–86). The very interior of the nucleus is a network of DNA, RNA, and proteins. It is the active transcriptions sites that are the targets of gene delivery.

The major barrier between the cytosolic and nucleoplasmic compartments is the hydrophobic barrier of the nuclear envelope. The overall lipid composition of the nuclear envelope membranes is dominated by phosphatidylcholine (PC), followed by phosphatidylinositol (PI) and phosphatidyleth-

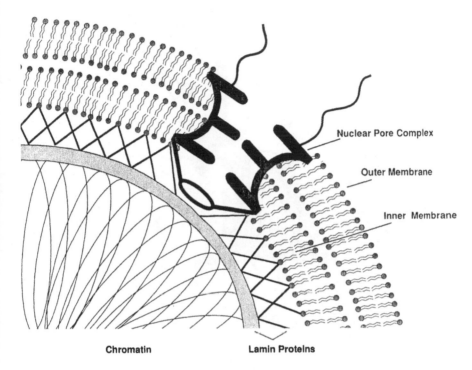

Figure 3 The nuclear envelope. The nuclear envelope is composed of a double bi-layer membrane, nuclear pore complexes, and lamin proteins.

anolamine (PE) in abundance (data are for rat liver nuclei; Ref. 87). The most striking difference between the nuclear envelope composition and that of the plasma membrane is the greater relative contribution of PE and sphingomyelin in the plasma membrane. Perhaps these differences are important when considering the different functions of both membrane systems and in designing a molecular system to perturb these individual membranes.

The inner and outer nuclear membranes, which would otherwise form a loose perinuclear compartment, are joined at nuclear pore complexes (NPCs) interspersed throughout the nuclear envelope (Fig. 4). These 125-MDa structures are composed of an estimated 100 proteins and span both membranes of the nuclear envelope (88,89). The NPC forms aqueous channels and supports both passive diffusion and active transport. It appears that the number of NPCs per nucleus (or NPCs/μm^2 nuclear membrane) corresponds to the trafficking needs of the cell and varies with cell type, cell cycle, and metabolic activity;

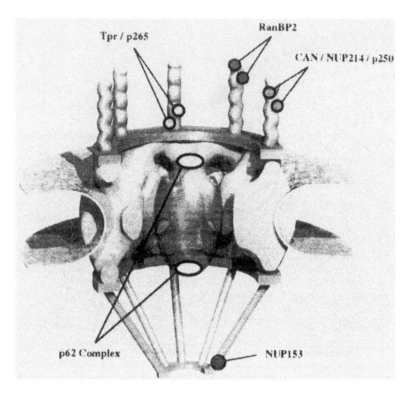

Figure 4 The nuclear pore complex. Consensus diagram of the nuclear pore complex. The NPC has a mass of approximately 125 MDa, consisting of an estimated 100 different proteins. The four major NPC components are (1) the spoke ring complex which provides a structural framework, (2) the central channel complex, (3) the cytoplasmic ring and associated filaments, and (4) the nuclear ring and associated basket. Structure of the spoke ring complex is suggested by negative stain/random conical tilt reconstructions; structure of the filaments and basket are confirmed by EM studies; structure of the central channel complex has yet to be established. (From Ref. 91, with permission.)

estimates of NPC density as low as $<1/\mu m^2$ for quiescent avian erythrocytes and up to $120/\mu m^2$ in tetrahymena (83). The structure and transport properties of NPCs are summarized in several recent reviews (83,88–95).

In summary, access into the nucleus is restricted as well as regulated by the nuclear envelope and the NPC. However, there are opportunities to access the nucleus through the nuclear pore complex or during telophase, when the nuclear envelope reassembles after cell division.

B. Nuclear Pore Complex

The nuclear pore complex consists of four basic elements—the spoke ring complex, the central transporter, cytoplasmic fibrils and the inner nuclear basket (Fig. 4) (for reviews see 88,91,94,96). The spoke-ring complex extends into the lipid bilayer and forms a barrellike structure that best displays the eightfold rotational symmetry characteristic of the NPC. The annulus of the spoke ring is thought to be composed of several of the aqueous diffusion channels, with a maximum size limit of 90 Å diameter. Active transport is proposed to take place through the central transporter within the spoke-ring complex. The central transporter pore sits within the spoke ring complex and is functionally observed to expand up to 260 Å in diameter (89,93) to transport large proteins across the nuclear envelope. Two large rings flank the nucleoplasmic and cytoplasmic sides of the NPC and lie coaxial with the spoke-ring complex. Fibrils attached to the cytoplasmic ring extend freely into cytoplasmic space, while fibrils on the nucleoplasmic side are arranged into a basketlike structure on the nuclear side. The NPC permits bidirectional transport of a wide size range of substrates, from 21-kDa histones to 2800 kDa ribosomal subunits (97). A molecule traversing through the NPC must travel 100 nm from cytoplasm to nuclear side. Studies examining trafficking of DNA molecules through the NPC have yet to be realized, so it is largely unknown how DNA of therapeutically relevant sizes (up to 10 kbp), exceeding the size of the NPC, is able to pass through the pore complex.

There are cellular conditions that can modulate the characteristics of nucleocytoplasmic transport. In addition to variations in nuclear pore complex surface density, the cell cycle appears to also modulate NPC active transport kinetics. Nuclei from proliferating cells exhibit increased active transport kinetics compared to nuclei from quiescent cells, despite equivalent nuclear surface-to-volume ratios (98–100). This may correspond to increased cell activity. The depletion of intracellular calcium stores, particularly from the endoplasmic reticulum lumen, results in the inhibition of diffusional transport (101, 102). The direct physiologic consequences of this effect is unknown; perhaps it is associated with a stress response. It is known that the phosphorylation activity of the cell increases during cell division released in increased phosphorylation of both nuclear and NPC proteins (103). Although these parameters have not been extensively exploited in gene delivery, the metabolic and functional characteristics of the target tissue are certainly opportunities to modulate and improve the efficiency and targeting of gene delivery. The addition of a metabolic activator such as PKC or butyric acid derivatives prior to DNA delivery has been shown to increase gene expression (103,210). Smith and co-workers administered sodium butyrate and tributyric acid to cells in cultures and found enhanced gene expression. This was attributed to the inhibition of histone deacetylases, resulting in an increased level of histone acetylation, which is associated with cellular activities involved in replication and transcription (105). Condensation of DNA with calcium ions could be used to create a musclelike environment for DNA trafficking or stimulate secondary messengers for phosphorylation of nuclear pore proteins. These alternatives are not yet explored, but they demonstrate that many avenues exist for the optimization of gene delivery.

C. Nuclear Pore Proteins

The NPC is composed of two major families of proteins. Proteins that extend through the membrane bilayer are called pore membrane proteins (POMs) and the proteins that sit upon this scaffold are termed nucleoporins (NUPs) (Table 2). Many NPC proteins are identified by their ability to bind the lectin wheat germ agglutinin (WGA), indicating the presence of O-linked N-acetyl glycosyl sugar modifications. With the use of WGA and antibodies, a few nucleoporins have been individually isolated and characterized. Several excellent reviews describe the different NPC proteins (88,94,95); here we highlight proteins of special interest to nucleoplasmic transport.

Table 2 Nuclear Pore Proteins and Characteristics

Nuclear pore protein	Characteristics	References
gp210	Pore membrane protein glycoprotein	114
CAN/Nup214/p250	Cytoplasmic ring/fibrils glycoprotein - XFXFG motif	91
Nup180	Cytoplasmic ring/fibrils no motif	106
RanBP2	Cytoplasmic ring/fibrils Zn^{2+} fingers binds Ran/TC4 GTPase	91
NUP159	Cytoplasmic ring/fibrils binds to NLS receptors, Imp-β	218
p62	Internal core complexes with p54, p58 direct interaction with mRNA	107, 108
Nup153	Zn^{2+} finger Motif binds DNA, in vitro nuclear localization	109
Nup145p/Nup116p/Nup100p	GLFG motif mRNA export, protein import	111

Nucleoporins are divided into two categories—those that display degenerate GLFG or XFXFG amino acid repeats indicative of O-linked glycosyl groups, and those that do not display this motif. Several proteins are identified with the cytoplasmic ring and fibrils, as shown in Figure 4 (91). CAN/Nup214/p250 binds WGA, indicative of the presence of the XFXFG repeat, while Nup180 and Tpr/p268 have neither characteristic (Table 2). It has not yet been determined what roles these proteins play in nucleocytoplasmic transport. The peripheral locations of the proteins suggest their participation in binding events that lead toward the NPC, although it is shown that antibodies to Nup180 do not inhibit protein transport (106). Recently the association between the NPC and nucleotide exhange proteins was localized to the cytoplasmic fibrils and confirmed by the identification of RanBP2, a Ran/TC4-binding protein (91). The active state of the protein, Ran-GTP is preferentially bound, and hydrolysis to Ran-GDP facilitates its release from the nucleoporin. RanBP2 is also O-linked glycosylated and displays 8 Zn^{2+} fingers, the function of which is unknown (91). Future studies regarding the components of the cytoplasmic fibril will verify the mechanism by which the nuclear transport substrate is concentrated at the NPC, a role that has long been suggested but only recently established by the characterization of NUP159. This yeast protein is localized to the cytoplasmic face of the NPC and has demonstrated binding to receptors that mediate protein nuclear import (refer to the section on Nuclear Localization Sequence Receptors).

One of the first isolated nucleoporins, p62 isolated from rat liver, displays the XFXFG motif. This protein tightly complexes with two other nucleoporins, p58 and p54 (107), and has been localized throughout the pore complex but mainly to the internal core of the NPC. Depletion of p62 and associated proteins from the nuclear envelope results in the loss of protein import function (107), and the direct interaction between mRNA and p62 implicates the nucleoprotein also functions in mRNA export (108). p62 also appears to play a structural role in the NPC, since the nucleoprotein is necessary for the in vitro formation of NPCs in annulate lamellae (94). Therefore p62 is involved with both import and export functions as well as providing structural support.

Nup153, containing a GLFG motif, contributes to the basket-like structure on the nucleoplasmic face. This protein displays Zn^{2+} finger domains which have been demonstrated in vitro to bind DNA reversibly (88,106,109). Nup153 interaction with cellular nucleic acids in situ has not yet been demonstrated; however, the property is intriguing when considering the transport and localization of imported nucleic acids.

A nucleoporin implicated in the transport of RNA is Nup145p. It exhibits an RNA-binding motif that would suggest a role in the binding and transport of RNA from the nucleus (110). The motif, an octapeptide RGYGCITF, is common to proteins that bind single-stranded nucleic acids, and this motif is shown through in vitro binding assays to preferentially bind poly(G) (111). The GLFG domain is not required for RNA export, but Nup145p mutants may be functionally complemented by Nup116p or Nup100p (111,112). Nup116 is essential for growth, for interaction with Kap95 (yeast nuclear localization sequence receptor, homolog for importin-β), and for mRNA export and protein import (113).

The transmembrane POMs, while distant from actual interaction with transport substrate, can nevertheless have an indirect impact on nucleocytoplasmic transport. One of the few identified POMs is gp210, which attaches the NPC to the nuclear envelope at the membrane interface between the inner and outer nuclear membranes (107,114). The lumenal domain of the gp210 is shown to disrupt protein import when bound by WGA or specific antibodies to this particular domain. This suggests that gp210 is important to the functionality of the NPC despite its remote location within the NPC. An additional important feature of the lumenal domain is the dominant sorting element within the transmembrane domain of the protein (114).

VIII. Studying Nucleocytoplasmic Transport

Microinjection, cell permeabilization, and isolated nuclei have been used to study the properties of nucleocytoplasmic transport. Microinjection allows the introduction of material into either the cytoplasmic or nuclear compartments in a living cell. This has been a powerful technique to study diffusion of macromolecules through the cytoplasm, the role of compartmentalization on expression of plasmid DNA, and the factors that regulate macromolecular transport into the nucleus. Digitonin has been effectively employed to permabilize the cell plasma membrane and introduce the substrate and supplemental cytosolic factors to replace proteins which escape while the membranes are compromised (developed by Adam and Gerace; see Ref. 115). This is a common method used to study the requirements for import of proteins into the nucleus. Isolated or reconstituted nuclei have also been used for studying nuclear transport in vitro and permit stringent control of the transport environment. More recently, genetic manipulation via deletion and complementation have been employed to determine the necessity of specific nuclear pore proteins for viability, growth, and transport.

Fluoresence and electron microscopy have been the principal tools for the detection and localization of transported macromolecules. Gold particles have been used to localize substrates and to assess the functional size of the nuclear pore complex and cytoplasmic motility. Fluoresence microscopy is more useful for measuring the kinetics of transport. Evaluation of the properties of the transport system typically includes temperature dependence, substrate with competing signal, energy inhibitors (NTP analogs), and known transport inhibitors such as WGA.

A. Nucleocytoplasmic Transport of Macromolecules

The nuclear pore complex is the gateway for the transport of many types of proteins and RNAs into the nucleus. These include replication enzymes, nucleolar proteins, steriod hormone receptor complexes, transcription factors, histones, messenger RNA, U snRNAs, and ribosomal RNA. Small molecules with molecular weights less than 60,000 to 70,000 Da can diffuse freely into the nucleus, while macromolecules larger than 90 Å in diameter require nuclear localizing sequences (NLS) to facilitate transport through the NPC. Indeed, some small proteins like histone H1 (21 kDa) also display NLS and bind to cytoplasmic receptors and are actively transported into the nucleus (116). Both proteins and RNAs (snRNPs, rRNAs) display bidirectionality with respect to transport. Following an introduction to nuclear localizing sequences, we will describe transport for each class of molecule.

Transit through the NPC requires energy in the form of GTP along with three distinct, consecutive events for nuclear entry of proteins containing an NLS (Fig. 5). The first step involves interaction of the NLS with a cytoplasmic protein. The second step requires receptor-mediated docking of the karyophilic protein to the nuclear pore, and this is followed by an energy-dependent translocation of the complex into the nucleus (for review see Ref. 117). Most NLS studies have focused on NPC-mediated transport of proteins; however, NLS have also been implicated in the transport of viral nucleic acids, mRNAs, and snRNPs.

B. Nuclear Localization Sequence (NLS)

The presence of an NLS in a protein sequence is the protein import signal that is involved with a complex series of protein-protein interactions that culminates in the energy-dependent expansion of the NPC central channel and translocation of the protein into the nucleus. Macromolecules with diameters greater than 90 Å that are found to enter the nucleus are more than likely to contain an NLS. An NLS sequence may be positioned along the surface-exposed residues in the protein sequence and a repertoire of sequences of different strengths and potencies exists. Nuclear targeting capability seems to be additive for proteins containing multiple NLS (Table 3).

The nuclear targeting sequence of SV40 large T-antigen is a paradigm and contains a cluster of basic amino acids (PKKKRKV) that confers the nuclear localizing activity. It was initially revealed that mutations of an internal lysine of the SV40 large T-antigen abolished nuclear accumulation (118,119).

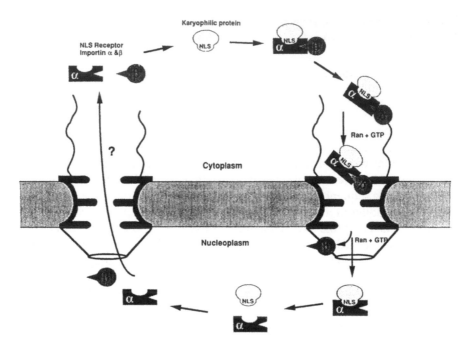

Figure 5 The NPC-mediated transport of karyophilic proteins into the nucleus. The initial event preceding the transport of protein through the nuclear pore complex is the binding of the karyophilic protein bearing a nuclear localization sequence (NLS) to the cytoplasmic NLS receptor complex importin-α/β. The NLS is recognized by the importin-α subunit which then permits binding of the importin-β subunit. This complex then binds to the NPC and, in the presence of Ran plus GTP, is transported into the cell nucleus. The importin-β may remain associated with the NPC, while the karyophilic protein-importin-α complex enters the nucleus. The mechanism of export of importin subunits has yet to be elucidated. (Adapted from Ref. 90.)

Investigation of other proteins for homologous NLS motifs have demonstrated that there are at least two classes of NLS. Those similar to the SV40 large T generally contain a short sequence of four to seven basic amino acids, and the second class is made up of a longer, bipartite motif consisting of two regions of basic amino acids separated by 10 less conserved amino acids. The ability of a sequence to confer nuclear uptake can be determined by the conjugation of the putative NLS to a reporter macromolecule (120) or the production of recombinant chimeric proteins (121) between a NLS sequence and a protein that otherwise localizes in the cytoplasm.

Nucleoplasmin, a major nuclear protein of *Xenopus* oocytes and embryos represents the most studied bipartite motif (122). Feldherr and colleagues

Table 3 Nuclear Localization Sequences

Protein	Sequence	Location (position)	References
SV-40 large T Ag	PPKKKRKV	N-term (132/708)	118, 119
Adenovirus E1A	SCKRPRP	C-term (239/243–289)	211
HIV *rev*	RRNRRRRW	Internal (45/116)	209
Mat α2 (yeast)	NKIPIKD	N-term (8/210)	214
Histone H2B (yeast)	GKKRSKA	Internal (35/131)	215
Xenopus nucleoplasmin	avkRPAATKKAGQAKKKkld	C-term (172/200)	216, 183
Xenopus N1/N2	LVRKKRKTEEESPLKDKDAKSKQ	C-term (553/589)	217
Rat glucocorticoid-R	RKTKKKIK	Internal (517/930)	219
Human lamin A	SVTKKRKLE	Internal (442/664)	220
Human estrogen-R	IRKDRRG	Internal (261/595)	75
Human c-*myc*	PAAKRVKL	C-term (327/439)	193
	RQRRNELKRSF	C-term (374/439)	193

Adapted from Ref. 83.

(123) demonstrated that the NLS of nucleoplasmin was sufficient to direct colloidal gold molecules with diameters of approximately 250 Å through the nuclear pore. Conjugation of a truncated nucleoplasmin molecule lacking the NLS was unable to duplicate these results. Consensus matching of NLS sequences with other nuclear proteins has revealed that the bipartite motif is more common that the SV40 large T motif (124). Several investigators have shown that a synthetic peptide containing the NLS can direct transport of large proteins including IgM into the cell nucleus (120,125,126).

C. Nuclear Localization Sequence Receptors

Recent investigations have identified cytosolic receptor proteins responsible for transport of NLS-containing proteins into the nucleus (Fig. 6) (127,128). These studies employing in vitro reconstitution of nuclear import showed that

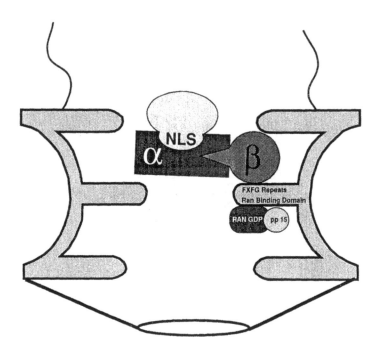

Figure 6 NPC interactions during protein transport. This model suggests that importin-β binds to the FXFG repeats found on various nuclear pore proteins. Some nucleoproteins contain Ran-binding domains, thus allowing close approximation of the NLS receptor and GTP hydrolysis machinery. The sequential steps of binding and release fueled by Ran/GTP and pp15 leads to transport of the karyophilic protein along the 100-nm length of the nuclear pore. (Adapted from Ref. 91.)

two cytosolic fractions were necessary and sufficient to reproduce nuclear protein import. Fraction A, associated with nuclear envelope docking activity, contains a protein heterodimer of importin-α and importin-β (128). Fraction B conferred nuclear pore translocation and has been shown to contain a GTP-binding protein called Ran/TC4, and an associated stimulatory factor called pp15. Görlich and colleagues have shown that Ran/TC4 and importin-α/β are sufficient to restore nuclear import. These cytosolic transport factors and their homologs have been the subject of extensive, ongoing investigations. Table 4 summarizes the transport factors identified to date.

The model for NPC receptor-mediated transport initially involves the binding of importin-α via its N-terminus subunit to importin-β (Fig. 5). The internal domain of importin-α consists of eight clusters of basic amino acids in the form of repeats implicated in NLS binding (128), and an unidentified region of importin-α is specifically recognized by importin-β. Importin-β then docks onto the NPC site containing degenerate amino acid repeats. The transport substrate is released from the NPC by GTP hydrolysis. One GTP hydrolysis event can account for 9 to 10 nm of distance movement. Yet, the transported protein must travel through the 100-nm-long pore channel before entry into the nucleus, and it is thought that multiple hydrolysis events are required for this to occur. It has been hypothesized (94,129) that a sequence of repetitive steps of binding and release along a track of repeat-containing nucleoporins is required for transport and that the affinity of karyophile lysines to nucleoporin repeats increases as it approaches the nuclear space.

There are several strategies to incorporate in the design of gene delivery systems given this information. The first is to incorporate proteins or peptides

Table 4 Nuclear Transport Factor Nomenclature

Nuclear transport factor	Homologues and synonyms	References
NLS receptor	Karyopherin α, importin 60, Rch1, hsrp1, NPI-1, m-importin, Srp1p *(Saccharomyces cerevisiae)*, pendulin, oho31 *(Drosophila)*	115, 190, 176, 127, 175, 177, 128, 173, 174, 178, 179, 180, 181
p97	Karyopherin β, importin-90	191, 192, 177, 128, 178
Ran	TC4	196, 194, 93
NTF2	B2, p10, pp15	182, 197
hsp70/hsc70		199, 93

From Ref. 221.

harboring nuclear targeting sequences with plasmid DNA molecules. These sequences can be either covalently linked to the DNA or present on DNA-binding proteins/peptides. The critical factor is for the NLS to remain with the plasmid DNA in the cytoplasm. Complexing the NLS protein with a lipid delivery vehicle would not be effective, as the lipid and plasmid DNA do not remain complexed during the transit of plasmid DNA to the nucleus. The second strategy is to use a more direct approach to targeting the nucleus, which would utilize the NLS receptor importin-α. The challenge when designing NLS or NLS receptor components is to ensure that the three-dimensional structure of the localizing motif is exposed to the critical cytoplasmic and nuclear targets.

D. Ribonucleic Acid Transport

The complexities of RNA transport are revealed in the distinct requirements for export and import of different RNA classes, dependent on both the proteins bound to the RNA and modifications made on the RNA itself (see Table 5). RNA transport mechanisms share certain similarities with protein transport; NLS peptides attenuate but do not completely shut down the transport of RNA. This suggests unique requirements for the transport of RNAs. RNA transport is saturable, temperature- and energy-dependent, indicating a carrier-mediated process. RNA transport is inhibited by antibodies to the NPC (92). RNA export is closely associated with transcription and processing events in the nucleus, and both export and import are mediated by proteins binding to the RNA (110). Most of the information on RNA transport has come from studies on mRNA export and U snRNA import.

The transport of RNA through the nuclear envelope requires RNA processing. Messenger RNA export is closely associated with the transcription and

Table 5 RNA Transport Requirements

RNA	Export requirements	Import requirements	References
mRNA	Intron spliced 5′ quanidine capped poly(A)+ tail protein: hnRNP A1	N/A	90, 110
U1 snRNA U2 snRNA	Monomethyl 5′ cap 3′ stem loop cap-binding protein	Trimethyl 5′ cap Sm proteins	97, 130, 206, 207, 208
U4 snRNA U5 snRNA	Monomethyl 5′ cap 3′ stem loop cap-binding protein	3′ stem loop Sm proteins	97, 207

processing machinery in the nucleus and unspliced mRNAs are not transported to the cytoplasm. The mRNA must be capped at the 5' end with a cyclic guanine, correctly and completely spliced, and elongated with a poly(A) tail prior to transport. Because RNA transport is dependent on correct processing and requires associated proteins, it has been shown that general trafficking of RNA follows a solid-state model (110,129).

The uridine-rich small nuclear RNAs (U snRNAs) often undergo both nuclear export and import. The export of U snRNA also requires a 5' cap with an additional methyl group (m7GpppG), stem loops at both ends of the nucleic acid strand, and an internal protein binding domain. Modifications for reimport of the snRNA typically includes the hypermethylation of the 5' cap, the presence of the 3' stem loop, and trimming of the 3' end of the nucleotide. All of the modifications are not essential for import of certain classes of snRNA, as depicted in Table 5. The requirement for the trimethyl cap is essential in *Xenopus* oocyte microinjection studies. However, its presence only enhanced import but was not essential in mammalian cells (130). This would imply the existence of multiple transport pathways, or the existence of specific proteins for different snRNA trafficking. Indeed, several methyl guanine cap-binding proteins have been described for the export of U1snRNA (131).

Specific proteins that bind RNA are also essential for nuclear transport. The hnRNP A1 shuttle protein has been determined to participate in mRNA trafficking across the nuclear membrane. A nuclear import/export signal of 38 amino acids has been identified for the hnRNP A1 protein (90), NQSSNFGPMKGGNFGGRSSGPYGGGGQYFAKPRNQGGY, but the specific mechanism by which hnRNP A1 mediates mRNA transport is not yet defined. The U snRNAs bind to their cognate proteins in the cytoplasm, the Sm proteins, and are reimported back into the nucleus, the site of their functional activity. It is likely that the Sm proteins contains NLS, yet the identification remains to be determined. The cytosol protein requirements for snRNA import have been noted to vary depending on the U snRNA in question and the nuclear system of choice (*Xenopus* oocyte vs. somatic) (130,132). U1 and U2 are similar in that they require binding of Sm proteins to an interior domain of the RNA, as well as hypermethylation of the 5' GpppG cap.

With the understanding that both nucleic acid modification and protein binding are required for nuclear import of RNA, perhaps such modifications could be incorporated into hybrid DNA-RNA sequences or into DNA-protein complexes. Perhaps the attachment of a 3' stem loop or the methylation of a hanging 5' end will permit nuclear transport via the snRNA reimport pathway. It may also be possible to design in RNA binding sites from Sm proteins to facilitate the protein requirement for RNA transport. It is not only possible to design in positive signals for transport, but perhaps it will also be possible to inhibit competing transport substrates with the administration of compounds that would temporarily interrupt the transport of a major substrate and permit the increased transport of the modified nucleic acid.

IX. Nuclear Transport of Viruses

The genome of a virus range from 2.5 to 200 kbp and can present a formidable barrier for movement through the cytoplasm and entry into the nucleus (Table 6 for properties of well studied viruses). Viruses have developed several efficient methods for entering the nucleus of the host cell. The nuclear entry mechanism depends on whether the genome is RNA or DNA as well as the types of accompanying viral proteins. Viral nucleic acid transport is an efficient process in which an estimated one of four viral particles in cell culture reaches the nucleus, and one out of 50 viral particles are infectious. Thus, for an efficient virus approximately 10 viral particles are required to enter the cell in order to cause productive infection. Mechanisms of viral entry into the nucleus have been elucidated for some viruses (reviewed in Ref. 133).

Viruses enter cells both through endocytic processes during which the viral capsid is uncoated from the nucleic acid core and via fusion with the plasma membrane. The nucleic acids, associated with virus-specific proteins (nucleoprotein capsid), are released into the cytoplasm, where they must make

Table 6 Virus Characteristics

Virus type	Nucleic acid	Size	Strategy	NLS	References
HIV	vRNA	10 kb	Preintegration complex consisting of MA protein, cDNA, RNA, integrase, reverse transcriptase	MA protein, *Vpr, Tat*	135
Influenza	mRNA	0.9–2.3 kb	4 ribonucleic acid-associated proteins (RNPs)	RNPs	148
SV40	DNA, circular	5 kbp	Compaction with octamer of cellular histones, NLS proteins	Vp3/2	149
Adenovirus	DNA, linear	36 kbp	DNA covalently linked to terminal protein and complexed with basic core protein V	Terminal protein, core protein V	198

their way to the nucleus. The influenza virus has approached the problem of size limits in nuclear pore gating by splitting its genome into eight smaller RNA segments, each of which can be actively transported through the nuclear pore. The genome of the adenovirus may facilitate its nuclear transport by its linear structure. Nuclear localization sequences are present within the viral proteins, or, as in the case of adenovirus, the nuclear localizing proteins are covalently linked to the 5' end of the DNA at the terminal dCMP. A variety of viral nucleic acids (HIV-1, influenza, SV40, and adenovirus; Table 6) are guided throught the NPC with assistance from at least one NLS-containing viral protein.

The HIV-1 matrix protein (MA), whose function was originally identified in the assembly of the viral capsid, has been recognized as the nuclear targeting component of the HIV-1 preintegration complex (134,135). The viral preintegration complex consists of RNA, cDNA, integrase, reverse transcriptase, and MA protein. The 17-kDa MA protein contains a putative nuclear localizing sequence (NLS) conferring nuclear transport activity to the HIV-1 preintegration complex, when attached to a secondary transport receptor (136) (Fig. 7). A stretch of basic amino acids on the N-terminus of MA matches the four residue consensus sequence (K-R/K-X-R/K) identified from the SV40 class of NLS (120). Mutations in this N-terminal region prevent replication of HIV-1 in nondividing cells (120), confirming the role of the native NLS sequence in nuclear transport. Some HIV-1 strains also contain an auxillary gene product, vpr, which is a redundant nuclear localizing protein and confers increased karyophilicity of the preintegration complex (137). Although a localization signal sequence has not been identified on *vpr*, an atypical one is suspected (135,136). Another protein that may serve a nuclear transport function is Tat, which has a nuclear localization signal sequence in addition to an RNA-binding domain (138). Transport of the HIV-1 pre-integration complex is ATP-dependent, inhibited by wheat germ agglutinin, and attenuated by competing with SV40 large T NLS or HIV MA (139). These results are consistent with a nuclear pore-mediated mechanism of entry.

Influenza virus is an RNA virus with a molecular weight of 5.9 to 6.3 MDa. Influenza nucleocapsids are composed of eight separate single-stranded RNAs, each coated with nucleoproteins and three viral polymerase subunits. The individual RNPs are 15 nm in diameter and 50 to 130 nm in length (140–142). It is noteworthy that the diameters of these particles are within the maximum transporter pore diameter observed for the nuclear pore complex. The transport activity is provided by an NLS located on each of the four RNP-associated proteins (143–145), although nucleoprotein (NP) is probably the most important of these proteins. Multiple transport signals have been shown to increase the rate of nuclear uptake of proteins (146,147), and the NP protein, present in high copy number (one NP per 20 nucleotides) in the RNP particle (140), may contribute to rapid uptake via an ATP-dependent mechanism.

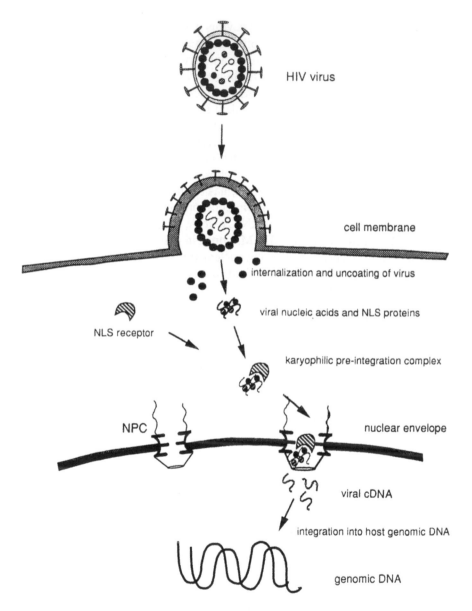

Figure 7 Nuclear import of karyophilic HIV preintegration complexes. Viral DNA is complexed with specific viral proteins which contain NLS motifs capable of directing transport of the viral nucleic acids through the NPC and into the nucleus. (Adapted from Ref. 139.)

Microinjection and electron microscopic studies indicate that the intact ribonucleoprotein complex entered the nucleus through the nuclear pore, and the time between penetration of the membrane and entry into nucleus was no longer than 10 min (148). Active transport of the RNP through the cytoplasm as well as the NPC is required to account for this observation.

Simian Virus 40 (SV40) is a DNA virus in the form of a minichromosome composed of cyclic double-stranded DNA. This virus is 5 kbp in length and is associated with about 21 octamers of cellular histones. Nuclear localizing sequences have been identified and confirmed on the Vp2 and Vp3 minor coat proteins (149). These proteins share the same amino acid sequence, with the Vp3 protein corresponding to the carboxy-terminal two-thirds of Vp2. Vp2 and Vp3 also contain a DNA-binding domain that is separate in location from the NLS. The transport of the SV40 genome has been demonstrated to be ATP-dependent but not inhibited by the lectin concanavalin A. Fusion of the viral envelope with the nucleus has been documented in vitro, and may be an alternative uptake pathway (150,151).

The adenovirus genome consists of a linear double-stranded DNA of 36 kbp in length and covalently linked to a 55-kDa terminal protein (TP) at each 5' end. The TP contains the motif RLPVRRRRRRVP, which has been confirmed as a nuclear localization sequence (152). Additionally, the TP protein has an affinity for the nuclear matrix (153,154). Adenovirus polymerase catalyzes the covalent linkage of the 5' terminal dCMP nucleotide to the β-hydroxy of a serine residue of the preterminal protein. The adenovirus genome is assembled into a chromatinlike virion core by association with two different basic proteins, protein V, VII and μ. Protein V contains an NLS based on homology with known NLSs.

Much has been learned from viruses with regard to nuclear pore entry. However, there is a paucity of information describing how these large structures traverse the cytoplasmic space. Microscopic analysis shows a diffuse as well as filamentous staining of virus in the cytoplasm, a pattern of staining indicative of cytoskeletal involvement in transport to the nucleus (148). Treatments designed to disrupt microtubules, microfilaments, and intermediate filaments, however, failed to inhibit ribonucleoprotein transport into the nucleus (148).

X. Assembly of the Nuclear Envelope

The nuclear envelope undergoes dramatic changes during mitosis (Fig. 8), and its disruption may allow colocalization of cytoplasmic plasmid DNA molecules and genomic DNA prior to nuclear reformation. This colocalization has been observed with oncoretroviruses capable of infecting only cells undergoing

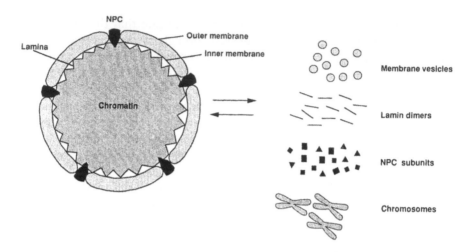

Figure 8 The reversible assembly of the nucleus during mitotis. The nuclear envelope disassembles into its various components at the onset of mitosis. Meanwhile, the chromosomes condense and segregate into the respective daughter cells. The nuclear envelope reassembly begins during the late stages of mitosis, an event associated with decondensation of chromosomes. (Left) Interphase nucleus; (right) Nuclear components during mitosis. (Adapted from Ref. 169.)

mitosis which lack an intact nuclear envelope. The viral cDNA is found integrated into the host's genomic DNA, indicating the presence of a process that leads to transport of viral nucleic acids to close proximity to genomic DNA.

Nuclear envelope disassembly and reassembly have been studied in cell-free systems derived from mammalian somatic cells or amphibian (*Xenopus*) eggs (for review see Ref. 155). *Xenopus* eggs, arrested in metaphase, are an ideal system to utilize for these studies since the isolated membrane fraction contains a vast amount of vesicles derived from the disassembled nuclear envelope, endoplasmic reticulum, and Golgi. Nuclear envelope assembly is studied by mixing cytosol, an ATP-generating system, and the membrane fraction with demembraned *Xenopus* sperm chromatin (156,157). Membrane vesicles are seen to associate with the chromatin and fuse to create a double lipid bilayer enveloping the chromatin. Perturbation of this system allows analysis of mechanisms involved in membrane disassembly.

Nuclear envelope disassembly begins in prophase where the nuclear membrane is seen to vesiculate (88,158–162), lamina proteins are phosphorylated leading to depolymerization (162), and the nuclear pore complex subunits are disassembled (163–165). This chain of events is thought to be initiated by a critical mitotic regulator MPF which orchestrates membrane receptor phosphorylation of key structural proteins by either activating a kinase or inhibiting a phosphatase. The molecular target of phosphorylation has not been elucidated, but as a result of phosphorylation a putative nuclear envelope receptor dissociates from the chromatin (166). Alternatively, dephosphorylation allows membrane vesicles to bind to the chromatin and initiate membrane reassembly. Fragmentation and vesiculation of the nuclear envelope, endoplasmic reticulum, and Golgi are believed to allow equal partitioning of these cellular components between daughter cells (155). Cells that are undergoing mitosis are biosynthetically quiescent; transport of vesicles between organelles ceases, and the cell is unresponsive to external stimuli.

Nuclear envelope assembly occurs in anaphase and telophase, and begins with dephosphorylation of the putative nuclear envelope receptor, which permits the sequential targeting of membrane components to the chromatin surface. Vesicles (80 to 200 nm in diameter) containing the lamin B receptor, an inner membrane protein, are the first components to initiate the assembly process (167–169). Experiments employing tryptic digestion have shown that vesicle chromatin association requires proteins on both the vesicle and the chromatin to interact for the priming of the spontaneous reassociation of nuclear envelope components (156,166). The fusion of membrane vesicles follows vesicle coating of chromatin, producing a double lipid bilayer structure. Nuclear membrane reassembly then proceeds by the incorporation of NPCs and reassembly of the lamina matrix. Vesicle binding to chromatin is independent of ATP (166); however, vesicle fusion requires the presence of both ATP and GTP (166,170). Vesicle fusion is inhibited by treatment of both cytoplasmic and membrane fractions with NEM, confirming the involvement of protein interaction in vesicle binding, and is inhibited by treatment with the calcium chelator BAPTA suggesting a role of calcium in the fusion process (169).

Forbes and co-workers (171) demonstrated that nuclear envelopelike structures can spontaneously form around bacteriophage and plasmid DNA which was microinjected into *Xenopus* eggs. The structures, formed within 90 min, included a double membrane similar to the inner and outer nuclear membrane, lamins, and nuclear pore complexes. Structures formed around plasmid DNA were much smaller and more fragile than those formed around both T4 and lambda bacteriophage DNA, suggesting differences in chromatin components available for membrane formation with *Xenopus* egg envelope components. Experiments with lambda phage showed that 50 to 200 nuclei were seen after cytoplasmic microinjection, indicating that small nuclei did not coalesce into one large nuclei during the 90-min time period observed. Whether more time is required for coalescence or transcription machinery is present in these structures is unknown. It is unlikely, however, that the small nuclei coalesce, since chromosomal fragments resulting from treatment of cells in culture with genotoxic chemicals remain separate from the main nuclei for up to 48 hours.

Chromosomes are guided through the cytoplasm by connection of microtubules emanating from the microtubule organizing center (MTOC) and terminating on the kinetochore attachment on the chromosome. This attachment ensures proper chromosome segregation during mitosis and close proximity of chromosomes during reformation of the nuclear envelope. Plasmid DNA lacks the kinetochore protein attachment site for microtubules, so it is unlikely that microtubules would associate with plasmid DNA and assist in the colocalization of plasmid with host DNA. Therefore, random events would be required to colocalize plasmid and genomic DNA during nuclear envelope reformation.

The question remains whether exogenous cytoplasmic DNA can become incorporated with genomic DNA and enclosed within the nuclear envelope following mitosis. Oncoretrovirus cDNA is found to integrate in host genomic DNA only in dividing cells which lack an intact nuclear envelope suggesting the presence of a virally based cDNA transport mechanism involved in colocalization of the cDNA with the host's genome. Colocalization may be a random process, yet this hypothesis fails to explain why given the high numbers of microinjected plasmid DNA molecules (1×10^5), transfection of cells in culture does not lead to 100% cells transfected (1). The situation for plasmid DNA is probably vastly different than for viruses based on the numbers of microinjected plasmids required for gene expression.

The efficiency of gene delivery may be improved by exploiting specific steps in membrane assembly. The first strategy involves the attachment of kinetechore proteins on plasmid molecules which would be accessible to microtubules and might result in colocalization of the plasmid with the chromosome. Another strategy would be to initiate the formation of the nuclear envelope around the cytoplasmic plasmid DNA. Here a critical mass of DNA molecules may be required for assembly of the nuclear envelope. Although the protein(s) that initiate the cascade of events required for nuclear assembly have yet to be identified, lamin B may be a protein that could be attached to plasmid DNA and serve as a nucleation site for incorporation of the plasmid into the nucleus during reassembly in telophase.

XI. Fusion with the Nuclear Membrane

Fusion of DNA/lipid/lamellar complexes with the nuclear membrane is another mechanism by which DNA may access the nucleus. The presence of viral vacuoles is observed close to the nuclear envelope, and markers engineered into the viral envelope have been observed in the nuclear envelope cisternae/lumen (150). Such a fusion process has been suggested to occur in certain liposome delivery systems (172), and various groups have observed electron dense particles in the vicinity of the nucleus after incubating cationic-lipid DNA complexes with cultured cells. However, the existence of a such a pathway is difficult to demonstrate since there are multiple membranes that a liposomal delivery system must traverse before entering the nucleus. When the liposome or lipid complex reaches the nuclear envelope, it must be able to fuse

Table 7 Strategies for Increasing Nuclear Delivery of DNA

Strategy	Mechanism	References
Cytoplasmic Trafficking		
Decrease transit of endosomes to lysosomes, with chloroquine	Prevent endosome fusion with lysosome, increase probability that DNA reaches the cytoplasm	187, 188
Disrupt microtubules	Alter vesicular trafficking; increase probability that DNA reaches the cytoplasm	55, 56, 80, 81, 188
Attach microtubule binding ligands to DNA	Permit DNA trafficking alone microtubules	73, 74, 75
Circumvent nuclear transport requirement	T7 polymerase regulated cytoplasmic expression	51, 59, 60, 61
Metabolic Manipulators/ Enhancers		
Protein kinase C activators	Unknown	104
Tributyrate	Acylate histones; direct mechanism is unknown	210
Osmotic shock	Increases permeability of endosomal and nuclear membranes	185, 186
Nucleocytoplasmic Trafficking		
Use low-MW DNA/hybrid DNA/modified nucleotide sequences to improve import	Permit diffusion through nuclear pore complexes	42
Utilize NLS nuclear targeting system	NLS peptide; nuclear protein	(see Table 3)
Utilize NLS nuclear transport system	Importin-β peptide to target nuclear pore complex	(see Table 4)
Utilize RNA targeting system	RNA targeting sequence to DNA	(see Table 5)
Attach lamin sequence	Increase nuclear incorporation of DNA during cell division	156, 166

across two bilayers since the envelope consists of a double membrane. The efficiency of such multiple fusions would be expected to be low and would not seem to be a promising choice to increase plasmid delivery into the nucleus. Cationic lipid-DNA complexes are proposed to fuse with the endosomal membrane (213,222,223). Hence a liposomal membrane would no longer surround the nucleic acid and could not participate in a pathway that requires multiple membrane fusions.

XII. Conclusion

Biology offers ample opportunites to circumvent factors that impede nucleic acid trafficking. In this review we have covered those biological systems that present potential avenues for increasing nucleic acid transport into the nucleus. Some of these systems have already been exploited by researchers in the field of gene therapy. In Table 7 we summarize successful interventions and propose new interventions based on the biology of macromolecular transport through the cytoplasm and into the nucleus. Previous successful pharmacologic interventions using chloroquine increased gene expression probably by decreasing fusion of endosomes with lysosomes and hence increased the probability that the gene would escape prior to delivery into the lysosomal compartment (Table 7). Other pharmacological interventions, such as the use of colchicine, appear to succeed in increasing the duration of gene expression in vivo (Table 7). The inclusion of proteins that are capable of binding to DNA and that are also transported into the nucleus have provided a fivefold increase in gene expression (Table 7). These examples support our belief that more sophisticated manipulations of these pathways offer great promise for enhancing gene expression from nonviral DNA delivery systems.

Acknowledgments

KEBM was supported by NIH DK46052 and the Biotechnology Research and Education Program. LSU was supported by BIH HL42368 and GM 26691.

References

1. Capecchi MR. High efficiency transformation by direct microinjection of DNA into cultured mammalian cells. Cell 1980; 22:479–488.
2. Bertling W, Hunger-Bertling K, Cline MJ. Intranuclear uptake and persistence of biologically active DNA after electroporation of mammalian cells. J Biochem Biophys Methods 1987; 14:223–232.
3. Tsong T, Electroporation of cell membranes. Biophys J 1991; 60:297–306.

4. Felgner P, Tsai Y, Sukhu L, et al. Improved cationic lipid formulations for in vivo gene therapy. Ann NY Acad Sci 1995; 772:126–139.

5. Gao X, Huang L, Cationic liposome-mediated gene transfer. Gene Therapy 1995; 2:710–722.

6. Chu C, Dijkstra J, Lai M, Hong K, Szoka FC Jr. Efficiency of cytoplasmic delivery by pH-sensitive liposomes to cells in culture. Pharm Res 1990; 7:824–834.

7. Kabanov A, Astafieva I, Maksimova I, Lukanidin E, Georgiev G, Kabanov V. Efficient transformation of mammalian cells using DNA interpolyelectrolyte complexes with carbon chain polycations. Bioconj Chem 1993; 4:448–454.

8. Kabanov A, Kabanov V. DNA complexes with polycations for the delivery of genetic material into cells. Bioconj Chem 1995; 6:7–20.

9. Haensler J, Szoka FC Jr. Polyamidoamine cascade polymers mediate efficient transfection of cells in culture. Bioconj Chem 1993; 4:372–379.

10. Behr J. Gene transfer with synthetic cationic amphiphiles: prospects for gene therapy. Bioconj Chem 1994; 5:382–389.

11. Boussif O, Lezoualc'h F, Zanta M, et al. A versatile vector for gene and oligonucleotide transfer into cells in culture and in vivo: polyethylenimine. Proc Natl Acad Sci USA 1995; 92:7297–7301.

12. Gao X, Huang L. Potentiation of cationic liposome-mediated gene delivery by polycations. Biochemistry 1996; 35:1027–1036.

13. Yaroslavov A, Sukhishvili S, Obolsky O, Yaroslavova E, Kabanov A, Kabanov V. DNA affinity to biological membranes is enhanced due to complexation with hydrophobized polycation. FEBS Lett 1996; 384:177–180.

14. Wolosewick JJ, Porter KR. Microtrabecular lattice of the cytoplasmic ground substance. Artifact or reality. J Cell Biol 1979; 82:114–139.

15. Schliwa M, Blerkhom JV. Structural interaction of cytoskeletal components. J Cell Biol 1981; 90:222–235.

16. Heuser J, Kirscher MW. Filament organization revealed in platinum replicas of freeze dried cytoskeletons. J Cell Biol 1980; 86:212–234.

17. Penman S, Fulton A, Capco D, Ze'ev AB, Wittleberger S, Tse CF. Cytoplasmic and nuclear architecture in cells and tissue: form, function, mode of assembly. Cold Spring Harbor Symp Quant Biol 1981; 46:1013–1028.

18. Porter KR. The cytomatrix: a short history of its study. J Cell Biol 1984; 99:3s–12s.

19. Fushimi K, Verkman A. Low viscosity in the aqueous domain of cell cytoplasm measured by picosecond polarization microfluorimetry. J Cell Biol 1991; 112:719–725.

20. Luby-Phelps K, Mujumdar S, Mujumdar R, Ernst L, Galbraith W, Waggoner A. A novel fluorescence ratiometric method confirms the low solvent viscosity of the cytoplasm. Biophys J 1993; 65:236–242.

21. Bicknese S, Periasamy N, Shohet S, Verkman A. Cytoplasmic viscosity near the cell plasma membrane: measurement by evanescent field frequency-domain microfluorimetry. Biophys J 1993; 65:1272–1282.

22. Hou L, Lanni F, Luby-Phelps K. Tracer diffusion in F-actin and Ficoll mixtures. Toward a model for cytoplasm. Biophys J 1990; 58:31–43.

23. Kao C, Sharon J. Chimeric antibodies with anti-dextran-derived complementarity-determining regions and anti-p-azophenylarsonate-derived framework regions. J Immunol 1993; 151:1968–1978.

24. Furukawa R, Arauz-Lara J, Ware B. Self-diffusion and probe diffusion in dilute and semidilute aqueous solutions of dextran. Macromolecules 1991; 24:599–605.

25. Murumatsu N, Minton A. Trace diffusion of globular proteins in concentrated protein solutions. Proc Natl Acad Sci USA 1988; 85:2984–2988.

26. Han J, Herzfeld J. Macromolecular diffusion in crowded solutions. Biophys J 1993; 65:1155–1161.

27. Garcia-Bustos J, Heitman J, Hall MN. Nuclear protein localization. Biochim Biophys Acta 1991; 1071:83–101.

28. Lang I, Peters R. Information and Energy Transduction in Biological Membranes. Nuclear Permeability: A Sensitive Indicator of Pore Complex Integrity. New York: Liss, 1984:377–386.

29. Lang I, Scholz M, Peters R. Molecular mobility and nucleocytoplasmic flux in hepatoma cells. J Cell Biol 1986; 102:1183–1190.

30. Peters R. Fluorescence microphotolysis to measure nucleocytoplasmic transport and intracellular mobility. Biochim Biophys Acta 1986; 864:305–509.

31. Luby-Phelps K, Taylor D, Lanni F. Probing the structure of cytoplasm. J Cell Biol 1986; 102:2015–2022.

32. Luby-Phelps K, Castle PE, Taylor DL, Lanni F. Hindered diffusion of inert tracer particles in the cytoplasm of mouse 3T3 cells. Proc Natl Acad Sci USA 1987; 84: 4910–4913.

33. Paine P. Diffusion between nucleus and cytoplasm. In: Nuclear Trafficking. New York: Academic Press, 1992:3–14.

34. Zelphati O, Szoka FC Jr. Intracellular distribution and mechanism of delivery of oligonucleotides mediated by cationic lipids. Pharm Res 1996; 13:1367–1372.

35. Loke SL, Stein CA, Zhang XH, et al. Characterization of oligonucleotide transport into living cells. Proc Natl Acad Sci USA 1989; 86:3474–3478.

36. Chin DJ, Green GA, Zon G, Szoka FC Jr, Straubinger RM. Rapid nuclear accumulation of injected oligodeoxyribonucleotides. New Biol 1990; 2:1091–1100.

37. Leonetti JP, Mechti N, Degols G, Gagnor C, Lebleu B. Intracellular distribution of microinjected antisense oligonucleotides. Proc Natl Acad Sci USA 1991; 88: 2702–2706.

38. Sixou S, Szoka FC Jr, Green G, Giusti B, Zon G, Chin D. Intracellular oligonucleotide hybridization detected by fluorescence resonance energy transfer (FRET). Nucleic Acids Res 1994; 22:662–668.

39. Leonetti JP, Degols G, Clarenc J-P, Mechti N, Lebleu B. Cell delivery and mechanisms of action of antisense oligonucleotides. Prog Nucleic Acid Res 1993; 44: 143–166.

40. Clarenc J-P, Lebleu B, Léonetti J-P. Characterization of the nuclear binding sites of oligodeoxyribonucleotides and their analogs. J Biol Chem 1993; 268: 5600–5604.

41. Geselowitz D, Neckers L. Analysis of oligonucleotide binding, internalization, intracellular trafficking utilizing a novel radiolabeled crosslinker. Antisense Res Dev 1992; 2:17–25.

42. Yoon K, Colestrauss A, Kmiec E. Targeted gene correction of episomal DNA in mammalian cells mediated by a chimeric RNA-center-dot-DNA oligonucleotide. Proc Natl Acad Sci USA 1996; 93:2071–2076.
43. Kunzelmann K, Legendre J-Y, Knoell D, Escobar L, Xu Z, Gruenert D. Gene targeting of CFTR DNA in CF epithelial cells. Gene Ther 1996; 3:859–867.
44. Brinster R, Chen H, Trumbauer M, Yagle M, Palmiter R. Factors affecting the efficiency of introducing foreign DNA into mice by microinjecting eggs. Proc Natl Acad Sci USA 1985; 82:4438–4442.
45. Mirzayans R, Aubin R, Paterson M. Differential expression and stability of foreign genes introduced into human fibroblasts by nuclear versus cytoplasmic microinjection. Mutat Res 1992; 281:115–122.
46. Thorburn A, Alberts A. Efficient expression of miniprep plasmid DNA after needle micro-injection into somatic cells. Biotechniques 1993; 14:358–359.
47. Dowty M, Williams P, Zhang G, Hagstrom J, Wolff J. Plasmid DNA entry into postmitotic nuclei of primary rat myotubes. Proc Natl Acad Sci USA 1995; 92: 4572–4576.
48. Wolff J, Malone R, Williams P, et al. Direct gene transfer into mouse muscle in vivo. Science 1990; 247:1465–1468.
49. Tsuchiya E, Shakuto S, Miyakawa T, Fukui S. Characterization of a DNA uptake reaction through the nuclear membrane of isolated yeast nuclei. J Bacteriol 1988; 170:547–551.
50. Loyter A, Scangos GA, Ruddle FH. Mechanisms of DNA uptake by mammalian cells: fate of exogenously added DNA monitored by the use of fluorescent dyes. Proc Natl Acad Sci USA 1982; 79:422–426.
51. Zabner J, Fasbender AJ, Moninger T, Poellinger KA, Welsh MJ. Cellular and molecular barriers to gene transfer by a cationic lipid. J Biol Chem 1995; 270: 18997–19007.
52. Felgner PL, Gadek TR, Holm M, et al. Lipofection: a highly efficient, lipid-mediated DNA-transfection procedure. Proc Natl Acad Sci USA 1987; 84:7413–7417.
53. Farhood H, Bottega R, Epand RM, Huang L. Effect of cationic cholesterol derivatives on gene transfer and protein kinase C activity. Biochim Biophys Acta 1992; 1111:239–246.
54. Thorsness P, Fox T. Escape of DNA from mitochondria to the nucleus in *Saccharomyces cerevisiae*. Nature 1990; 346:376–379.
55. Chowdhury N, Wu C, Wu G, Yerneni P, Bommineni V, Chowdhury J. Fate of DNA targeted to the liver by asialoglycoprotein receptor-mediated endocytosis in vivo. Prolonged persistence in cytoplasmic vesicles after partial hepatectomy. J Biol Chem 1993; 268:11265–11271.
56. Findeis M, Wu C, Wu G. Ligand-based carrier systems for delivery of DNA to hepatocytes. Methods Enzymol 1994; 247:341–351.
57. Chiou H, Tangco MV, Levine S, et al. Enhanced resistance to nuclease degradation of nucleic acids complexed to asialoglycoprotein-polylysine carriers. Nucleic Acids Res 1994; 22:5439–5446.
58. Kaneda Y, Iwai K, Uchida T. Increased expression of DNA cointroduced with nuclear protein in adult rat liver. Science 1989; 243:375–378.

59. Elroy-Stein O, Fuerst T, Moss B. Cap-independent translation of mRNA conferred by encephalomyocarditis virus 5' sequence improves the performance of the vaccinia virus/bacteriophage T7 hybrid expression system. Proc Natl Acad Sci USA 1989; 86:6126–6130.

60. Gao X, Huang L. Cytoplasmic expression of a reporter gene by co-delivery of T7 RNA polymerase and T7 promoter sequence with cationic liposomes. Nucleic Acids Res 1993; 21:2867–2872.

61. Rose J, Buonocore L, Whitt M. A new cationic liposome reagent mediating nearly quantitative transfection of animal cells. Biotechniques 1991; 10:520–525.

62. Cole NB, Lippincott-Schwartz J. Organization of organelles and membrane traffic by microtubules. Curr Opin Cell Biol 1995; 7:55–64.

63. Langford G.M. Actin- and microtuble-dependent organelle motors: interrelationships between the two motility systems. Curr Opin Cell Biol 1995; 7:82–88.

64. Atkinson SJ, Doberstein SK, Pollard TD. Moving off the beaten tracks. Curr Biol 1992; 2:326–328.

65. Vallee RB, Shpetner H. Motor proteins of cytoplasmic microtubules. Annu Rev Biochem 1990; 59:909–932.

66. Walker RA, Sheetz MP. Cytoplasmic microtubule-associated motors. Annu Rev Biochem 1993; 62:429–451.

67. Cole DG, Scholey JM. Structural variations among the kinesins. Trends Cell Biol 1995; 5:259–262.

68. Brady S, Sperry A. Biochemical and functional diversity of microtubule motors in the nervous system. Curr Opin Neurobiol 1995; 5:551–558.

69. Jellali A, Metz-Boutique MH, Surgucheva I, et al. Structural and biochemical properties of kinesin heavy chain associated with rat brain mitochondria. Cell Motil Cytoskeleton 1994; 28:79–93.

70. Feiguin F, Ferreira A, Kosik KS, Caceres A. Kinesin-mediated organelle translocation revealed by specific cellular manipulations. J Cell Biol 1994; 127:1021–1039.

71. Hollenbeck PJ, Swanson JA. Radial extension of macrophage tubular lysosomes supported by kinesin. Nature 1990; 127:1021–1039.

72. Lippincott-Schwartz J, Coel NB, Marotta A, Conrad PA, Bloom GS. Kinesin is the motor for microtubule-mediated Golgi-to-ER membrane traffic. J Cell Biol 1995; 128:293–306

73. Burkhardt JK. In search of membrane receptors for microtubule-based motors—is kinectin a kinesin receptor? Cell Biol 1996; 6:127–131.

74. Yu H, Nicchitta CV, Kumar J, Becker M, Toyoshima I, Sheetz MP. Characterization of kinectin, a kinesin-binding protein: primary sequence and N-terminal topogenic signal analysis. Mol Biol Cell 1995; 6:171–183.

75. Kumar J, Yu H, Sheetz MP. Kinectin, an essential anchor for kinesin-driven vesicle motility. Science 1995; 267:1834–1837.

76. Rickard JE, Kreis TE. Binding of pp170 to microtubules is regulated by phosphorylation. J Biol Chem 1991; 266:17597–17605.

77. Cheney R.E, Mooseker M.S, Unconventional myosins. Curr Opin Cell Biol 1992; 4:27–35.

78. Finer J, Simmons R, Spudich J. Single myosin molecule mechanics: piconewton forces and nanometre steps. Nature 1994; 368:113–119.
79. Chowdhury N, Hays R, Bommineni V, et al. Microtubular disruption prolongs the expression of human bilirubin-uridinediphosphoglucuronate-glucuronosyltransferase-1 gene transferred into Gunn rat livers. J Biol Chem 1996; 271:2341–2346.
80. Wu G, Chowdhury J, Bommineni V, Basu S, Wu C, Chowdhury N. Fate of DNA targeted to hepatocytes by asialoglycoprotein polylysine conjugates. Proc Assoc Amer Physicians 1995; 107:211–217.
81. Plourde R, Phillips A, Wu C, et al. A hepatocyte-targeted conjugate capable of delivering biologically active colchicine in vitro. Bioconj Chem 1996; 7:131–137.
82. Goldberg MW, Allen TD. Structural and functional organization of the nuclear envelope. Curr Opin Cell Biol 1995; 7:301–309.
83. Miller M, Park MK, Hanover JA. Nuclear pore complex: structure, function and regulation. Physiol Rev 1991; 71:909–949.
84. Taniura H, Glass C, Gerace L. A chromatin binding site in the tail domain of nuclear lamins that interacts with core histones. J Cell Biol 1995; 131:33–44.
85. Glass J, Gerace L. Lamins A and C bind and assemble at the surface of mitotic chromosomes. J Cell Biol 1990; 111:1047–1057.
86. Paddy M, Belmont A, Saumweber H, Agard D, Sedat J. Interphase nuclear envelope lamins form a discontinuous network that interacts with only a fraction of the chromatin in the nuclear periphery. Cell 1990; 62:89–106.
87. Gennis R. Introduction: the structure and composition of biomembranes. In: Biomembranes—Molecular Structure and Function. New York: Springer-Verlag, 1989:1–35.
88. Davis LI. The nuclear pore complex. Annu Rev Biochem 1995; 64:865–896.
89. Hinshaw JE. Architecture of the nuclear pore complex and its involvement in nucleocytoplasmic transport. Biochem Pharmacol 1994; 47:15–20.
90. Görlich D, Mattaj IW. Nucleocytoplasmic transport. Science 1996; 271:1513–1518.
91. Panté N, Aebi U. Exploring nuclear pore complex structure and function in molecular detail. J Cell Sci 1995; suppl 19:1–11.
92. Csermely P, Schnaider T, Szanto I. Signalling and transport through the nuclear membrane. Biochim Biophys Acta 1995; 1241:425–451.
93. Melchior F, Gerace L. Mechanisms of nuclear protein import. Curr Opin Cell Biol 1995; 7:310–318.
94. Hunt EC. The nuclear pore complex. FEBS Lett 1993; 325:76–80.
95. Newmeyer DD. The nuclear pore complex and nucleocytoplasmic transport. Curr Opin Cell Biol 1993; 5:395–407.
96. Hurt EC. Importins/karyopherins meet nucleoporins. Cell 1996; 84:509–515.
97. Michaud N, Goldfarb D. Microinjected U snRNAs are imported to oocyte nuclei via the nuclear pore complex by three distinguishable targeting pathways. J Cell Biol 1992; 116:851–861.
98. Feldherr CM, Akin D. The permeability of the nuclear envelope in dividing and nondividing cell cultures. J Cell Biol 1990; 111:1–8.
99. Feldherr CM, Akin D. Signal-mediated nuclear transport in proliferating and growth-arrested BALB/c 3T3 cells. J Cell Biol 1991; 115:933–939.

100. Feldherr CM, Akin D. Regulation of nuclear transport in proliferating and quiescent cells. Exp Cell Res 1993; 205:179–186.

101. Greber UF, Gerace L. Depletion of calcium from the lumen of endoplasmic reticulum reversibly inhibits passive diffusion and signal-mediated transport into the nucleus. J Cell Biol 1995; 128:5–14.

102. Stehno-Bittel L, Perez-Terzic C, Clapham DE. Diffusion across the nuclear envelope inhibited by depletion of the nuclear Ca^{2+} store. Science 1995; 270:1835–1838.

103. Macaulay C, Meier E, Forbes D. Differential mitotic phosphorylation of proteins of the nuclear pore complex. J Biol Chem 1995; 270:254–262.

104. Reston J, Gould-Fogerite S, Mannino R. Potentiation of DNA mediated gene transfer in NIH3T3 cells by activators of protein kinase C. Biochim Biophys Acta 1991; 1088:270–276.

105. Kruh J. Effects of sodium butyrate, a new pharmacological agent, on cells in culture. Mol Cell Biochem 1982; 42:65–82.

106. Wilken N, Kossner U, Senécal J-L, Scheer U, Dabauvalle M-C. Nup180, a novel nuclear pore complex protein localizing to the cytoplasmic ring and associated fibrils. J Cell Biol 1993; 123:1345–1354.

107. Finlay DR, Meier E, Bradley P, Horecka J, Forbes DJ. A complex of nuclear pore proteins required for pore function. J Cell Biol 1991; 114:169–183.

108. Dargemont C, Schmidt-Zachmann MS, Kühn LC. Direct interaction of nucleoporin p62 with mRNA during its export from the nucleus. J Cell Sci 1995; 108:257–263.

109. Sukegawa J, Blobel G. A nuclear pore complex protein that contains zinc finger motifs, binds DNA, faces the nucleoplasm. Cell 1993; 72:29–38.

110. Zapp ML. The ins and outs of RNA nucleocytoplasmic transport. Curr Opin Genet Dev 1995; 5:229–233.

111. Fabre E, Boelens WC, Wimmer C, Mattaj IW, Hurt EC. Nup145p is required for nuclear export of mRNA and binds homopolymeric RNA in vitro via a novel conserved motif. Cell 1994; 78:275–289.

112. Wente S, Blobel G. NUP145 encodes a novel yeast glycine-leucine-phenylalanine-glycine (GLFG) nucleoporin required for nuclear envelope structure. J Cell Biol 1994; 125:955–969.

113. Iovine MK, Watkins JL, Wente SR. The GLFG repetitive region of the nucleoporin Nup116p interacts with Kap95p, essential yeast nuclear import factor. J Cell Biol 1995; 131:1699–1713.

114. Wozniak RW, Blobel G. The single transmembrane segment of gp210 is sufficient for sorting to the pore membrane domain of the nuclear envelope. J Cell Biol 1992; 119:1441–1449.

115. Adam S, Gerace L. Cytosolic proteins that specifically bind nuclear location signals are receptors for nuclear import. Cell 1989; 66:837–847.

116. Breeuwer M, Goldfarb D. Facilitated nuclear transport of histone H1 and other small nucleophilic proteins. Cell 1990; 60:999–1008.

117. Dingwall C, Laskey R. The nuclear membrane. Science 1992; 258:942–947.

118. Kalderon D, Roberts BL, Richardson WD, Smith AE. A short amino acid sequence able to specify nuclear location. Cell 1984; 39:499–509.

119. Lanford R, Butel J. Construction and characterization of an SV40 mutant defective in nuclear transport of T antigen. Cell 1984; 37:801–813.
120. Chelsky D, Ralph R, Jonak G. Sequence requirements for synthetic peptide-mediated translocation to the nucleus. Mol Cell Biol 1989; 9:2487–2492.
121. Guiochon-Mantel A, Delabre K, Lescop P, Milgrom E. Nuclear localization signals also mediate the outward movement of proteins from the nucleus. Proc Natl Acad Sci USA 1994; 91:7179–7183.
122. Robbins J, Dilworth SM, Laskey RA, Dingwall C. Two interdependent basic domains in nucleoplasmin nuclear targeting sequence: Identification of a class of bipartite nuclear targeting sequence. Cell 1991; 64:615–623.
123. Feldherr C, Kallenbach E, Schultz N. Movement of a karyophilic protein through the nuclear pores of oocytes. J Cell Biol 1984; 99:2216–2222.
124. Dingwall C, Laskey R. Nuclear targeting sequences-a consensus? Trends in Biochemical Sciences 1991; 16:478–481.
125. Lanford RE, Feldherr CM, White RG, Dunham RG, Kanda P. Comparison of diverse transport signals in synthetic peptide-induced nuclear transport. Exp Cell Res 1990; 186:32–38.
126. Yoneda Y, Semba T, Kaneda Y, et al. A long synthetic peptide containing a nuclear localization signal and its flanking sequences of SV40 T-antigen directs the transport of IgM into the nucleus efficiently. Exp Cell Res 1992; 201:313–320.
127. Görlich D, Prehn S, Laskey RA, Hartmann E. Isolation of a protein that is essential for the first step of nuclear protein import. Cell 1994; 79:767–768.
128. Görlich D, Kostka S, Kraft R, et al. Two different subunits of importin cooperate to recognize nuclear localization signals and bind them to the nuclear envelope. Curr Biol 1995; 5:383–392.
129. Agutter P. Intracellular structure and nucleocytoplasmic transport. Int Rev Cytol 1995; 162B:183–224.
130. Marshallsay C, Lührmann R. In vitro nuclear import of snRNPs: cytosolic factors mediate m_3G-cap dependence of U1 and U2 snRNP transport. EMBO J 1994; 13:222–231.
131. Citovsky V, Zambryski P. Transport of nucleic acids through membrane channels: Snaking through small holes. Ann Rev Microbiol 1993; 47:167–197.
132. Baserga S, Gilmore-Hebert M, Yang X. Distinct molecular signals for nuclear import of the nucleolar snRNA, U3. Genes Dev 1992; 6:1120–1130.
133. Goldfarb D. Viruses: the Trojan horses of the cell. Trends Cell Biol 1996; 6: 8–15.
134. Gallay P, Swingler S, Song J, Bushman F, Trono D. HIV nuclear import is governed by the phosphotyrosine-mediated binding of matrix to the core domain of integrase. Cell 1995; 83:569–576.
135. Stevenson M, Portals of entry: uncovering HIV nuclear transport pathways. Trends Cell Biol 1996; 6:9–15.
136. Bukrinsky M, Haggerty S, Dempsey MP, et al. A nuclear targeting signal within HIV–1 matrix protein that governs infection of non-dividing cells. Nature 1993; 365:666–669.

137. Heinzinger NK, Bukrinsky M, Haggerty S, et al. The Vpr protein of human immunodeficiency virus type 1 influences nuclear localizaiton of viral nucleic acids in nondividing cells. Proc Natl Acad Sci USA 1994; 91:7311–7315.
138. Fawell S, Seery J, Daikh Y, Moore C, Chen L.L, Pepinsky B, Barsoum J, Tat-mediated delivery of heterologous proteins into cells. Proc Natl Acad Sci USA 1994; 91:664–668.
139. Gulizia J, Dempsey MP, Sharova N, et al. Reduced nuclear import of human immunodeficiency virus type 1 preintegration complexes in the presence of a prototypic nuclear targeting signal. J Virol 1994; 68:2021–2025.
140. Compans RW, Content J, Duesberg PH. Structure of the ribonucleoprotein of influenza virus. J Virol 1972; 10:795–800.
141. Heggeness MH, Smith PR, Ulmanen I, Krug RM, Choppin PW. Studies on the helical nucleocapsid of influenza virus. Virology 1982; 118:466–470.
142. Jennings PA, Finch JT, Winter G, Robertson JS. Influenza virus ribonucleoprotein guide sequence rearrangements in influenza viral RNA? Cell 1983; 34:619–627.
143. Jones I, Reay P, Philpott K. Nuclear location of all three influenza polymerase proteins and a nuclear signal in polymerase PB2. EMBO J 1986; 5:2371–2376.
144. Lin BC, Lai CJ. The influenza virus nucleoprotein synthesized from cloned DNA in a simian virus 40 vector is detected in the nucleus. J Virol 1983; 45:434–438.
145. Smith GL, Levin JZ, Palese P, Moss B. Synthesis and cellular location of the ten influenza polypeptides individually expressed by recombinant vaccinia viruses. Virology 1987; 160:336–345.
146. Roberts BL, Richarson WD, Smith AE. The effect of protein context on nuclear location signal function. Cell 1987; 50:465–501.
147. Dworetzky SI, Lanford RE, Feldherr CM. The effects of variations in the number and sequence of targeting signals on nuclear uptake. J Cell Biol 1988; 107:1279–1287.
148. Martin K, Helenius A. Transport of incoming influcenza virus nucleocapsids into the nucleus. J Virol 1991; 65:232–244.
149. Clever J, Yamada M, Kasmatsu H. Import of simian virus 40 virions through nuclear pore complexes. Proc Natl Acad Sci USA 1991; 88:7333–7337.
150. Nishimura T, Kawai N, Kawai M, Notake K, Ichihara I. Fusion of SV40-induced endocytotic vacuoles with the nuclear membrane. Cell Struct Funct 1986; 11:135–141.
151. Nishimura T, Kawai N, Ichihara I. Interaction of endocytotic vacuoles with the inner nuclear membrane in simian virus 40 entry into CV–1 cell nucleus. Cell Struct Funct 1991; 16:441–445.
152. Zhao L, Padmanabhan R. Nuclear transport of adenovirus DNA polymerase is facilitated by interaction with preterminal protein. Cell 1988; 55:1005–1015.
153. Schaack J, Ho W, Freimuth P, Shenk T. Adenovirus terminal protein mediates both nuclear matrix association and efficient transcription of adenovirus DNA. Genes Dev 1990; 4:1197–1208.
154. Fredman J, Engler J. Adenovirus precursor to terminal protein interacts with the nuclear matrix in vivo and in vitro. J Virol 1993; 67:3384–3395.
155. Warren G. Membrane partitioning during cell division. Annu Rev of Biochem 1993; 62:323–348.

156. Wilson K, Newport J. A trypsin-sensitive receptor on membrane vesicles is required for nuclear envelope formation in vitro. J Cell Biol 1988; 107:57–68.
157. Newmeyer D, Wilson K, Egg extracts for nuclear import and nuclear assembly reactions. Methods Cell Biol 1991; 36:607–634.
158. Robbins E, Gonatas NK. The ultrastructure of a mammalian cell during the mitotic cycle. J Cell Biol 1964; 21:429–463.
159. Zatsepina OV, Polyakov VY, Chentsov YS. Some structural aspects of the fate of the nuclear envelope during mitosis. Cytobiologie 1977; 16:130–144.
160. Zeligs GW, Wollman SH. Mitosis in rat thyroid epithelial cells in vivo. I. Ultrastructural changes in cytoplasmic organelles during the mitotic cycle. J Ultrastruct Res 1979. 66:53–77.
161. Gerace L, Blum A, Blobel G. Immunocytochemical localization of the major polypeptides to the nuclear pore complex-lamina fraction. Interphase and mitotic distribution. J Cell Biol 1978; 79:546–566.
162. Gerace L, Blobel G. The nuclear envelope lamina is reversibly depolymerized during mitosis. Cell 1980; 19:277–287.
163. Gerace L, Blobel G. Nuclear lamina and the structural organization of the nuclear envelope. Cold Spring Harbor Symp Quant Biol 1982; 46(pt 2):967–978.
164. Davis LI, Blobel G. Identification and characterization of a nuclear pore complex protein. Cell 1986; 45:699–709.
165. Snow CM, Senior A, Gerace L. Monoclonal antibodies identify a group of nuclear pore complex glycoproteins. J Cell Biol 1987; 104:1143–1156.
166. Newport J, Dunphy W. Characterization of the membrane binding and fusion events during nuclear envelope assembly using purified components. J Cell Biol 1992; 116:295–306.
167. Chaudhary N, Courvalin J-C. Stepwise reassembly of the nuclear envelope at the end of mitosis. J Cell Biol 1993; 122:295–306.
168. Foisner R, Gerace L. Integral membrane proteins of the nuclear envelope interact with lamins and chromosomes, binding is modulated by mitotic phosphorylation. Cell 1993; 73:1267–1279.
169. Sullivan K, Wilson K. A new role for IP$_3$ receptors: Ca^{2+} release during nuclear vesicle fusion. Cell Calcium 1994; 16:314–321.
170. Boman Al, Delannoy MR, Wilson KL. GTP hydrolysis is required for vesicle fusion during nuclear envelope assembly in vitro. J Cell Biol 1992; 116:281–294.
171. Forbes D, Kirschner M, Newport J. Spontaneous formation of nucleus-like structures around bacteriophage DNA microinjected into *Xenopus* eggs. Cell 1983; 34:13–23.
172. Nicolau C, Cudd A. Liposomes as carriers of DNA. Crit Rev Ther Drug Carrier Syst 1989; 6:239–271.
173. Imamoto N, Tachibana T, Matsubae M, Yoneda Y. A karyophilic protein forms a stable complex with cytoplasmic components prior to nuclear pore binding. J Biol Chem 1995; 270:8559–8565.
174. Imamoto N, Shimamoto T, Takao T, et al. In vivo evidence for involvement of a 58 kDa component of nuclear pore-targeting complex in nuclear protein import. EMBO J 1995; 14:3617–3626.

175. Adam S. The importance of importin. Trends Cell Biol 1995; 5:189–191.
176. Belanger K, Ma MK, Wei S, Davis L. Genetic and physical interactions between Srp1p and nuclear pore complex proteins Nup1p and Nup2p. J Cell Biol 1994; 126:619–630.
177. Enenkel C, Blobel G, Rexach M. Identification of a yeast karyopherin heterodimer that targets import substrate to mammalian nuclear pore complexes. J Biol Chem 1995; 270:16499–16502.
178. Moroianu J, Blobel G, Radu A. Previously identified protein of uncertain function is karyopherin alpha and together with karyopherin beta docks import substrate at nuclear pore complexes. Proc Natl Acad Sci USA 1995; 92:2008–2011.
179. Moroianu J, Hijikata M, Blobel G, Radu A. Mammalian karyopherin alpha 1 beta and alpha 2 beta heterodimers: alpha 1 or alpha 2 subunit binds nuclear localization signal and beta subunit interacts with peptide repeat-containing nucleoporins. Proc Natl Acad Sci USA 1995; 92:6532–6536.
180. Weis K, Mattaj I, Lamond A. Identification of hSRP1 alpha as a functional receptor for nuclear localization sequences. Science 1995; 268:1049–1053.
181. Yano R, Oakes M, Yamaghishi M, Dodd J, Nomura M. Cloning and characterization of SRP1, a suppressor of temperature-sensitive RNA polymerase I mutations, in *Saccharomyces cerevisiae*. Mol Cell Biol 1992; 12:5640–5651.
182. Moore MS, Blobel G. A G protein involved in nucleocytoplasmic transport: the role of Ran. TIBS 1994; 19:211–215.
183. Dingwall C, Robbins J, Dilworth SM, Roberts B, Richardson WD. The nucleoplasmin nuclear location sequence is larger and more complex than that of SV40 large T antigen. J Cell Biol 1988; 107:841–850.
184. Dang CV, Lee WMF. Completion of DNA replication is monitored by a feedback system that controls the initiation of mitosis in vitro: studies in *Xenopus*. Cell 1990; 61:811–823.
185. Takai T, Ohmori H. DNA transfection of mouse lymphoid cells by the combination of DEAE-dextran-mediated DNA uptake and osmotic shock procedure. Biochim Biophys Acta 1990; 1048:105–109.
186. Zauner W, Lichler A, Schmidt W, Sinski A, Wagner E. Glycerol enhancement of ligand-polylysine/DNA transfection. Biotechniques 1996; 20:905–913.
187. Luthman H, Magnusson G. High efficiency polyoma DNA transfection of chloroquine treated cells. Nucleic Acids Res 1983; 11:1295–1308.
188. Fraley R, Straubinger R, Rule G, Springe E, Papahadjopoulos D. Liposome-mediated delivery of deoxyribonucleic acid to cells: enhanced efficiency of delivery related to lipid composition and incubation conditions. Biochemistry 1981; 20: 6978–6987.
189. Goodsell D. Inside a living cell. TIBS 1991; 16:203–206.
190. Adam SA, Gerace L. Cytosolic proteins that specifically bind nuclear location signals are receptors for nuclear import. Cell 1991; 66:837–847.
191. Adam EJ, Adam SA. Identification of cytosolic factors required for nuclear location sequence-mediated binding to the nuclear envelope. J Cell Biol 1994; 125: 547–555.

192. Chi NC, Adam EJ, Adam SA. Sequence and characterization of cytoplasmic nuclear protein import factor p97. J Cell Biol 1995; 130:265–274.
193. Dang CV, Lee WM. Identification of the human c-myc protein nuclear translocation signal. Mol Cell Biol 1988; 8:4048–4054.
194. Melchior F, Weber K, Gerke V. A functional homologue of the RNA1 gene product in *Schizosaccharomyces pombe*: purification, biochemical characterization, identification of a leucine-rich repeat motif. Mol Biol Cell 1993; 4:569–581.
195. Melchior F, Sweet DJ, Gerace L. Analysis of Ran/TC4 function in nuclear protein import. Methods Enzymol 1995; 257:279–291.
196. Moore MS, Blobel G. The GTP-binding protein Ran/TC4 is required for protein import into the nucleus. Nature 1993; 365:661–663.
197. Paschal BM, Gerace L. Identification of NTF2, a cytosolic factor for nuclear import that interacts with nuclear pore complex protein p62. J Cell Biol 1995; 129:925–937.
198. Russell WC, Kemp GD. Role of adenovirus structural components in the regulation of adenovirus infection. Curr Top Microbiol Immunol 1995; 199:81–98.
199. Yang J, DeFranco DB. Differential roles of heat shock protein 70 in the in vitro nuclear import of glucocorticoid receptor and simian virus 40 large tumor antigen. Mol Cell Biol 1994; 14:5088–5098.
200. Wojcieszyn JW, Schlegel RA, Wu E-S, Jacobson KA. Diffusion of injected macromolecules within the cytoplasm of living cells. Proc Natl Acad Sci USA 1981; 78:4407–4410.
201. Tanner JE. Intracellular diffusion of water. Arch Biochem Biophys 1983; 224:416–428.
202. Horowitz SB. The permeability of the amphibian oocyte nucleus, in situ. J Cell Biol 1972; 54b:609–625.
203. Horowitz SB, Moore LC. The nuclear permeability, intracellular distribution, diffusion of inulin in the amphibian oocyte. J Cell Biol 1974. 60:405–415.
204. Paine PL, Moore LC, Horowitz SB. Nuclear envelope permeability. Nature (Lond) 1975; 254:109–114.
205. Luby-Phelps K, Lanni F, Taylor DL. Behavior of a fluorescent analogue of calmodulin in living 3T3 cells. J Cell Biol 1985; 101:1245–1256.
206. Hamm J, Darzynkiewicz E, Tahara SM, Mattaj IW. The trimethylguanosine cap structure of U1 snRNA is a component of a bipartite nuclear targeting signal. Cell 1990; 62:569–577.
207. Izaurralde E, Stepinski J, Darzynkiewicz E, Mattaj IW. A cap binding protein that may mediate nuclear export of RNA polymerase II-transcribed RNAs. J Cell Biol 1992; 118:1287–1295.
208. Terns MP, Dahlberg JE, Lund E. Multiple cis-acting signals for export of pre-U1 snRNA from nucleus. Genes Dev 1993; 7:1898–1908.
209. Malim M, Tiley LS, McCarn DF, Rusche JR, Hauber J, Cullen BR. HIV–1 structural gene expression requires binding of the rev trans-activator to its RNA target. Cell 1990; 60:675–683.

210. Smith J, German J, Hickman M. Incorporation of tributyrin enhances the expression of a reporter gene in primary and immortalized cell lines. Biotechniques 1995; 18:852–5.

211. Lyons RH, Ferguson BQ, Rosenberg M. Penta-peptide nuclear localization signal in adenovirus E1A. Mol Cell Biol 1987; 7:2451–2456.

212. Cochrane AW, Perkins A, Rosen CA. Identification of sequences important in the nucleolar localization of human immunodeficiency virus rev: relevance of nucleolar localization to function. J Virol 1990; 64:881–885.

213. Legendre JY, Szoka FC Jr. Delivery of plasmid DNA into mammalian cell lines using pH-sensitive liposomes: comparison with cationic liposomes. Pharm Res 1992; 9:1235–1242.

214. Hall MN, Hereford L, Herskowitz I. Targeting of *E. coli* ß-galactosidase to the nucleus in yeast. Cell 1984; 36:1057–1065.

215. Moreland RB, Langevin GL, Singer RH, Garcea RL, Hereford LM. Amino acid sequences that determine the nuclear localization of yeast histone 2B. Mol Cell Biol 1987; 7:4048–4057.

216. Burglin TR, Robertis EMD. The nuclear migration signal of *Xenopus laevis* nucleoplasmin. EMBO J 1987; 6:2617–2625.

217. Kleinschmidt A, Seiter A. Identifcation of domains involved in nuclear uptake and histone binding of protein N1 *Xenopus laevis*. EMBO J 1988; 7:1605–1614.

218. Kraemer DM, Strambio-de-Castillia C, Blobel G, Rout MP. The essential yeast nucleoporin NUP159 is located on the cytoplasmic side of the nuclear pore complex and serves in karyopherin-mediated binding of transport substrate. J Biol Chem 1995; 270:19017–19021.

219. Picard D, Yamamoto KR. Two signals mediate hormone-dependent nuclear localizaiton of the glucocorticoid receptor. EMBO J 1987; 6:3333–3340.

220. Loewinger L, McKeon F. Mutations in the nuclear lamin proteins resulting in their aberrant assembly in the cytoplasm. EMBO J 1988; 7:2301–2309.

221. Sweet DJ, Gerace L. Taking from the cytoplasm and giving to the pore: soluble transport factors in nuclear protein transport. Trends Cell Biol 1995; 5:444–447.

222. Zelphati O, Szoka FC Jr. Mechanism of oligonucleotide release from cationic liposomes. Proc Natl Acad Sci USA 1996; 93:11493–11498.

223. Xu Y, Szoka FC Jr. Mechanism of DNA release from cationic liposome/DNA complexes used in cell transfection. Biochemistry 1996; 35:5616–5623.

9

Delivery of Genes Through the Lung Circulation

DAVID M. RODMAN

University of Colorado Health
  Sciences Center
Denver, Colorado

ELIZABETH G. NABEL

University of Michigan
Ann Arbor, Michigan

I. Why Deliver Genes Via the Vasculature?

The majority of effort to date in lung gene transfer has centered on the problem of delivery of genes via the airway. The reason for the airway focus is twofold: first, by delivering genes via the airway, transgene is restricted almost entirely to the lung; second, the principal focus of gene therapy in the lung has been on correcting epithelial disorders, such as the ion transport defect of cystic fibrosis. However, there are many reasons why delivering genes via the pulmonary circulation might be useful. The most obvious of these reasons is to affect vascular function. Diseases such as pulmonary hypertension, persistent pulmonary hypertension of the neonate, chronic thromboembolic disease, and vasculitis could be targets for pulmonary vascular gene therapy. In addition to use in a therapeutic paradigm, gene transfer to the pulmonary circulation could also be a valuable tool in understanding pulmonary vascular biology. In the field of oncology, the pulmonary circulation may be used as access to primary or metastatic lung tumors (1). While speculative at present, the pulmonary endothelium could be used as a "factory" to produce and secrete proteins into the circulation, which could then affect cell and organ function at a distant site.

Finally, and potentially most important, the pulmonary and/or bronchial circulation may provide an indirect route for delivery of genes to the airway.

There are three general strategies that can be used to deliver transgene to the pulmonary circulation (Fig. 1 on page 187): implantation of previously transduced endothelial cells; systemic injection of lipid/DNA complexes; and catheter or surgical isolation and perfusion of the lung circulation with either viral or nonviral vectors. Each technique has unique advantages and disadvantages, and the current literature utilizing any of these techniques is quite limited. Implantation of transduced cells has the advantage of ability to use retroviral vectors, which are too inefficient in nonreplicating cells to be useful for in vivo vascular gene transfer but which allow incorporation of transgene into the host cell genome and thus stable transfection (2–4). However, this technique is very inefficient and has largely been replaced by direct in vivo gene transfer strategies. Systemic lipid delivery has the advantage of ease of use, but the disadvantage of inefficiency and lack of complete lung selectivity (5,6). Directed approaches, using either surgical isolation or catheter delivery, offer the best approach for local delivery, but can be limited by the invasiveness of the technique and the relatively small area of pulmonary circulation subtended by the catheter (7,8).

II. Endothelial Cell Implantation

Ex vivo transduction of endothelial cells, with subsequent implantation in the microcirculation was studied by Messina and co-workers (9). They cultured canine jugular vein endothelium, transducing these cells in culture with a murine amphotrophic retroviral vector containing genes coding for ß-galactosidase and neomycin resistance. The stably transfected endothelial cell population was then selected with the neomycin analog G418. A second group of cells were ^{125}I-labeled. These cells were then injected into the femoral artery of a group of syngenetic rats, and organ localization and transgene expression studied up to 28 days later using scintillation counting and X-gal staining, respectively. ^{125}I activity 1 hour postinjection was greatest in the injected leg. However, despite the fact that the cells were injected into the femoral artery, ^{125}I activity in lung was still 20% of leg activity. By 24 hours, ^{125}I activity had declined by approximately 50% in both hindlimb and lungs, while it had doubled in liver. This general pattern was maintained for 28 days, with measurable activity in lungs throughout the period of observation. X-gal staining confirmed incorporation of transduced endothelium into skeletal muscle capillaries. However, no microscopic analysis of lung tissue was performed.

Bernstein and co-workers (10) transduced a spontaneously immortalized rat pulmonary endothelial cell line with a retroviral vector containing a human

growth hormone fusion gene and the neomycin resistance gene, selecting transfected cells with G418. These cells were then injected either intraperitoneally,
subcutaneously, or intravenously into nude mice. Serum growth hormone levels were measured from 1 to 15 days postinjection. All three methods of implantation resulted in measurable increases in growth hormone levels for the
first 2 days after injection, though levels returned to baseline by 15 days in all
three groups. Intravenous injection was most efficient, requiring 5% to 7% of
the numbers of cells needed in the other two groups to achieve similar levels
of maximal activity. These investigators did not determine into which organs
the transduced pulmonary endothelium implanted.

These two studies demonstrate the feasibility of producing stably transduced endothelial cells which can be implanted into the microcirculation. The
study by Messina and co-workers (9) is provocative in that cells injected intra-
arterially in the leg found their way into the lung, where they presumably were
incorporated into the lung capillary network. One would predict that direct
infusion of transduced endothelium into the pulmonary artery might result in
substantially greater lung activity. Obviously, for this technique to become
useful for pulmonary circulatory gene transfer, a great deal of additional work
would be required, such as characterizing the location and number of transfected endothelial cells in the lung circulation. However, given the relatively
slow turnover of endothelial cells in vivo, ex vivo retroviral transfection and
infusion could be a viable technique to produce long-term expression of transgene in pulmonary endothelial cells.

III. Systemic Injection of Lipid/DNA Complexes

Polycationic lipids have been shown to increase the transfection efficiency of
naked plasmid DNA by 100- to 1000-fold or more (11). Several studies of
systemic injection of lipid/DNA complexes have demonstrated lung expression. Brigham and co-workers (12) first reported the ability of systematically
administered lipid:DNA complexes to transduce the lung. In that study an
expression plasmid utilizing the chloramphenocol acetyltransferase (CAT) reporter gene under control of the simian virus 40 (SV40) early promoter was
complexed to DOTMA:DOPE (Lipofectin) liposomes in 1:5 ratio (w/w). Tail
vein injection of complexes containing either 15 or 30 μg DNA resulted in
dose-dependent expression of CAT in lung for up to 7 days, with no activity
detected in other organs. While this study suggested organ selectivity using
DOTMA:DOPE, it is likely that the use of a plasmid containing the relatively
weak SV40 promoter contributed to a lack of detectable CAT activity in other
organs.

Zhu and co-workers (5), using DOTMA:DOPE liposomes complexed
to the CAT reporter gene under control of the human cytomegalovirus (CMV)

immediate early promoter and enhancer, studied organ expression after tail vein injection in mice. They found a dose-dependent increase in lung CAT activity over the dose range of 50 to 150 μg DNA in 1:8 ratio with lipid. In contrast to the initial report by Brigham, expression was seen not only in lung but in a variety of other organs. However, lung expression exceeded that seen in other organs at all doses studied, with a three- to fourfold relative selectivity of lung over spleen and heart. Thus, use of the stronger CMV promoter resulted in generally higher expression in a variety of organs, including lung. Still, some organ selectivity was seen, as CAT activity was absent in liver. While maximal CAT activity was seen after 2 days, in animals transduced with the lac-Z reporter gene significant numbers of X-gal positive cells were detected in lung up to 9 weeks after injection, suggesting that relatively long-term expression is possible using systemic lipid:DNA complexes.

Canonico et al. (13) compared intravenous to intratracheal DOTMA: DOPE liposomes in adult rabbits using the α1-antitrypsin gene under control of the CMV promoter. Distribution of intravenous radiolabeled lipid was found to be diffusely distributed in liver, and selectively taken up in the endothelium of conduit pulmonary arteries near bifurcations. Five hundred micrograms of plasmid DNA in 1:5 ratio with lipid was used for transfection experiments. Northern analysis detected messenger RNA for α1-antitrypsin in both lung and liver for up to 7 days with either route of administration. mRNA levels were greater in lungs of animals given intratracheal complexes and in the livers of animals given an intravenous vector. In rabbits receiving intravenous complexes, protein expression was detected by immunohistochemical staining with a sheep polyclonal antibody to α1-antitrypsin. One day after injection α1-antitrypsin was seen in conduit artery endothelium, mainly at bifurcations. By day 7, greater staining intensity was seen, with expression not only in endothelium but also in a small percentage of subjacent vascular smooth muscle cells as well as airway epithelial cells. Low-level expression was also detected in alveolar macrophages. In contrast, in rabbits given intratracheal complexes, expression was seen in both airway and alveolar epithelial cells and alveolar macrophages, with no expression seen in the pulmonary vasculature. This study was unique in that a potentially functional eukaryotic protein, rather than a reporter gene, was used though no functional consequence of α1-antitrypsin overexpression was documented.

The search for improved liposome vectors for systemic gene delivery has been vigorous. Thierry and co-workers (6) tested dioctadecylamindoglyclsperidine (DLS) liposomes complexed to a plasmid containing the luciferase reporter gene under control of either the Rous sarcoma virus (RSV) or CMV immediate early promoter in mice. They found that luciferase activity 4 days after injection could be detected in lung, spleen, and liver. However, there was a dose-dependent effect on organ selectivity. In lung, expression increased throughout the dose range from 25 to 300 μg of plasmid DNA. In contrast, in spleen expression increased in the dose range from 25 to 100 μg, then decreased to undetectable levels at 300 μg DNA. Liver expression was maximal at 50 μg DNA, with no further increased detected at the higher dose range. This effect of higher DNA concentration, which the authors attributed to greater cationic charge of the complexes, resulted in relative lung selectivity at all doses greater than 100 μg of DNA. Lung luciferase activity was 13-fold greater than activity in the other organs studied at DNA doses of 200 μg or greater. The duration of expression was comparable in all organs, with small amounts of luciferase activity detected up to 3 months after injection. This study also compared a variety of plasmid constructs that utilized either the RSV or CMV promoter. They found that use of the RSV promoter resulted in more lung-selective expression than the CMV promoter. However, owing to the greater activity of the CMV promoter, expression in all tissues, including lung, was greater using CMV promoter constructs. This apparent lung selectivity of the RSV promoter in systemic lipid:DNA complexes is similar to results by Brigham et al. (12) using the SV40 promoter. Conclusions about lung selectivity of these constitutively active viral promoters should be judged cautiously, as it is likely that by using relatively weak promoters expression in other organs may still be present, but fall below detectable levels.

The factors that confer relative lung selectivity of lipid vectors remain uncertain. As noted above, it is possible that more cationic complexes have greater lung selectivity, though this remains unproven (8). The cationic moiety of most lipid vectors reported to date is either quaternary ammonium or a primary, secondary, or tertiary amine. Recently a novel series of lipids based on an imidazolinium moiety have been reported to have enhanced in vivo efficiency, and appear to have lung selectivity (14). The most active of this series is DOTIM (1-[2-(9(Z)-octadecenoyloxy)ethyl]-2-(9(Z)-heptadecenyl)-3-(2-hydroxyethyl)imidazolinium chloride). Enhanced gene transfer was seen when 1:1 complexes with cholesterol, rather than dioleoylphospharidylethanoliamine (DOPE), were studied in vitro, distinguishing this lipid from DOTMA and most other lipids studied to date. The optimum DNA:lipid ratio for DOTIM in vivo was found to be 0.1 μg/nmol. Using 60 μg of plasmid containing the CAT gene under control of the CMV immediate early promoter, DOTIM: Chol injected by tail vein in mice resulted in detectable CAT activity in lung, spleen, heart, liver, and kidney. Lung expression was 8-, 20-, and 35-fold greater than that seen in spleen, liver, and heart, respectively. Our own experience with this lipid suggests that pulmonary vascular expression is restricted to the endothelium, and while there is an initial 60% reduction of maximum lung CAT activity between days 2 and 7 after transfection, expression levels of approximately 20% of maximum are maintained for up to 60 days (15).

These studies suggest that systemically administered lipid:DNA complexes are expressed in the lung circulation. The predominant cell type expressing transgene is the endothelial cell, although, depending on the lipid and complex design, other cell types including vascular smooth muscle cells and alveolar epithelial cells may express transgene as well. While the overall efficiency of systemic lipid in delivering transgene is relatively low, this technique may prove useful for the study of genes of interest which are highly active, particularly when expressed in plasmids with strong regulatory elements, such as the CMV immediate early promoter and enhancer. However, systemic injection of lipid:DNA complexes has little likelihood of translation to human therapy, as the amount of lipid and DNA required could be prohibitive, and possibly toxic. Therefore, methods of directed delivery, possibly utilizing lung-avid lipids, will need to be developed.

IV. Directed Delivery to the Lung Circulation

Directed delivery offers several potential advantages, including the ability to restrict gene delivery to the lung and the opportunity to use viral vectors. At the present time, replication-defective adenovirus is the most efficient vector available (8). However, systemic injection of adenovirus results primarily in gene expression in liver, with little if any expression in lung (16). LeMarchand and co-workers (7) circumvented this problem by using a preparation in which the flow to one lung lobe could be isolated and perfused with adenovirus. They surgically isolated the pulmonary artery and veins to the right upper lobe of sheep, inserted a catheter into the cranial segment, and infused 1 to 3 × 10^{11} plaque-forming units (pfu) of adenovirus expressing the lac-Z reporter gene. After 15 min flow was restored, surgical wounds were closed, and animals were allowed to recover for 1 to 4 days. Expression was detected in pulmonary arterial, capillary, and venous endothelium. No expression was seen in vascular smooth muscle. In addition, expression was seen in alveolar, bronchial, and submucosal gland epithelium. Transgene expression was not detected in non-transfected left lung, though no attempt was made to determine if adenovirus transduced other organs, such as the liver.

A similar approach was used by Schachtner and co-workers (17) in rats. They attempted to deliver transgene with a percutaneous catheter placed with fluoroscopic guidance in a pulmonary capillary wedge position. Using a replication-defective adenovirus expressing lac-Z, they infused 6 × 10^8 to 3 × 10^{10} pfu over 15 min, but failed to detect transgene expression in lung. However, expression was seen in liver. Similar to LeMarchand's study in sheep, these investigators then surgically isolated and occluded the blood supply to the left lung, inserted a catheter into the pulmonary artery, and infused 6 × 10^9 to

2×10^{10} pfu of adenovirus for 20 min, after which flow was reestablished, wounds were closed, and animals were allowed to recover for 48 to 72 hours. Twenty percent of rats demonstrated X-gal-positive lung staining after the surgical approach. Three percent to 5% of lung cells were X-gal-positive, with 90% of positive cells being alveolar epithelial cells. Less than 4% of positive cells were endothelial cells in this study. The reasons for difference in cell types expressing transgene between sheep and rats are unclear, and may be related to differences in techniques or species. Both studies employed a vector utilizing a strong constitutive promoter.

Catheter-mediated gene transfer has many advantages over surgical techniques, including potential applicability to human gene transfer. For that reason, our group tested a catheter-based gene delivery system utilizing both

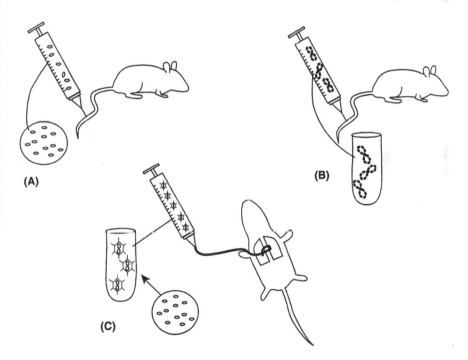

Figure 1 Three methods of pulmonary vascular gene transfer. (A) Endothelial cells in tissue culture are stably transfected using a retrovirus, selected for antibiotic resistance, then infused, after which they implant in the lung microcirculation. (B) Lipid/DNA complexes are injected systemically, resulting in pulmonary endothelial cell transduction. (C) Replication-defective adenovirus is injected by right heart catheter. Surgical isolation and perfusion of a lobar pulmonary artery are also possible.

lipid and adenoviral vectors (8). Using pigs we placed a 7 Fr end-hole balloon-tipped catheter (Medi-Tech) into the pulmonary capillary wedge position in the left lower lobe using fluoroscopic and hemodynamic guidance. The balloon was left inflated and dye was instilled to confirm arrest of blood flow. The pulmonary artery was then flushed with sterile saline, after which vector, either lipid:DNA complexes (5 μg DNA 1:2 ratio with DOSPA:DOPE, Lipofectamine) or replication-defective adenovirus (1.2×10^9 pfu), was instilled. Both expression cassettes utilized the human placental alkaline phosphatase (hpAP) reporter gene under control of the RSV promoter. After 20 min, flow was restored by deflating the balloon, a guide wire was inserted to identify the location of transgene insertion, and the catheter was removed. Animals were sacrificed either 5, 14, or 28 days later for assessment of transgene expression.

Figure 2 shows representative examples of histochemical staining of lung for hpAP activity. Gene expression was seen in pulmonary artery endothelium and in epithelial cells of the alveolar septum. Little expression was seen in vascular smooth muscle, and none in epithelial cells of conducting airways. Expression efficiency 5 days after transfection was compared between lipid- and adenovirus-transfected animals. Variable expression was seen in the lipid-transduced animals, with a mean of $1.9 \pm 1.5\%$ of cells in vascular structures and alveoli alkaline phosphatase positive. Adenoviral-transduced animals also showed some variability, though the relative number of transduced cells was greater ($22.6 \pm 7.6\%$). In both groups of animals, hpAP expression was diminished but detectable 14 days after transfection, and absent 28 days after transfection. An important question when using viral vectors is the degree of inflammation caused by the vector itself. We therefore systematically evaluated the degree of inflammation in control, lipid-, and adenovirus-transduced lungs. While a mild perivascular and peribronchiolar inflammatory response was detected in some animals, there did not appear to be significant differences between control animals and those receiving either vector. Additionally, toxicity was not detected in other organs. This study demonstrates the feasibility and lack of toxicity of catheter-mediated gene delivery to the pulmonary circulation using either lipid or adenoviral vectors.

V. Summary

Gene delivery to the pulmonary circulation is feasible. Using present technology, isolation of the lung circulation with either a catheter-based or a surgical approach yields the highest expression. Both lipid and adenoviral vectors yield lung expression, though adenovirus produces higher efficiency. While systemic delivery utilizing lipid:DNA complexes appears to result primarily in endothelial cell expression, the higher levels of expression achieved with directed

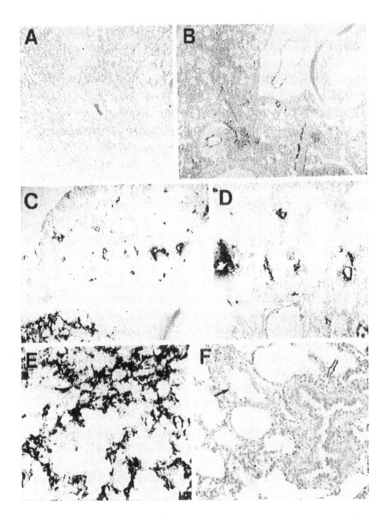

Figure 2 Expression of human placental alkaline phosphatase protein in nontransfected and transfected porcine pulmonary tissue in vivo. Alkaline phosphatase protein was not detected in nontransfected right lung (A), whereas alkaline phosphatase staining was present in cells lining vascular structures and in the interstitium of lung after liposomal transfection (B) and adenoviral infection (C, D, and E). Alkaline phosphatase staining was observed in arteries/arterioles and capillaries (B, C, and D, black arrows) and in alveolar septa (B and E, black arrows) but was not observed in bronchi (B, white arrow). Panel D is a higher power magnification of panel C. Panel F is a representative section of formalin-fixed, paraffin-embedded tissue showing a mild interstitial and peribronchial inflammatory cell infiltrate after adenoviral infection (white arrow); black arrow indicates adjacent vascular structure. Original magnifications ×25 (A and B), ×100 (C), ×200 (D and E), and ×400 (F).

approaches result in expression in a variety of other lung cells, including alveolar epithelial cells. Of note, little expression in vascular smooth muscle cells has been detected in any study to date. The present studies still represent early stages of development of this technology. While a variety of techniques result in detectable reporter gene activity, it remains to be proven if these techniques provide sufficient efficiency to produce biological effects from genes of interest. Challenges for the future include development of more efficient vectors and improved cell targeting. Despite these limitations, pulmonary vascular gene transfer holds promise as a useful investigative and therapeutic tool.

References

1. Nabel G, Yang Z, Muller D, et al. Safety and toxicity of catheter gene delivery to the pulmonary vasculature in a patient with metastatic melanoma. Human Gene Ther 1994; 5:1089–1094.
2. Miller A. Retrovirus packaging cells. Human Gene Ther 1990; 1:5–14.
3. Wilson, JM, Birinyi K, Salomon N, et al. Implantation of vascular grafts lined with genetically modified endothelial cells. Science 1989; 244:1344–1346.
4. Nabel G, Plautz G, Boyse D, Stanley JC, Nabel GJ. Recombinant gene expression in vivo within endothelial cells of the arterial wall. Science 1989; 244:1342–1344.
5. Zhu N, Liggitt D, Yong L, Debs R. Systemic gene expression after intravenous DNA delivery into adult mice. Science 1993; 261:209–212.
6. Thierry AR, Lunardi-Iskandar Y, Dryant JL, et al. Systemic gene therapy: biodistribution and long-term expression of a transgene in mice. Proc Natl Acad Sci USA 1995; 92:9742–9746.
7. LeMarchand P, Jones M, Danel C, Yamada I, Mastrangeli A, Crystal RG. In vivo adenovirus-mediated gene transfer to lungs via pulmonary artery. J Appl Physiol 1994; 76(6):2840–2845.
8. Muller DWM, Gordon D, San H, et al. Catheter-mediated pulmonary vascular gene transfer and expression. Circ Res 1994; 75:1039–1049.
9. Messina LM, Podrazik RM, Whitehill TA, et al. Adhesion and incorporation of lacZ-transduced endothelial cells into the intact capillary wall in the rat. Proc Natl Acad Sci USA 1992; 89:12018–12022.
10. Bernstein SC, Skoskiewicz MJ, Jones R, Zapol WM, Russell S. Recombinant gene expression in pulmonary vascular endothelial cells: polarized secretion in vivo. FASEB J 1990; 4:2665–2670.
11. Logan JJ, Bebok Z, Walker LC, et al. Cationic lipids for reporter gene and CFTR transfer to rat pulmonary epithelium. Gene Ther 1995; 2:38–49.
12. Brigham KL, Meyrick BO, Christman B, Magnuson M, King G, Berry LC Jr. Rapid communication: in vivo transfection of murine lungs with a functioning prokaryotic gene using a liposome vehicle. Am J Med Sci 1989; 296:278–281.
13. Cononico AE, Conary JT, Meyrick BO, Brigham KL. Aerosol and intravenous transfection of human α1-antitrypsin gene to lungs of rabbits. Am J Respir Cell Mol Biol 1994; 10:24–29.

14. Solodin I, Brown CS, Bruno MS, et al. A novel series of amphiphilic imidazolinium compounds for in vitro and in vivo gene delivery. Biochemistry 1995; 34:13537–13544.
15. Rodman DM, Tyler RC, Gorman C, et al. Cationic liposome-mediated gene transfer to lungs of normal and pulmonary hypertensive mice in vivo. FASEB J 1995; 9:A:857. Abstract.
16. Perricaudet LD, Makeh I, Perricaudet M, Briand P. Widespread long-term gene transfer to mouse skeletal muscles and heart. J Clin Invest 1992; 90:626–630.
17. Schachtner SK, Rome JJ, Hoyt RF Jr, Newman KD, Virmani R, Deichek DA. In vivo adenovirus-mediated gene transfer via the pulmonary artery of rats. Circ Res 1995; 76:701–709.

10

Targeting Gene Expression to the Lung

JEFFREY A. WHITSETT and STEPHAN W. GLASSER

Children's Hospital Medical Center
Cincinnati, Ohio

I. Introduction

Bioactive compounds can be selectively delivered to or activated in specific tissues on the basis of their physical, chemical, and metabolic properties. Such considerations have provided the fundamental principles underlying modern medical pharmacology. These principles also apply to the field of gene therapy. The feasibility of delivering cDNAs or genes to somatic cells for therapy of genetic or acquired diseases has developed rapidly, necessitating the design of vectors that selectively deliver and express genes in target tissues. In contrast to other organs, gene transfer to the lung is facilitated by its unique accessibility, via the trachea, offering a remarkable opportunity to bring gene transfer vectors in direct physical contact with respiratory epithelial cells. In addition, both pulmonary and bronchial arteries provide blood supply to the lung, offering the ability to deliver gene transfer vectors to distinct sites within the pulmonary vasculature. However, successful gene therapy for many diseases will likely require the ability to efficiently deliver and precisely regulate the expression of therapeutic DNAs in target cells.

Consideration of the cells requiring the introduction of therapeutic DNA for correction of metabolic or acquired lung diseases is highly dependent upon knowledge of the pathophysiology of the disorder. Clarification of the role of therapeutic genes in cellular metabolism and the potential consequences of aberrant expression of an introduced gene will be needed to complement scientific advances in the development of gene transfer vectors. While the introduction of DNA to the one cell may correct a metabolic abnormality, inappropriate levels of the therapeutic protein in nontarget cells may be detrimental. Therefore the level and duration of expression of introduced genes may be of clinical importance. Ultimately, the means to efficiently deliver DNA to appropriate cellular targets, wherein precise expression of the introduced gene is controlled, will be required to effect therapeutic interventions for a variety of diseases. Ideally, the transferred gene should contain all of the cis-regulatory information necessary to control the level of the therapeutic mRNA as precisely as that of the endogenous gene.

II. Cellular Targets in the Lung

The lung is a complex organ containing more than 40 cell types, each with distinct morphological and biochemical properties. The respiratory epithelium consists of distinct populations of columnar cells lining the conducting and peripheral airways. These tubules lead to the alveolar gas exchange region, lined by cuboidal (Type II) and squamous (Type I) epithelial cells. In the human lung, tracheobronchial glands are also lined by a complex and highly differentiated respiratory epithelium containing several distinct cell types. The lung parenchyma is richly supplied by capillaries from both bronchial and pulmonary circulations. In the human lung, more than a dozen distinct epithelial cell types are readily discerned by morphologic criteria alone; their location and metabolic properties change during development and in response to injury or other stimuli. The unique accessibility of these cells, by instillation or inhalation, makes the lung an ideal target for gene therapy. However, the heterogeneity of the cellular targets may complicate therapies that require the introduction of DNA into distinct subsets of respiratory epithelial cells.

Gene transfer to respiratory epithelial cells may be considered for the introduction of cDNAs or genes directing the synthesis of proteins that are secreted systemically, or alternatively, for the delivery of genes encoding cytotoxic or immunostimulatory proteins that can be used to ablate specific cell types for treatment of pulmonary cancer. The large surface area of the lung and its rich vascular supply provide an extensive layer of cells that can be transfected with cDNAs that direct the secretion of polypeptides that are active at distant sites. On the other hand, gene therapy may be used to correct disorders caused by mutations in genes resulting in metabolic abnormalities that alter the function of specific subsets of cells in the lung. For example, correction of hereditary SP-B deficiency and cystic fibrosis will require complementation of gene defects in subsets of respiratory epithelial cells. Such diverse therapeutic goals may require the design of vectors that will allow targeting specific cell types and allow the precise control of the expression of the introduced gene.

A. Vector Considerations

A wide variety of vectors are being developed for gene transfer including viruses, liposomes, and protein-DNA conjugates. Recombinant viral vectors, including adenoviruses, retroviruses, and adeno-associated viruses, enter target cells based on presence of viral proteins that interact with cell surfaces. The efficiency of viral entry is often highly species and cell type-specific—e.g., ecotropic and amphotropic retroviruses, which bind to distinct cell transport proteins on the surface of the cell. The abundance or absence of specific cell surface receptors may strongly influence the ability of a viral vector to transfect the target cell and, in turn, may be influenced by cell differentiation and proliferation that alters the efficiency of viral entry and transport of DNA to the nucleus.

B. Targeting Specific Cell Surfaces in the Lung

Respiratory epithelial cells of the conducting and peripheral airways display a variety of morphological and biochemical features. Distinct subsets of receptors, surface glycoproteins, glycoconjugates, and adhesion molecules on the surface of respiratory epithelial cells may be utilized by viruses or nonviral liposomal-ligand complexes to facilitate entry of DNA into target cells. Not surprisingly, a number of protein DNA complexes have been developed to target respiratory epithelial cells based upon the biology of protein transport in the lung. In particular, surfactant proteins A, B, and C are expressed and taken up at high rates by subsets of respiratory epithelial cells in the lung (1). Surfactant protein A, B, and C conjugated to polylysine have been utilized in vitro to transfect DNA into respiratory epithelial cells with some cell selectivity (2). The polymeric IgA system has also been utilized in vivo to express genes in target cells in the respiratory tract utilizing DNA-protein conjugates based on polymeric IgA receptor system (3,4). Polymeric IgA receptors normally transfer IgA from the systemic circulation to respiratory epithelial cells. After proteolysis, active IgA is secreted into the airway where it serves host defense functions in the lung. Polymeric IgA-DNA conjugates efficiently and selectively deliver genes to respiratory epithelial cells of tracheobronchial glands and to subsets of upper airway epithelial cells after systemic injection (4). Binding to the endogenous polymeric IgA receptor enhances the efficiency and specificity of cellular uptake by the lung. Similar conjugates have also been used for delivery of genes to the liver—e.g., the asialoglyoprotein-DNA conjugate system used by Wu et al. (5,6) to enhance targeting of DNA to hepatocytes. Strategies based upon carbohydrate binding specificity of lectins on the cell surface or, alternatively, the use of lectin-DNA conjugates, have been used to transfer DNA into pulmonary cells. Lectin-DNA conjugates and lipospermine or lipopolyamine-DNA conjugates form condensed particles that enter cells on the basis carbohydrate or lectin specificity have been considered for gene therapy (7). The activity of such conjugates can be further enhanced by coadministration of the DNA conjugates with viral proteins that enhance DNA uptake, escape from lysosomes, and transfer to the nucleus (6,8). Cationic and anionic lipids have been developed that increase the uptake and activity of plasmid DNA vectors. Recently, cationic lipid-DNA particles expressing the CFTR (cystic fibrosis transmembrane conductance regulator) have been administered to the respiratory epithelium for correction of cystic fibrosis (9). While cationic lipid-DNA mixtures do not achieve precise organ or cellular targeting, their efficiency in the lung may be influenced by their size, chemical composition, and site of administration. Manipulation of chemical or structural composition of DNA-protein conjugates can further alter the selectivity in gene transfer. However, vector delivery to each tissue will be highly dependent upon the site of injection and the anatomy of the vascular or tracheal-bronchial systems that may influence the concentrations of vector delivered to cellular targets. As with all vectors, the site and volume of administration, concentration, dwell time, and intrinsic properties of the vector may greatly influence the selectivity and efficiency of gene transfer.

In spite of the development and testing of a variety of gene transfer vectors, precise cellular specificity has been difficult to achieve. In contrast, endogenous genes are often expressed in precise temporal-spatial patterns that are determined by both transcriptional and posttranscriptional mechanisms that precisely regulate mRNA levels and protein production. Such transcriptional and posttranscriptional controls can be readily introduced into the DNA, whether viral or nonviral, and may be used to control the levels of exogenous mRNAs within specific cell types in the lung.

III. Transcriptional Control of Gene Expression

While therapy of genetic disease may require the precise temporal, spatial, and humoral control of the introduced gene, the level of expression needed to correct most genetic diseases will likely be less than that normally expressed by affected tissues. Heterozygote carriers generally have little or no adverse effects from their mutant allele, even when accompanied by a 50% reduction in gene expression related to the loss of a normal gene copy. In some disorders, very low levels of expression, perhaps limited to few cells, may suffice to correct physiologic abnormalities—e.g., adenosine deaminase deficiency. Nevertheless, the ability to precisely regulate levels of gene expression in a physiological relevant manner and to increase or decrease expression of the delivered gene without reintroducing the vector would be useful for the therapy of a number of genetic or acquired disorders. Such control also would be critical if inappropriate levels of gene expression prove detrimental to cellular function. Under normal conditions, this control is mediated intercellularly, primarily at transcriptional and posttranscriptional levels, influencing the synthesis, processing, stability, or translation of mRNAs. These activities can be introduced in cis with transferred DNAs, to control gene transcription or mRNA stability.

A. Identification and Characterization of Genes Expressed in the Respiratory Epithelium

The lung bud envaginates into the splanchnic mesenchyme from epithelial cells lining the foregut endoderm on day 9 (in the mouse) or the 7th or 8th week of gestation (in the human). Thereafter the conducting airway and gas exchange areas are produced by dichotomous branching. Ultimately, more than a dozen distinct epithelial cell types are derived from the common set of progenitor cells that produce the cells with the distinct morphological, biochemical, and functional properties characteristic of the mature airway. Cis-acting elements controlling lung-specific gene expression have been isolated from genes encoding proteins that are expressed selectively in various subsets of respiratory epithelial cells. Such cis-acting regulatory elements have been isolated from the 5' regions of the genes encoding surfactant proteins SP-A, SP-B, SP-C, and CCSP (Clara cell secretory protein) and TTF-1 (thyroid transcription factor-1). Each promoter-enhancer controls gene expression in respiratory epithelial cells in vitro and, in transgenic mice, in vivo. The cis-acting elements and transacting factors that control transcription of these genes provide the reagents that will be useful in the design of gene therapy vectors that will express genes selectively in the respiratory epithelium.

B. Combinatorial Interactions Among Nuclear Transcription Proteins That Determine Lung Epithelial Cell Selectivity

Surfactant proteins and CCSP are expressed in overlapping yet also distinct subsets of respiratory epithelial cells in vivo. While SP-A and SP-B are synthesized in epithelial cells in tracheal-bronchial glands and in bronchiolar and alveolar epithelial cells (10,11), SP-C is expressed only by Type II epithelial cells (12,13). In contrast, CCSP is expressed in the nonciliated epithelial cells in the conducting airways and is excluded from the alveolus (14) (Fig. 1). In spite of these distinctions, these respiratory epithelial cells share common progenitors in the foregut endoderm and also share in the utilization of several

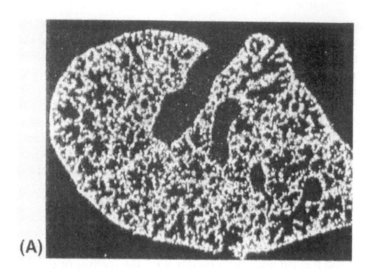

(A)

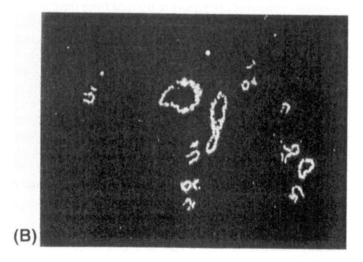

(B)

Figure 1   In situ hybridization analysis of gene expression in the lung. Serial sections
of fetal day 18 mouse lungs were analyzed using riboprobes complementary to the
mRNA for SP-C (A), CCSP (B), SP-A (C), SP-B (D). SP-C is detected only in alveolar
epithelial cells. CCSP is detected along the tracheobronchiolar axis; SP-B expression
is detected in distal airway and alveolar epithelial cells. SP-A is detected in alveolar
epithelial cells (and postnatally in distal airway epithelial cells). (Kindly provided by
Susan E. Wert.)

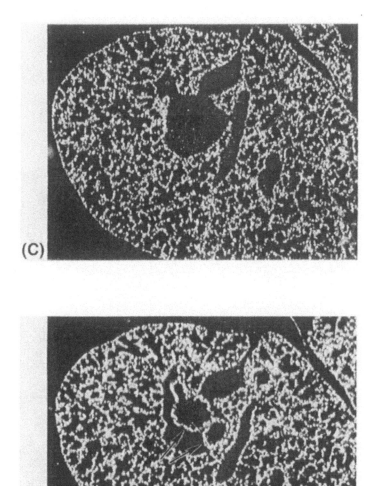

transcription factors that control cell differentiation and gene expression. Elucidation of the cis-acting elements and transacting factors controlling expression of these distinct genes demonstrates that each gene shares common regulatory elements and that distinctions in control are likely provided by the interactions of shared cis-acting sites and their transacting factors. The finding

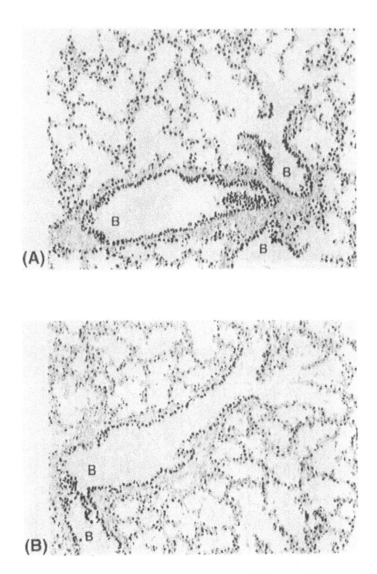

Figure 2 Distribution of the transcription factors TTF-1 and HNF-3β in the lung. Expression of the transcription factors TTF-1 and HNF-3β was determined by immunocytochemical staining of fetal day 18 mouse lung using antibodies specific to either TTF-1 (A) or HNF-3β (B). TTF-1 and HNF-3β is detected in conducting and alveolar epithelial cells in patterns that are consistent with CCSP and SP gene expression. (Kindly provided by Lan Zhou.)

that thyroid transcription factor-1 (TTF-1) and various members of the fork-head family of transcription factors, particularly hepatocyte nuclear factor-3 (HNF-3), and AP-1 family members activate lung specific transcription has now provided the initial framework from which to build cis-acting elements capable of controlling gene expression in DNA-transfer vectors in subsets of respiratory cells (15–18) (Fig. 2).

C. Cis-Acting Elements Controlling SP-B Gene Transcription

A map of the 5' regulatory region and regulatory motifs controlling SP-B transcription is provided by Figure 3. A region as small as –218 bp from the start

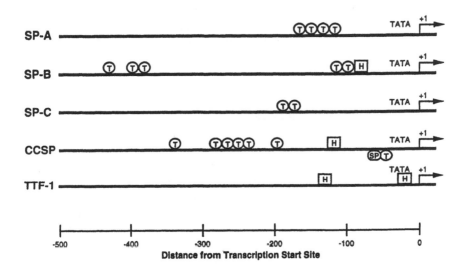

Figure 3 Summary of nuclear protein-binding sites associated with genes expressed selectively in the lung. The location of tissue restricted TTF-1 (T), HNF-3 (H), and ubiquitous SP-1, SP-3 (SP) transcription factor binding sites is shown relative to the basal promoter (TATA) and start of transcription (+1 at arrow) for the surfactant protein, Clara cell secretory and thyroid transcription factor-1 genes. The scale below the promoter binding sites is in base pairs. Only transcription factor-DNA interactions that have been demonstrated by direct DNA binding experiments are shown. DNase footprint analysis suggests that numerous other factors are associated with these promoter regions. The array of TTF-1 and HNF-3 binding sites is unique for each gene, suggesting that combinatorial interactions (that may include related family members for each transcription factor) may mediate the distinct patterns of expression of each gene. The figure represents data from murine (SP-A and SP-C), human (SP-B and TTF-1), and rat CCSP (below line); and murine CCSP (above line) genes.

of transcription of the SP-B gene provides all of the elements required to produce lung-specific gene expression in transgenic mice. The level of gene expression is markedly increased by a second region that serves as a position independent enhancer located –330 to –440 from the start of transcription (19). The proximal (–70 to –110) and distal (–330 to –440) elements have been mapped by DNase footprinting, electromobility shift assay (EMSA), and site-specific mutagenesis to determine DNA-protein interactions that confer lung cell specific gene transcription (19,20).

At present, the mechanisms controlling the expression of SP-B are best understood. SP-B is expressed only in the respiratory epithelium, beginning as early as day 12 or 13 of gestation in the mouse or 10 to 12 weeks in the human, and is found in subsets of nonciliated respiratory epithelial cells in the conducting airways and in Type II epithelial cells in the alveolus of the mature lung (10,21). SP-B is required for respiratory function after birth and mutations in the SP-B gene cause lethal respiratory distress in humans and in SP-B gene targeted transgenic mice (22,23). The SP-B gene is controlled by both transcriptional and posttranscriptional mechanisms and is strongly influenced by the developmental and humoral signals.

IV. Proximal SP-B Enhancer

The proximal enhancer region is located in close proximity to the start of transcription and is critical to SP-B gene expression. Site specific mutagenesis of elements in this region completely block transcriptional activity of promoter constructs containing this region. Two sites, termed SP-B (F1) and SP-B (F2), and a distinct site located in the first intron, are critical to the function of the proximal promoter element (19). SP-B (F1) contains a consensus element (CAAG) that binds thyroid transcription factor-1 and serves as a critical lung cell selective activator of this gene. The TTF-1 binding site is located in close proximity to SP-B (F2) a cis-active element that binds hepatocyte nuclear factor-3 (HNF-3). Mutations in either the TTF-1 or HNF-3 sites in the proximal element markedly decreased the function of the SP-B promoter, suggesting that interactions among these sites are critical to the formation of a transcription complex that influences the basal transcriptional apparatus of the gene.

AP-1 family members bind to a cis-acting element located near a TTF-1 binding site in the first intron of the mouse gene. Its occupation and function are critical to SP-B gene expression but do not suffice to confer lung cell specificity per se. Together the SP-B (F1) and SP-B (F2) serve to confer lung cell-specific gene expression in the context of the basal SP-B promoter. The element does not activate gene expression from other basal promoter elements (20). In contrast, an upstream SP-B enhancer, located approximately –350 from the start of transcription of the human SP-B gene, contains TTF-1 sites (two or three in the human or mouse gene, respectively) that markedly enhance gene expression and serve as an enhancer of viral promoters. TTF-1 is critical to the activity of the distal element; mutations blocking TTF-1 binding to this site markedly inhibited enhancer activity (24). The distal SP-B enhancer acts in an orientation-independent fashion and can activate a minimal promoter from the herpes simplex thymidine kinase gene, producing lung cell-selective control of gene expression after transfection of respiratory epithelial cells in vitro. Such elements can therefore be introduced into gene transfer vectors to activate gene expression in a respiratory epithelial cell selective manner.

Genomic DNA flanking both the SP-C promoter and CCSP promoter were also used to direct expression of a chloramphenicol acetyl transferase (CAT) reporter gene in the respiratory epithelial cells. Approximately 3.7 kb of the human SP-C gene and 2.4 kb of the rat CCSP gene were sufficient to selectively direct CAT expression to the lungs of transgenic mice (14,25,26). With both of these promoter constructs, CAT activity was observed in the lungs of multiple independent founder lines, indicating that these cis-acting sequences were sufficient to determine pulmonary specific expression. In situ hybridization analysis demonstrated that the pattern of transgene expression was similar to that of the endogenous murine CCSP and SP-C genes. CCSP-CAT expression was detected in the tracheal and bronchiolar epithelium, consistent with the distribution of Clara cells. In contrast, mRNA derived from 3.7 SP-C-CAT transgene was detected only in distal bronchiolar epithelium and in Type II cells throughout the distal pulmonary parenchyma. The pattern of expression of a transgene derived from the mouse SP-C gene was identical to that of the endogenous SP-C that is expressed only in alveolar Type II cells. These data demonstrate that the CCSP and SP-C promoter constructs can be used to express therapeutic genes in vivo. The feasibility of this approach has been recently demonstrated by the finding that expression GM-CSF in the respiratory epithelium under control of the SP-C promoter fully corrected the alveolar proteinosis in the GM-CSF knockout mouse (27). The 3.7 SP-C promoter has also been used to direct expression of the human CFTR in the distal airway of transgenic mice (28).

The SP-A, SP-C, and CCSP promoter regions also contain TTF binding motifs associated with DNase footprint sites. Four clustered TTF-1 binding sites were identified in the –256 bp region of the murine SP-A promoter. Site-specific mutation of the TTF-1 binding sites in the SP-A gene were used in gel shift experiments to identify DNA binding sites with distinct affinities for TTF-1 (29). Likewise, two TTF-1 binding sites were identified in the murine SP-C promoter (–186 to –163 bp) and were used to express genes in a heterologous promoter construct. Mutation of either of the TTF-1 sites in the SP-C gene eliminated stimulation by TTF-1 (30). Multiple TTF-1 binding sites have been identified in the –344 bp region of the murine CCSP promoter, where they serve critical roles in transcriptional control of the CCSP genes (31,32) (Fig. 2).

The number, location, and affinity of TTF-1 binding sites in the surfactant and CCSP genes are highly heterogenous. The distribution of TTF-1 protein and mRNA in the developing and mature lung is consistent with the sites of SP-A, -B, -C, and CCSP expression (Fig. 2). The critical role of TTF-1 in lung development is further supported by the null mutation of the TTF-1 gene, which inhibited development of distal lung parenchyma (33).

Thus transcriptional elements containing HNF-3, AP-1, and TTF-1 binding sites are found in the promoter-enhancer regions of the surfactant protein and CCSP genes. Distinctions in the location and number of sites, the affinity for TTF-1, HNF, and AP-1 proteins, and their interactions with other enhancer elements likely provide the subtle controls that mediate the distinct patterns of gene expression observed for the surfactant proteins and CCSP. Combinatorial interactions among this set of nuclear transcription factors and cis-active elements in each of the promoters provide an elegant and complex system by which gene expression is controlled by humoral, temporal, spatial, and developmental influences.

While these studies have provided fundamental insights into the control of respiratory epithelial cell gene expression, it is highly likely that other transcription factors, and their cognate elements, remain to be identified. Such factors are likely to function in combination with known elements to provide the full complexity of gene expression that is determined by the 5′ regions of each of these genes. This detailed information is derived from analysis of the promoters that provide the building blocks that will be required for the construction of viral or plasmid vectors conferring highly controlled lung epithelial cell selective gene expression in gene transfer vectors.

V. Posttranscriptional Control of Gene Expression

Experience in the study of numerous genes demonstrated 5′ and 3′ elements that alter the level of steady-state mRNA stabilizing or destabilizing mRNAs in target tissues. Such elements have been identified in early response genes, and are mediated by interactions of proteins with AUUA repeats that destabilize the mRNAs. Similarly, elements controlling mRNA stability are contained within the 3′ regions of surfactant protein genes that stabilize surfactant protein mRNAs under basal conditions or destabilize surfactant protein mRNAs after exposure to various cytokines. For example, TNF-α and TPA decrease SP-B mRNA stability. This effect is determined by the interaction of proteins with elements located in the 3′-untranslated region of the human SP-B mRNA

(34,35). Such elements provide cis-acting control of mRNA stability that can be introduced into gene transfer vectors. Similar strategies can be used to provide control of mRNA stability after exposure to hormones or other pharmacologic agents.

VI. Summary

Advances in our understanding of vector biology, gene transcription, and post-transcriptional control of gene expression provide the reagents with which to design vectors capable of precisely controlling gene expression in subsets of target cells. At present, these studies have elucidated mechanisms controlling transcription in respiratory epithelial cells and do not provide insight into control of gene expression in other cell types in the lung. Nevertheless, these studies have provided the framework that will be highly useful in future analysis of the elements controlling gene expression in other cells in the lung. Identification of genes expressed selectively in various cellular compartments of the lung and the elucidation of the cis-active and trans-acting factors controlling such genes, will provide the regulatory elements that can be used to engineer gene transfer vectors in the future. Ultimately, success of gene therapy may depend upon a more complete knowledge of the stochastic, temporal-spatial, and cellular requirements of gene expression required to achieve amelioration of acquired and genetic diseases in the lung. This knowledge, when taken into context with advancements in viral, plasmid, or chromosomal-based gene delivery systems, may some day revolutionize therapies for acquired and genetic pulmonary disease.

Acknowledgments

Supported by HL51832 and the Cystic Fibrosis Foundation.

References

1. Horowitz AD, Moussavian B, Whitsett JA. Roles of SP-A, SP-B, and SP-C in modulation of lipid uptake by pulmonary epithelial cells in vitro. Am J Physiol (Lung Cell Mol Physiol 14) 1996; 270:L69–L79.
2. Baatz JE, Bruno MD, Ciraolo PJ, et al. Utilization of modified surfactant-associated protein B for delivery of DNA to airway cells in culture. Proc Natl Acad Sci USA 1994; 91:2547–2551.
3. Ferkol T, Kaetzel CS, Davis PB. Gene transfer into respiratory epithelial cells by targeting polymeric immunoglobulin receptor. J Clin Invest 1993; 92:2394–2400.
4. Ferkol T, Perales JC, Eckman E, Kaetzel CS, Hanson RW, Davis PB. Gene transfer into the airway epithelium of animals by targeting the polymeric immunoglobulin receptor. J Clin Invest 1995; 95:493–502

5. Wu GY, Wu CH. Specific inhibition of Hepatitis-B viral gene expression in vitro by targeted antisense oligonucleotides. J Biol Chem 1992; 267(18):12436–12439.

6. Wu Gy, Zhan P, Sze LL, Rosenberg AR, Wu CH. Incorporation of adenovirus into a ligand-based DNA carrier system results in retention of original receptor specificity and enhances targeted gene expression. J Biol Chem 1994; 269:11542–11546.

7. Yin W, Cheng PW. Lectin conjugate-directed gene transfer to airway epithelial cells. Biochem Biophys Res Commun 1994; 205:826–833.

8. Cristiano RJ, Smith LC, Kay MA, Brinkley BR, Woo SLC. Hepatic gene therapy: efficient gene delivery and expression in primary hepatocytes utilizing a conjugated adenovirus-DNA complex. Proc Natl Acad Sci USA 1993; 90:11548–11552.

9. Caplen NJ, Alton EWFW, Middleton PG, et al. Liposome-mediated CFTR gene transfer to the nasal epithelium of patients with cystic fibrosis. Nature Med 1995; 1:39–46.

10. Stahlman MT, Gray ME, Whitsett JA. The ontogeny and distribution of surfactant protein-B in human fetuses and newborns. J Histochem Cytochem 1992; 40(10): 1471–1480.

11. Wohlford-Lenane CL, Snyder JM. Localization of surfactant-associated proteins SP-A and SP-B mRNA in rabbit fetal lung tissue by in situ hybridization. Am J Respir Cell Mol Biol 1992; 7:335–343.

12. Wohlford-Lenane CL, Durham PL, Snyder JM. Localization of surfactant-associated protein C (SP-C) mRNA in fetal rabbit lung tissue by in situ hybridization. Am J Respir Cell Mol Biol 1992; 6:225–234.

13. Wert SE, Glasser SW, Korfhagen TR, Whitsett JA. Transcriptional elements from the human SP-C gene direct expression in the primordial respiratory epithelium of transgenic mice. Dev Biol 1993; 156:426–443.

14. Stripp BR, Sawaya PL, Luse DS, et al. Cis-acting elements that confer lung epithelial cell expression of the CC10 gene. J Biol Chem 1992; 267:14703–14712.

15. Lazzaro D, Price M, De Felice M, DiLauro R. The transcription factor TTF-1 is expressed at the onset of thyroid and lung morphogenesis and in restricted regions of the foetal brain. Development 1991; 113:1093–1104.

16. Ikeda K, Clark JC, Shaw-White JR, Stahlman MT, Boutell CJ, Whitsett JA. Gene structure and expression of human thyroid transcription factor-1 in respiratory epitheial cells. J Biol Chem 1995; 270:8108–8114.

17. Clevidence DE, Overdier DG, Peterson RS, et al. Members of the HNF-3/forkhead family of transcription factors exhibit distinct cellular expression patterns in lung and regulate the surfactant protein B promoter. Dev Biol 1994; 166:195–209.

18. Zhou L, Lim L, Costa RH, Whitsett JA. Thyroid transcription factor-1, hepatocyte nuclear factor-3β, surfactant protein B, C, and Clara cell secretory protein in developing mouse lung. J Histochem Cytochem 1996; 44:1183–1193.

19. Bohinski RJ, Huffman JA, Whitsett JA, Lattier DL. Cis-active elements controlling lung cell-specific expression of human pulmonary surfactant protein-B gene. J Biol Chem 1993; 268(15):11160–11166.

20. Bohinski RJ, DiLauro R, Whitsett JA. The lung-specific surfactant protein B gene promoter is a target for thyroid transcription factor 1 and hepatocyte nuclear factor 3, indicating common factors for organ-specific gene expression along the foregut axis. Mol Cell Biol 1994; 14:5671–5681.
21. D'Amore-Bruno MA, Wikenheiser KA, Carter JE, Clark JC, Whitsett JA. Sequence, ontogeny and cellular localization of murine surfactant protein SP-B mRNA. Am J Physiol: Lung Cell Mol Physiol 1992; 262:L40–L47.
22. Nogee LM, Garnier G, Dietz HC, et al. A mutation in the surfactant protein B gene responsible for fatal neonatal respiratory disease in multiple kindreds. J Clin Invest 1994; 93:1860–1863.
23. Clark JC, Wert SE, Bachurski CJ, et al. Targeted disruption of the surfactant protein B gene disrupts surfactant homeostasis, causing respiratory failure in newborn mice. Proc Natl Acad Sci USA 1995; 92:7794–7798.
24. Yan C, Sever Z, Whitsett JA. Upstream enhancer activity in the human surfactant protein B gene is mediated by thyroid transcription factor 1. J Biol Chem 1995; 270:24852–24857.
25. Glasser SW, Korfhagen TR, Bruno MD, Dey C, Whitsett JA. Structure and expression of the pulmonary surfactant protein-SP-C gene in the mouse. J Biol Chem 1990; 265(35):21986–21991.
26. Glasser SW, Korfhagen TR, Wert SE, et al. Genetic element from human surfactant protein SP-C gene confers bronchiolar-alveolar cell specificity in transgenic mice. Am J Physiol: Lung Cell Mol Physiol 1991; 261(5):L349–L356.
27. Huffman JA, Hull WM, Dranoff G, Mulligan RC, Whitsett JA. Pulmonary epithelial cell expression of GM-CSF corrects the alveolar proteinosis in GM-CSF-deficient mice. J Clin Invest 1996; 97:649–655.
28. Whitsett JA, Dey CR, Stripp BR, et al. Human cystic fibrosis transmembrane conductance regulator directed to respiratory epithelial cells of transgenic mice. Nature Genet 1992; 2:13–20.
29. Bruno MD, Bohinski RJ, Huelsman KM, Whitsett JA, Korfhagen TR. Lung cell specific expression of the murine surfactant protein A gene is mediated by interactions between the SP-A promoter and thyroid transcription factor-1. J Biol Chem 1995; 270:6531–6536.
30. Kelly SE, Bachurski CJ, Burhans MS, Glaser SW. Transcription of the lung-specific surfactant protein C gene is mediated by thyroid transcription factor-1. J Biol Chem 1996; 271:6881–6888.
31. Ray MK, Chen C-Y, Schwartz RJ, DeMayo FJ. Transcriptional regulation of a mouse Clara cell-specific protein (mCC10) gene by the NKx transcription factor family members thyroid transcription factor 1 and cardiac muscle-specific homeobox protein (CSX). Mol Cell Biol 1996; 16:2056–2064.
32. Toonen RGF, Gowan S, Bingle CD. The lung enriched transcription factor TTF-1 and the ubiquitously expressed proteins Sp1 and Sp3 interact with elements located in the minimal promoter of the rat Clara cell secretory protein gene. Biochem J 1996; 316:467–473.
33. Kimura S, Hara Y, Pineau T, et al. The T/ebp null mouse: thyroid specific enhancer-binding protein is essential for the proganogenesis of the thyroid, lung, ventral forebrain, and pituitary. Gene Dev 1996; 10:60–69.

34. Pryhuber GS, O'Reilly MA, Clark JC, Hull WM, Fink I, Whitsett JA. Phorbol ester inhibits surfactant protein SP-A and SP-B expression. J Biol Chem 1990; 265: 20822–20828.
35. Pryhuber GS, Church SL, Kroft T, Panchal A, Whitsett JA. 3′-Untranslated region of SP-B mRNA mediated inhibitory effects of TPA and TNF-α on SP-B expression. Am J Physiol (Lung Cell Mol Physiol 11) 1994; 267:L16–L24.

11

Strategies to Accomplish Targeted Gene Delivery Employing Tropism-Modified Adenoviral Vectors

JOANNE T. DOUGLAS and DAVID T. CURIEL

University of Alabama at Birmingham
Birmingham, Alabama

I. Limitations of Adenoviral Vectors for Pulmonary Gene Therapy Applications

A number of gene therapy approaches for both inherited and acquired diseases of the lung have been proposed (1). These approaches have generally involved direct in vivo gene transfer to pulmonary cellular targets. One vector system that has demonstrated a high level of efficiency in this delivery context has been recombinant adenoviral vectors (2,3). To date, all adenoviral vectors employed for gene therapy have been of serotypes 2 (Ad2) and 5 (Ad5), which are highly homologous members of subgroup C and share a common cellular receptor (4). Direct in vivo gene transfer has been achieved in a broad array of pulmonary cellular targets after delivery of the adenoviral vector via the airway, vascular, or lymphatic routes (5,6). Thus, the unique routing options allowing access to pulmonary parenchyma, along with the broad cellular tropism of the recombinant adenovirus, have permitted the genetic modification of diverse pulmonary targets in various approaches to gene therapy for diseases of the lung.

However, adenoviral vectors suffer from the disadvantage that the widespread distribution of the cellular receptor for serotypes 2 and 5 precludes the targeting of specific cell types. This feature would result in a decrease in the efficiency of transduction of lung cells by adenoviral vectors, as the number of virus particles available for delivery to the target cells would be decreased by sequestration by nontarget cells. Furthermore, this would allow ectopic expression of the delivered transgene, with possibly deleterious consequences. Therefore a means must be developed to redirect the tropism of the adenoviral vector specifically to lung cells. The ability to accomplish targeted gene delivery to specific cells of the pulmonary system would thus greatly enhance the various current gene therapy schemas for lung disorders.

It should also be mentioned that another recognized problem with the use of existing adenoviral vectors deleted only in the E1 early region of the genome is that the low-level expression of late adenovirus gene products in transduced cells triggers an immune response in the host. This is manifested as an inflammatory immune attack on the transduced cells, which leads to transient expression of the transgene (7,8). In addition, immune recognition of the adenoviral vector precludes repeated gene transfer with the same vector agent (9–12). However, promising advances are being reported by investigators who are trying to overcome this immune response problem by: (1) genetic strategies to decrease the replication of adenovirus by making additional deletions in other early regions of the viral genome (13–15); and (2) strategies to modify the host immune response to prevent an attack on the transduced cells and/or vector (16,17).

Therefore, there are two potential problems associated with employing adenoviral vectors for gene therapy for diseases of the lung: (1) the lack of lung cell-specific tropism; and (2) vector-induced immune responses. While other investigators are developing promising strategies to address the second issue, the focus of this chapter is the development of a targeted adenoviral vector for lung cell-specific gene delivery.

II. Tropism-Modified Viral Vectors Can Be Constructed to Achieve Targeted, Cell-Specific Gene Delivery

Attempts to modify the tropism of adenoviral vectors should be considered with regard to strategies that have been utilized to modify the cell-binding specificity of other viral vectors. It is generally recognized that the ability to accomplish targeted, cell-specific transduction offers many advantages for gene therapy (18). Consequently, strategies to modify the tropism of viral agents employed for gene transfer are being actively developed. Most work to date has focused on altering retroviral vectors to allow cell-specific transduction. The host and tissue specificity of a retroviral vector are defined by the cell surface receptors which the viral envelope glycoprotein is able to recognize and bind (19) (Fig. 1A). Hence, targeting of retroviral vectors has been attempted by introducing alternate envelope glycoproteins or by modifying the retroviral envelope glycoprotein to confer new cell-binding activity or tropism.

One approach to the modification of the target cell specificity of retroviruses has involved the construction of "pseudotypes," in which the retroviral genome is encapsidated by the envelope protein of another virus (Fig. 1B). In early work, superinfection of cells producing Rous sarcoma virus (RSV), an avian retrovirus, with vesicular stomatitis virus (VSV) was shown to produce pseudotyped particles containing the genome of one virus within the envelope of the other (20). The host range of the pseudotype was defined by the virus

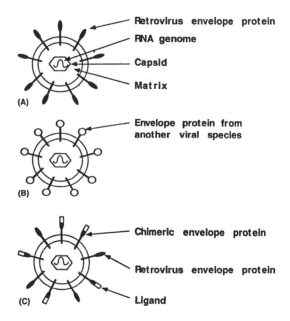

Figure 1 Strategies to modify the tropism of retroviral vectors. (A) Schematic diagram of the retrovirus virion. The RNA genome is contained within the nucleocapsid, which is surrounded by the envelope. Infection results from the specific interaction of the retrovirus envelope glycoproteins with cellular receptors. (B) Pseudotyped retrovirus vector with envelope protein from another species of virus. The tropism of the vector is defined by the host range of the virus supplying the envelope glycoproteins. (C) Pseudotyped retrovirus vector with hybrid envelope. The tropism of the vector is modified by chimeric proteins consisting of a ligand fused to part of the retrovirus envelope protein, permitting the specific targeting of novel cell types by means of ligand-receptor interactions. (From Ref. 18.)

providing the envelope protein. In later studies, Dong et al. (21) showed that a chimeric protein comprising the signal peptide of the RSV envelope glycoprotein fused in frame to the influenza virus hemagglutinin glycoprotein could be incorporated into RSV virions in place of the RSV envelope glycoprotein. In this case, the influenza virus hemagglutinin conferred a broadened host range on the pseudotyped RSV and redirected the normal route of entry of the retrovirus.

Similar pseudotyping experiments have been performed with ecotropic Moloney murine leukemia virus (MoMLV). Preliminary work demonstrated that it was possible to assemble virions in which the genome of ecotropic MoMLV was encapsidated by the envelope protein of a related murine amphotropic retrovirus (22). The resultant virions were able to infect a wider range of host cells than those normally susceptible to MoMLV. Further work by this group resulted in the formation of pseudotyped MoMLV virions with alternative envelope glycoproteins derived from gibbon ape leukemia virus, human T-cell leukemia virus, or vesicular stomatitis virus (22,23). Once again, this approach was shown to be of utility in extending the host range of MoMLV to cells that are normally resistant to infection by this retrovirus. However, the targeting specificity of pseudotyped retrovirions is still limited to cells defined by the host range of the virus supplying the envelope glycoproteins.

Cell-specific targeting of retroviral vectors has also been approached by genetic modifications of the envelope designed to introduce a cell-binding domain not dictated by a second virus. This has been accomplished by fusing the retroviral envelope protein with either an antibody fragment that recognizes a cell-specific antigen, or a ligand with a cognate receptor on the target tissues (Fig. 1C). Preliminary experiments to construct retroviral particles displaying single-chain antibodies (sFvs) on the surface involved hapten model systems. In one approach, Russell et al. (24) fused an sFv directed against the hapten 4-hydroxy-5-iodo-3-nitrophenacetyl caproate (NIP) close to the amino terminal of the envelope glycoprotein of an ecotropic murine retrovirus, MoMLV, by fusing the respective genes. They hypothesized that the sFv would be folded and displayed as a separate domain at the N-terminus of the envelope fusion protein. Chu et al. (25) developed a similar system by fusing a gene coding for an sFv against the hapten 2,4-dinitrophenol (DNP) to 3' regions of the envelope gene of spleen necrosis virus (SNV). In both cases, coexpression of the fusion gene with a retroviral plasmid in ecotropic packaging cells led to the recovery of particles displaying the sFv on the surface, conferring a novel binding specificity for the respective haptens. Furthermore, the modified retroviral particles were shown to be competent for infection, although direct infection of cells targeted by the sFv was not demonstrated. Having established the proof of concept in this manner, Chu and Dornburg (26) repeated their work using an sFv directed against an antigen expressed on the surface of various human carcinoma cells. They showed that SNV particles possessing the sFv-envelope fusion protein could infect human cells expressing the target antigen, albeit at low efficiency. Viral infectivity was blocked in competition experiments employing the parental monoclonal antibody from which the sFv was derived, thus confirming that the infection was mediated by the antibody moiety. In a similar manner, Somia et al. (27) have generated a targeted retroviral vector by fusing an sFv directed against the low density lipoprotein receptor (LDLR) on to the envelope of ecotropic MoMLV vector. The recombinant retroviral vector containing the chimeric envelope was shown to infect human cells specifically through the LDLR. This body of work therefore supports the strategy of retroviral targeting to specific cell surface receptors by means of chimeric sFv-envelope proteins.

Retroviral retargeting has also been achieved by genetic modifications to the envelope which introduce ligands directed to specific cell receptors. In experiments with avian leukosis virus (ALV), Valsesia-Wittmann et al. (28) substituted small segments of the variable regions of the receptor-binding domain of the envelope glycoprotein with a 16 amino acid ligand for cell surface integrin receptors. Some of these minimally modified envelope proteins were correctly processed and incorporated into virions. Moreover, the chimeric envelope proteins permitted the infection of target mammalian cells expressing the integrin receptor but resistant to wild-type ALV. Other groups have made larger modifications in fusing a ligand to the retroviral envelope protein for cell-specific targeting. The first success was reported by Kasahara et al. (29), who deleted a portion of the gene encoding the amino-terminal 150 amino acids of the envelope protein of ecotropic MoMLV and replaced it with sequences encoding the same number of amino acids from the blood protein erythropoietin (EPO). Retroviruses coated with this chimeric envelope protein in addition to wild-type ecotropic envelope protein were then generated from a producer cell line. Incorporation of the chimeric EPO-envelope proteins into intact virions was demonstrated by the fact these viruses were recognized by both anti-EPO and antienvelope antisera. The viruses carrying the EPO-envelope hybrid were shown to infect an EPO receptor-containing murine cell line with a greater than sixfold increase in efficiency over the parental wild-type retrovirus. The increase in infection was abolished by the addition of soluble EPO peptide at the time of infection, demonstrating that it was mediated by specific targeting of the EPO receptor by the EPO-envelope hybrid. The modified retrovirus also became specifically infectious for human cells carrying the EPO receptor while failing to infect other human cells. The same group subsequently applied this strategy for the retroviral targeting of human breast cancer cells via the peptide ligand heregulin (30). They showed that the modified ecotropic murine MoMLV viruses could cross species to specifically infect human breast cancer cells which overexpress receptors for heregulin, while cell lines with low or basal levels of these receptors were not infected.

Other work has revealed that not all cell surface receptors are suitable for targeting by retroviral vectors with genetically modified envelope proteins. Cosset et al. (31) generated ecotropic MoMLV vectors with chimeric envelope proteins carrying at their amino terminus polypeptides able to bind to either the Ram-1 phosphate transporter (the first 208 amino acids of amphotropic murine leukemia virus, MLV-A, surface protein) or the epidermal growth factor (EGF) receptor (the 53 amino acids of EGF). While specific virion binding was achieved in both cases, only Ram-1 permitted viral infection of the target cells, albeit at lower efficiency than virions carrying wild-type MLV-A envelopes. The infectivity of virions targeted to EGFR was blocked at a postbinding step, probably as a result of trafficking to lysosomes. Thus only some cellular receptors permit productive infection of tropism-modified retroviruses. Furthermore, the spacing between the envelope protein and the ligand dictates the efficiency of viral transduction: the incorporation of interdomain amino acid spacers between the Ram-1 binding domain and the MoMLV envelope can result in a significant increase in gene transfer by the retroviral vector (32).

In the instances described above, the demonstration of infection of human cells through the receptor targeted by the sFv or ligand fused to the retroviral envelope protein indicates that it is possible to employ genetic methods to engineer recombinant viral vectors with modified tropism. Neda et al. (33) employed an alternative approach to achieve redirection of the target cell specificity of the ecotropic murine retrovirus MoMLV. They demonstrated that chemical attachment of lactose to the surface of purified MoMLV virions permitted specific infection of human hepatocyte cells via asialoglycoprotein receptors unique to these cells. This chemical modification therefore conferred upon the retrovirus a novel targeting specificity based on receptor recognition of carbohydrate. This form of chemical modification of the retroviral surface would be limited to ligands which could be chemically coupled to the virion proteins.

Roux et al. (34) used an immunological strategy to accomplish retroviral targeting by modification of purified virions. Biotinylated monoclonal antibodies directed against the envelope protein of a murine ecotropic retrovirus on one side, and against specific membrane markers of nonpermissive cells on the other, were linked via a streptavidin bridge. The bifunctional complex was used to link the retrovirus to the target cell, allowing the successful, although inefficient, infection of human cells by means of major histocompatibility complex class I and class II antigens. This demonstrated that the molecular bridging approach permits ecotropic retroviruses to infect normally nonpermissive cells, overcoming the lack of specific retroviral receptors by targeting other receptors. Subsequent molecular bridging studies indicated that the receptors for epidermal growth factor and insulin were also capable of routing the ecotropic murine viruses through alternate pathways to mediate the infection of normally resistant human cells (35). Furthermore, biotinylated proteins could substitute for the anti-cell receptor antibody to mediate infection by the bispecific complex, albeit at a lower efficiency, thus extending the versatility of the approach. However, as also shown by Cosset et al. (31), not all cell membrane molecules were suitable for allowing retroviral infection of human cells: targeting of the transferrin, high density lipoprotein and galactose receptors did not permit infection (35). These findings demonstrate that the nature of the receptor, the specific entry pathway and the composition of the cell-retrovirus linker are important in determining the efficiency of cell targeting.

Thus, the capacity to achieve targeted, cell-specific gene delivery by modifications of viral tropism has been established by these studies which employed genetic, chemical and immunological approaches. It is thus logical to pursue strategies to modify the tropism of adenoviral vectors, which are of great utility for in vivo gene delivery. This would theoretically allow the derivation of a vector exhibiting both in vivo efficacy and cell-specificity, which are desirable features for pulmonary gene therapy applications.

III. The Pathway of Adenoviral Entry Is Partially Characterized

It is important to base strategies to modify adenoviral tropism on an understanding of the biology of viral infection of target cells (Fig. 2). Three distinct, sequential steps are required for viral infection: (1) attachment of the adenovirus to specific receptors on the surface of the target cell; (2) internalization of the virus; and (3) transfer of the viral genome to the nucleus where it can be expressed (36). Thus any attempt to modify the tropism of an adenoviral vector must preserve its ability to perform these three functions efficiently.

Two separate cellular receptors are required for adenoviral entry, one mediating attachment and the other mediating internalization (37,38). The initial high-affinity binding of Ad2 and Ad5 to an as yet unidentified primary cellular receptor occurs via the fiber capsid protein, as demonstrated by the ability of soluble fiber or antifiber antibodies to inhibit productive infection (37,39). The fiber protein is a homotrimer of 62 kDa subunits (40,41) that protrudes from each of the twelve vertices of the icosahedral viral particle where it is attached noncovalently to the penton base (42–44). The amino-terminal tail of the fiber is separated from the carboxy-terminal knob domain by a long, rodlike shaft comprising a 15-amino acid residue motif repeated 22 times in Ad2 and Ad5 (45). It has recently been demonstrated by Henry et al. (46) that Ad5 infection of HeLa cells can be blocked both by a recombinant

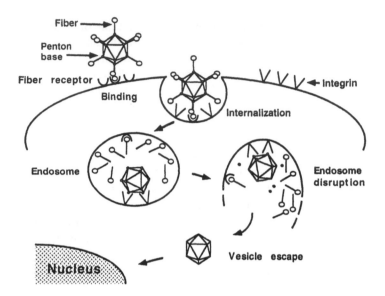

Figure 2 Pathway of adenoviral entry. Adenoviral attachment to cells is accomplished by the high affinity binding of the knob domain of the fiber to an as-yet-unidentified membrane surface receptor. Internalization of the virion by receptor-mediated endocytosis is mediated by the interaction of Arg-Gly-Asp (RGD) peptide sequences in the penton base with secondary host cell receptors, α_v integrins. After internalization, the adenovirus is localized within endosomes. Acidification of the endosomes induces hydrophobic alterations in the capsid proteins, allowing their interaction with the cell vesicle membrane. The virion then localizes to the nuclear pore, and its genome is translocated to the nucleus of the host cell. (Figure courtesy of Sharon Michael.)

adenovirus type 5 trimeric knob expressed in *E. coli* and by antiknob antibody (46). Similarly, Ad2 infection can be blocked by recombinant serotype 2 knob produced in a baculovirus expression system (47); thus the knob is both necessary and sufficient for virion binding to host cells.

Following attachment, the next step in infection by adenovirus serotypes 2 and 5 is internalization of the virion by receptor-mediated endocytosis (48–51). Several lines of evidence have indicated that this process is mediated by the interaction of Arg-Gly-Asp (RGD) peptide sequences in the penton base with secondary host cell receptors, identified as integrins $\alpha_v\beta_3$ and $\alpha_v\beta_5$. Wickham et al. (38) have demonstrated that Ad2 infection is inhibited both by soluble Ad2 penton base and by synthetic peptides containing the RGD motif, but not RGE (Arg-Gly-Glu). In a separate study, Bai et al. (52) showed that mutations of the RGD sequence in the Ad2 penton base abolish cell adhesion and significantly decrease the efficiency of virus infection. The RGD motif is present in a number of molecules, including fibronectin and vitronectin, in which it mediates cellular adhesion via cell surface integrins. A role for $\alpha_v\beta_3$ and $\alpha_v\beta_5$ integrins in the internalization of Ad2 was demonstrated by Wickham et al. (38), who employed function-blocking anti-α_v antibodies to inhibit Ad2 internalization without affecting cellular attachment. Furthermore, Ad2 was shown to bind to cultured cells lacking α_v integrins, but failed to be internalized. However, transfection of these α_v-deficient cells with cDNA encoding α_v resulted in the expression of $\alpha_v\beta_3$ and $\alpha_v\beta_5$ integrins and permitted virus internalization and infection. Huang et al. (53) demonstrated that specific cell subsets such as T lymphocytes and alveolar macrophages are not efficiently infected by adenoviruses because they express very low levels of the $\alpha_v\beta_3$ and $\alpha_v\beta_5$ integrin receptors. These resistant cells can become susceptible to adenoviral infection upon induction of cell differentiation or cell activation when they express these two integrins (53). Similarly, the resistance of terminally differentiated muscle cells to adenoviral infection has been shown to correlate with a paucity of appropriate integrin receptors (54).

Hence, adenovirus attachment and internalization are two distinct, sequential steps. The high-affinity binding of the fiber protein to the primary cellular receptor is followed by the lower-affinity binding of the penton base to the secondary receptor, an α_v integrin, to mediate internalization. After internalization, the virus is localized within the cellular vesicle system, initially in clathrin-coated vesicles and then in cell endosomes (48–51). Acidification of the endosomes allows the virions to escape and enter the cytosol (49). This step has been hypothesized to occur via a pH-induced alteration in the hydrophobicity of the adenoviral capsid proteins which allows their interaction with the cell vesicle membrane (36). The virion then localizes to the nuclear pore, and its genome is translocated to the nucleus of the host cell (51). This understanding of the adenoviral entry pathway should facilitate attempts to modify the tropism of adenoviral vectors to permit the targeting of specific cell types.

IV. Adenoviral Cellular Binding and Internalization Can Be Uncoupled From Subsequent Steps in Infection

Modification of the tropism of the adenoviral vector so that it recognizes and binds to a novel receptor on specific target cells requires that the vector still be able to accomplish the distal steps of internalization and gene transfer. Data derived in our laboratory suggest that this goal may be achievable in the context of the adenovirus. We have explored methods to accomplish gene transfer via the receptor-mediated endocytosis pathway employing molecular conjugate vectors (55). The basic design of a molecular conjugate vector consists of plasmid

DNA attached to a macromolecular ligand which can be internalized by the cell type of interest. To accomplish this, a molecular conjugate vector contains two distinct functional domains: (1) a DNA-binding domain which is composed of a polycation such as polylysine; and (2) a ligand domain which binds to a specific cell surface receptor. Previous investigations employing this vector system had found that the efficiency of gene transfer was idiosyncratic due to entrapment of the conjugate-DNA complex in endosomes after internalization (56). To overcome this limitation, we incorporated a replication-deficient adenovirus into the conjugate design to capitalize on its ability to accomplish endosome disruption (57). It could be shown that incorporation of the adenovirus into the vector configuration dramatically augmented the gene transfer efficiency of the vector because the complex could avoid entrapment in the cell vesicle system (58,59). However, the introduction of the adenovirus into the system undermined one of the theoretical attributes of this vector: the ability to accomplish targeted, cell-specific gene delivery based upon the incorporated ligand domain. To overcome this limitation, we derived an antifiber antibody that could specifically block adenoviral binding and entry into target cells (55). It was hypothesized that coating the complex with antibody would block adenoviral binding, thus permitting targeted gene delivery exclusively via the ligand domain. In these studies, it was shown that the use of antibody-coated, binding-incompetent adenovirus did not decrease the overall levels of gene expression observed (Fig. 3). Thus, despite entry via an alternate internalization pathway, fiber binding was not required for the adenovirus to mediate endosomal vesicle disruption: in spite of being routed through a nonadenoviral binding pathway, the virus accomplished efficient postinternalization entry events. Hence, this study demonstrated that the processes of adenoviral binding and subsequent entry steps are not functionally linked. It should therefore be possible to reroute recombinant adenoviral vectors through heterologous cellular entry pathways in a similar manner while retaining their desirable downstream entry properties. This finding thus provides the impetus for our further exploration of adenoviral retargeting schemas.

V. Toward a Tropism-Modified Adenoviral Vector for Cell-Specific Gene Delivery

Successful adenoviral entry requires two distinct, sequential steps: binding of the adenoviral capsid to the cell surface, which is determined by the knob domain of the fiber protein, followed by a secondary interaction between the penton base and cellular integrins. The initial interaction between the fiber and its cognate receptor thus plays the major role in determining adenoviral tropism. This suggests that modifications of the fiber to permit the recognition

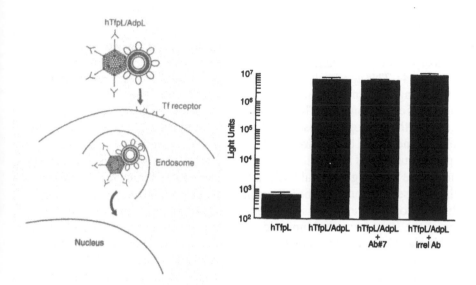

Figure 3 Effect of antifiber antibody on the ability of adenovirus-polylysine-DNA complexes to mediate gene transfer. Complexes were prepared containing adenovirus conjugated to human transferrin-polylysine-DNA. Prior to incorporation, the adenovirus was treated with either a neutralizing antifiber antibody or an irrelevant antibody (PY203). Complexes were delivered to HeLa cells, and cell lysates were evaluated for expression of the luciferase reporter gene. (From Ref. 59.)

of alternative specific cell surface molecules will be of primary importance in the development of targeted adenoviral vectors for gene therapy. However, it might also be necessary in certain instances to modify the penton base to permit viral internalization by the target cells.

A. Redirection of Adenoviral Cell-Binding Specificity by Modifications to the Fiber Protein

As described for retargeted retroviral vectors, adenoviral tropism could be altered by modifications of the cell-binding protein, the fiber, to permit the targeting of specific cell surface receptor molecules. In order to achieve this goal, there are two considerations. In the first instance, a new binding specificity must be introduced into the adenoviral fiber to allow recognition of cell surface markers characterizing the target cells. In addition, the ubiquity of the receptor for Ad2 and Ad5 means that ablation of endogenous viral tropism must be achieved to restrict gene transfer exclusively to the target cells.

Redirection of Adenoviral Cell-Binding Specificity by Genetic Methods

We have reported a strategy to genetically modify the adenovirus fiber protein by incorporating heterologous ligands at the carboxy terminus of the molecule (60) (Fig. 4). This approach capitalizes on the knowledge that the endogenous cell-binding domain of the adenovirus is localized within the carboxy terminal knob of the fiber protein. The aim was thus to localize the novel cell-binding ligand in the analogous position, thereby accomplishing two goals: (1) the novel ligand would be localized in the region of the endogenous ligand, likely a propitious site as relates to interaction with the cognate cellular receptor; and (2) the novel ligand would be removed from other adenoviral capsid proteins, whose function might be important in distal, postbinding entry functions. The addition of heterologous peptides to the fiber protein required consideration of the strict structural limitations of the fiber quaternary configuration. In this regard, the fiber protein is synthesized as a monomer which undergoes trimerization in the cytoplasm by virtue of intramolecular, noncovalent interactions, initiated at the carboxy terminus (61,62). The trimer is then transported to the nucleus (61), where viral particles are assembled. In this process, the amino terminus of the trimeric fiber associates with the penton base; fiber monomers cannot bind to the penton base (63). Thus, it is necessary to make modifications to the knob domain of the fiber without impairing trimer

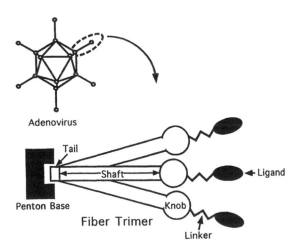

Figure 4   Strategy to incorporate heterologous peptide ligand in the knob domain of adenoviral fiber. This strategy involves the genetic modification of the fiber gene to generate a fiber chimera consisting of mature trimeric fiber, peptide linker, and an added physiologic ligand at the knob domain. (From Ref. 60.)

formation and preventing incorporation of chimeric fiber molecules into the mature adenoviral capsid. In addition to these considerations, it is important to achieve a final quaternary configuration whereby the incorporated ligand is localized on the exterior of the mature fiber trimer and is thus accessible for target cell binding. With these considerations in mind, a strategy was undertaken to create fiber-ligand fusion proteins by genetically incorporating into the fiber gene heterologous sequences encoding peptides with physiologic ligand function.

To create adenoviral fiber-peptide ligand fusion proteins, several cloning maneuvers were performed on the 3' end of the adenovirus type 5 fiber gene, corresponding to the carboxy terminus of the fiber protein (Fig. 5). First, site-directed mutagenesis was performed to introduce a unique *Bam*H1 restriction site in this region of the gene to facilitate cloning of the test ligand. Second, a flexible 10 amino acid [Pro(SerAla)₄Pro] hinge region (linker) was introduced

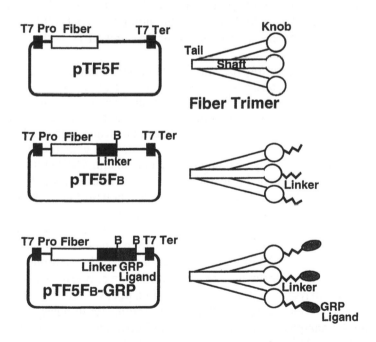

Figure 5 Expression vectors and schematic diagram of genetically modified fiber proteins. Three different expression vectors driven by the T7 promoter were used to transiently express the Ad5 fiber protein variants: pTF5F, which expresses the wild-type Ad5 fiber protein; pTF5FB, which expresses the modified fiber protein plus the linker; and pTF5FB-GRP, which expresses the modified fiber protein plus both the linker and GRP ligand. (From Ref. 60.)

between the carboxy terminus of the fiber protein and the peptide ligand to minimize any possible steric constraints and to present the test ligand extended away from the body of the fiber protein. Finally, we chose the terminal decapeptide of the gastrin-releasing peptide (GRP) as the initial test ligand owing to its small size and its ability to be internalized into its target cell by a receptor-mediated endocytosis pathway (64). An autocrine growth loop involving GRP has been implicated in the growth deregulation of small cell lung cancers in which the cell surface GRP receptors are overexpressed (65), making GRP a suitable ligand for lung cell-specific targeting.

The fiber fusion proteins were expressed in HeLa cells using a vaccinia virus expression system. HeLa cells were infected with a recombinant vaccinia virus expressing T7 RNA polymerase prior to lipofection with plasmids pTF5FB and pTF5FB-GRP, which contain fusion genes driven by the T7 promoter. To confirm the presence of a fiber-GRP fusion product, cell lysates prepared from vaccinia-infected HeLa cells which had been transfected with either pTF5FB or pTF5FB-GRP were immunoprecipitated with either a monoclonal anti-fiber antibody (4D2, 2A6, or AF7A) or a rabbit antihuman GRP antibody. 4D2 recognizes fiber monomers and trimers, while 2A6 and AF7A both recognize fiber trimers only.

As shown in Figure 6A, the anti-GRP antibody immunoprecipitates a protein that is of a size consistent with that of mature fiber and is also recognized by an antifiber antibody, 4D2. This immunoprecipitation product was identical in size to the product of immunoprecipitated fiber-GRP fusion protein with an antifiber antibody that recognizes fiber trimers only, AF7A. In contrast, when HeLa cell lysates were transfected with a construct encoding the fiber with the linker only, and no GRP ligand, an anti-GRP antibody was unable to precipitate a protein of similar size. This result indicated that the GRP peptide was indeed synthesized as part of a fusion protein with the adenovirus fiber. To confirm this result, HeLa cell lysates that were transfected with a plasmid containing the fiber-GRP fusion gene were first immunoprecipitated with different antifiber antibodies and then subjected to immunoblot analysis with an anti-GRP antibody. Figure 6B indicates that immunoprecipitated fiber-GRP fusion proteins were recognized by an anti-GRP antibody. In contrast, when cell lysates containing the modified fiber with the linker and no ligand were immunoprecipitated with different antifiber antibodies, subjected to immunoblot analysis and probed with an anti-GRP antibody, no antibody-specific bands could be detected. Thus, these studies demonstrate that the fusion gene construct we have derived is capable of directing the expression of a fiber-GRP fusion protein.

Correct fiber protein folding is absolutely required for incorporation of the fiber protein into the vertices of nascent adenovirus capsids. Since our aim is to construct a recombinant adenovirus with a genetically modified fiber, it

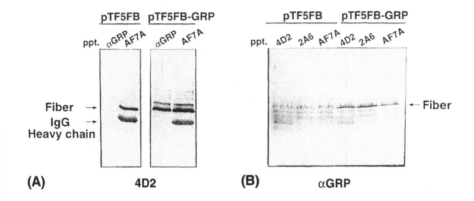

Figure 6 Evaluation of adenoviral fiber protein variants. (A) HeLa cell lysates transfected with plasmids encoding either the fiber variant containing the linker only (pTF5FB) or the fiber-GRP fusion protein (pTF5FB-GRP) were immunoprecipitated with either an anti-GRP polyclonal antibody (αGRP) or an antifiber antibody (AF7A) that recognizes Ad5 fiber trimers only. The precipitates were boiled, separated by SDS-PAGE, and transferred to solid membranes. The blots were then probed with an anti-fiber antibody (4D2) that recognizes both fiber monomers and trimers. The fiber-GRP fusion protein immunoprecipitated with an anti-GRP antibody was also recognized by an anti-fiber antibody, indicating the presence of a fiber-GRP fusion product. (B) HeLa cell lysates that had been transfected with pTF5FB and pTF5FB-GRP were immunoprecipitated with three antifiber antibodies: 4D2, which recognizes both fiber monomers and trimers; and 2A6 and AF7A, which recognize fiber trimers only. The precipitates were treated as before, and the blots were probed with an anti-GRP polyclonal antibody. The fiber-GRP fusion protein immunoprecipitated with antifiber antibodies was recognized by an anti-GRP antibody, indicating the presence of a fiber-GRP fusion product. (From Ref. 60.)

was first important to determine whether addition of exogenous peptides to the carboxy terminus of fiber still allowed proper fiber protein folding into the native quaternary configuration. To determine the quaternary structure of the fiber-GRP fusion protein, HeLa cell lysates transfected with pTF5F, pTF5FB, or pTF5FB-GRP were subjected to immunoblot analysis and probed with an antifiber antibody 4D2. On native gel analysis, boiled fiber protein migrates as a monomer, whereas in unboiled samples fiber migrates as a trimer. Figure 7 shows that boiled fiber-GRP fusion protein migrated in the monomeric form of the protein whereas unboiled fiber-GRP fusion protein migrated as a trimer. This result indicates that it is possible to add exogenous sequences to the carboxy terminus of the fiber protein, at least as large as 22 amino acids, without perturbing the quaternary structure of the protein. This result

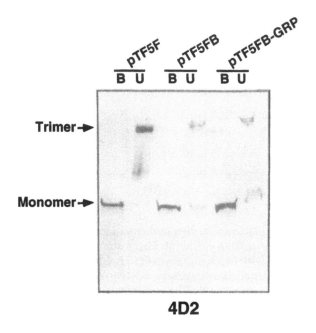

4D2

Figure 7 Determination of the quaternary structure of the fiber-GRP fusion protein. To determine whether the fiber-GRP fusion protein was able to form a trimer, as required for association with the penton base during the assembly of viral capsids, an immunoblot assay was performed. Boiled (B) and unboiled (U) lysates of cells transfected with plasmids encoding: wild-type fiber (pTF5F); fiber plus linker (pTF5FB); or fiber plus linker-GRP (pTF5FB-GRP) were separated by SDS-PAGE and transferred to a membrane. The blot was probed with antifiber antibody 4D2 which recognizes both fiber monomers and trimers. The result indicates that the fiber-GRP fusion protein trimerized. (From Ref. 60.)

demonstrates that the derived fiber-ligand fusion gene retains the requisite quaternary configuration characteristics required for its incorporation into assembled adenoviral capsids.

We have shown that it is possible to add short peptide sequences to the carboxy terminus of the adenovirus fiber without impairing either the biosynthesis or the proper folding of the protein. It was then necessary to determine whether the GRP ligand in the trimeric fiber-GRP fusion protein was exposed in its quaternary structure, which would suggest that it would be accessible to its cellular receptor. To answer this question, an immunoblot assay was performed in which boiled and unboiled HeLa cell lysates from cells transfected with pTF5FB or pTF5FB-GRP were probed with either 4D2, which recognizes

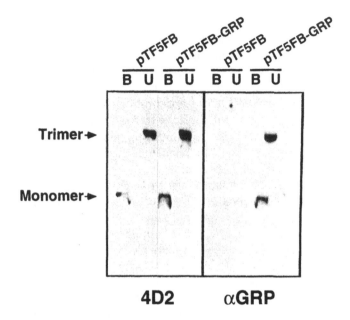

Figure 8 Accessibility of the GRP ligand in the fiber-GRP trimer. To determine whether the GRP ligand in the native form of the fiber-GRP fusion protein was accessible to binding, an immunoblot assay was performed. Boiled (B) and unboiled (U) HeLa cell lysates that had been transfected with the expression vector encoding either the modified fiber plus the linker (pTF5FB) or the fiber-GRP fusion protein were separated by SDS-PAGE and transferred to a membrane. The blot was probed with either antifiber antibody 4D2, which recognizes both fiber monomers and trimers, or with an anti-GRP antibody (αGRP). The results indicate that the GRP ligand in the fiber-GRP fusion protein is accessible to binding by an anti-GRP antibody in both the monomeric and trimeric forms of the protein. (From Ref. 60.)

both fiber trimers and monomers, or anti-GRP antibody. As shown in Figure 8, when boiled and unboiled lysates containing the modified fiber with the linker only (pTF5FB) were subjected to immunoblot analysis and probed with 4D2, a monomer and trimer band were detected for the boiled and unboiled samples, respectively. When the same samples were probed with anti-GRP antibody, no bands could be detected, owing to the absence of ligand in this construct. When boiled and unboiled cell lysates containing the fiber-GRP fusion protein (pTF5FB-GRP) were subjected to the same type of analysis and probed with 4D2, a monomer and trimer band were detected, respectively. When the same samples were probed with an anti-GRP antibody, a monomer band could be detected in the boiled sample and a trimer band could be detected in the unboiled sample. These results indicate that the GRP ligand in the fiber-GRP fusion protein is also exposed and accessible to binding in the native or trimeric form of the protein.

These preliminary studies demonstrated several key properties of the fiber-GRP fusion protein: (1) the protein retains its native configuration; (2) the protein retains its native biosynthesis profile; and (3) the protein presents the added ligand in an exterior, surface-exposed localization. These studies thus demonstrate the feasibility of introducing heterologous peptide ligands into the cell-binding domain of the adenoviral fiber protein in a manner consistent with the ultimate derivation of targeted adenoviral particles.

With a view to being able to target a variety of lung cells in a specific manner, alternative ligands will be employed in the construction of fiber fusion proteins. In this regard, the neuropeptide somatostatin and insulinlike growth factor I (IGF-I) have, like GRP, been reported to serve as autocrine growth factors for small cell lung cancers (66) and are therefore suitable ligands for targeting these cell types. Other ligands will be investigated for the purpose of targeting lung cell subsets not associated with malignant transformation. For example, we have shown that the binding patterns of lectins directed against the cell surface glycocalyx can be exploited to target molecular conjugate vectors to specific cells (67).

Thus we will explore a range of ligands, ranging in size from small peptides to larger molecules. In each case, it will be necessary to determine whether the incorporation of the ligand is compatible with trimerization of the chimeric fiber, which is required for association with the penton base in the assembly of viral particles. This analysis will provide information regarding the size constraints of the ligand that can be employed in this manner. The binding profile of purified chimeric fiber-ligand proteins will then be investigated to determine whether the ligand retains the ability to bind to its novel target receptor. In a further study, we will determine whether the fiber-ligand fusion protein retains the ability to recognize the native adenoviral receptor, which is ubiquitously expressed. If native binding of the fiber to its receptor is not ablated by the incorporation of the ligand, it will be necessary to pursue an alternative strategy to achieve this end, which is required to target the fiber-ligand chimeric protein specifically to receptors dictated by the ligand. In this regard, mutagenesis of the knob domain of the adenovirus fiber to ablate its ability to recognize its native receptor, while preserving its trimeric quaternary structure, would be of great utility. This approach will be facilitated by the work of Xia et al. (68), who have resolved the crystal structure of the Ad5 knob, identifying amino acid residues that form putative receptor-binding sites and that are therefore candidates for mutagenesis.

Thus we aim to identify recombinant fiber-ligand fusion proteins that (1) bind to the cognate cellular receptor for the ligand, and (2) fail to recognize the native adenoviral receptor. The next step will be the incorporation of the modified fibers into adenoviral particles to generate vectors targeted to specific cells. A number of approaches have been employed to generate adenoviral vectors by in vivo homologous recombination between noninfectious adenoviral genome constructs to yield an infectious viral genome capable of propagating progeny virions. Techniques reported to date have included the use of overlapping linear DNA constructs (69–71) and the use of plasmids (72). In the latter instance, a two-plasmid system involving recombination between a shuttle plasmid, containing foreign gene sequences, and a rescue plasmid, providing the required viral functions, has been widely employed (72). At present we are actively developing a strategy to generate adenoviral vectors with modified fibers by homologous recombination between a fiber-deleted rescue plasmid and a shuttle plasmid containing the fiber-ligand fusion gene. The tropism-modified adenoviral vectors will be evaluated for their ability to accomplish the three steps necessary for targeted expression of the transgene: (1) attachment to the target cells; (2) internalization; and (3) gene transfer.

Redirection of Adenoviral Cell-Binding Specificity by Immunological Methods

In an alternative approach to the generation of a targeted adenoviral vector, we hypothesized that the twin goals of ablation of native binding and introduction of novel binding could be achieved by employing a neutralizing anti-knob antibody chemically conjugated to a cell-specific ligand (Fig. 9). Therefore we generated a panel of monoclonal antibodies (mAbs) directed against the Ad5 knob. One mAb (designated 1D6.14), which possessed the ability to (1) recognize trimeric Ad5 knob in an enzyme-linked immunosorbent assay (ELISA), and (2) block infection of HeLa cells by recombinant adenoviruses, was chosen for further study. The Fab fragment of this mAb was generated by papain digestion, thereby removing the potential for the formation of large complexes by crosslinking of adenoviral particles by the parent immunoglobulin molecule. Functional analysis confirmed that the Fab fragment of 1D6.14 possessed the ability to neutralize adenovirus infection. For initial proof of concept we decided to target the folate receptor, which is overexpressed in several malignant cell lines, including approximately 50% of lung carcinomas (73). In addition, several authors have demonstrated that folate conjugates can be employed to deliver macromolecules, including DNA and protein, specifically to folate receptor-bearing cells (74–76). Therefore we conjugated folate to the Fab fragment of the neutralizing antiknob antibody 1D6.14. This conjugate was incubated with a recombinant adenoviral vector carrying the luciferase reporter gene prior to infection of target cells overexpressing the folate receptor. Whereas the unconjugated Fab fragment was shown to inhibit Ad

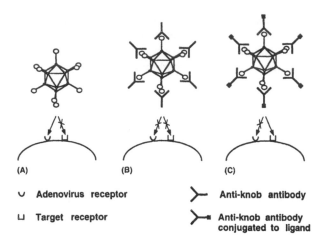

Figure 9 Strategy for immunological retargeting of adenoviral vector. (A) Adenoviral attachment to cells is accomplished by the high-affinity binding of the knob domain of the fiber to an as-yet-unidentified membrane surface receptor. (B) When complexed with a neutralizing antibody directed against the knob domain, the adenovirus is unable to bind to its cellular receptor. (C) Conjugation of a cell-specific ligand to the neutralizing antibody is hypothesized to permit binding to a novel target receptor on the cell surface. (Figure courtesy of Meizhen Feng and Victor Krasnykh.)

infection, the Fab-folate conjugate permitted infection by the adenoviral vector. This infection was shown to be directed specifically to the folate receptor, as demonstrated by competition experiments with free folate which ablated the ability of the adenovirus to infect the target cells when complexed with the Fab-folate conjugate. This work provides the initial validation of the concept that a conjugate between a fragment of a neutralizing antiknob antibody and a cell-specific ligand can retarget adenoviral binding. The range of lung cell lines that can be targeted by this immunological modification of adenoviral tropism will be expanded by conjugating other ligands to the neutralizing antibody. In this regard, this immunological strategy for modification of adenoviral tropism might not be subject to the constraints on the size of the ligand that are inherent in the genetic strategy to construct trimeric fiber-ligand chimeras.

B. Redirection of Adenoviral Internalization by Genetic Modifications to the Penton Base

Successful gene therapy is dependent on efficient gene transfer to the target cells. In the case of adenovirus-mediated gene transfer, this means that the vectors must be efficiently internalized by the target cells. Thus in addition to employing modifications to the fiber to permit adenoviral retargeting, it might also be necessary in certain instances to modify the RGD motif of the penton base to permit viral internalization by alternative secondary receptors on the target cells.

Preclinical and preliminary clinical studies in mice and humans suggest that some airway epithelia, such as the columnar epithelia of the nasal cavity, are resistant to adenoviral-mediated transduction, as demonstrated by inefficient transfer of the cDNA encoding the cystic fibrosis transmembrane regulator (77). A similar finding was reported by Dupuit et al. (78), who used a recombinant adenoviral vector carrying the *LacZ* reporter gene to infect different culture models of regenerating human nasal polyp surface epithelium. They found that transduced cells were not detected in normal pseudostratified areas of the ex vivo nasal polyp tissue, but were found in areas of regenerating tissue. In a separate study, Goldman and Wilson (79) employed a human bronchial xenograft model to study the biology of gene transfer to the human proximal pulmonary conducting airway by adenoviral vectors. These investigators demonstrated that the state of differentiation of the epithelia played a significant role in the efficiency of adenoviral-mediated gene transfer. Undifferentiated epithelia, which express high levels of $\alpha_v\beta_5$ integrins, were readily infected with recombinant adenoviruses. In contrast, pseudostratified epithelia, which do not express $\alpha_v\beta_5$ in differentiated columnar cells, were relatively resistant to adenovirus infection. Moreover, the low level of gene transfer was not inhibited by an RGD peptide or anti-$\alpha_v\beta_5$ antibody, suggesting that it is independent of α_v integrin binding to the penton base.

Thus, the fully differentiated pseudostratified epithelium of the proximal pulmonary conducting airway is relatively refractory to adenovirus infection because of the absence of $\alpha_v\beta_5$ integrins. This suggests that it will be necessary to exploit alternative pathways of viral entry to achieve efficient adenovirus-mediated gene transfer to these cells. In this regard, Wickham et al. (80) have demonstrated that wild-type adenovirus serotype 2 penton base which normally binds to $\alpha_v\beta_3$ and $\alpha_v\beta_5$ integrins to mediate viral internalization can be targeted to new receptors (80). This was achieved by substituting the wild-type RGD motif with peptide motifs specific for other integrins. Sequence comparisons between different adenovirus serotypes indicate that the RGD sequence is located in a hypervariable region of the penton base (52). Moreover, analysis of the secondary structure of this domain predicts that the RGD motif is positioned at the apex between two alpha helices (52,81). Wickham et al. (80) thus rationalized that the RGD region was an ideal site to accommodate potentially large peptide sequences to redirect the specificity of the penton base for cellular receptors. To test this concept, they constructed a chimeric penton base in which the RGD motif was replaced with a peptide motif specific

for integrin $\alpha_4\beta_1$ which is expressed at high levels on lymphocytes and monocytes but is absent from epithelial and endothelial cells. In a second chimera,
the wild-type sequences flanking the RGD motif were mutated, ablating its
interaction with $\alpha_v\beta_5$ while retaining its specificity for $\alpha_v\beta_3$. The chimeric penton bases were produced in insect cells using a baculovirus expression system.
These genetic modifications to the penton base were shown to redirect binding
to the novel target receptors while preventing recognition of the wild-type
secondary adenovirus receptor. However, the authors did not demonstrate
whether the chimeric penton bases were capable of assuming the authentic
pentameric quaternary structure required for assembly of adenoviral particles.
Nevertheless, these results provide an encouraging demonstration that it will
be possible to improve the efficiency of adenovirus infection of normally resistant lung cells, and hence the efficiency of adenoviral-mediated gene transfer, by modifications to the penton base.

VI. Conclusion

The development of a targeted adenoviral vector able to accomplish lung cell-
specific gene delivery would be of great value for gene therapy for diseases of
the lung. Attempts to modify adenoviral tropism must be based on an understanding of the viral entry pathway and preserve the ability of the vector to
perform the three steps of cellular attachment, internalization, and gene transfer. In this regard, the primary interaction between the adenoviral fiber and
its cellular receptor plays the major role in the determination of adenoviral
tropism. Therefore we are exploring genetic and immunological strategies to
modify the binding specificity of the fiber to permit the targeting of novel
receptors. Following binding, viral internalization is mediated by a secondary
interaction between the adenoviral penton base and α_v integrins on the cell
surface. Since an absence of appropriate integrin receptors on certain cells is
correlated with a low efficiency of adenoviral infection, strategies to modify
the penton base to permit internalization by alternative pathways will be of
utility in these instances. Thus a combination of approaches to modify the
receptor specificities of the adenovirus fiber and penton base should permit
the generation of tropism-modified adenoviral vectors capable of targeted,
high-efficiency infection of specific cell types.

Acknowledgments

The authors acknowledge their co-workers in the University of Alabama at
Birmingham Gene Therapy Program: Meizhen Feng, Victor Krasnykh, Galina
Mikheeva, Sharon Michael, Ryan Miller, and Buck Rogers. We appreciate the

financial support provided by grants from the National Institutes of Health, R01 5025505, US Army, DAMD 17-94-J-4398, and the Muscular Dystrophy Association.

References

1. Curiel D, Pilewski J, Albeda S. Gene therapy approaches for inherited and acquired lung diseases. Am J Respir Cell Mol Biol 1996; 14:1–18.
2. Siegfried W. Perspectives in gene therapy with recombinant adenoviruses. Exp Clin Endocrinol 1993; 101:7–11.
3. Wilson J. Gene therapy for cystic fibrosis: challenges and future directions. J Clin Invest 1995; 96:2547–2554.
4. Trapnell B, Gorziglia M. Gene therapy using adenoviral vectors. Curr Opin Biotech 1994; 5:617–625.
5. Engelhardt J, Yang Y, Stratford-Perricaudet L, et al. Direct gene transfer of human CFTR into human bronchial epithelia of xenografts with E1-deleted adenoviruses. Nature Genet 1993; 4:27–33.
6. Huard J, Lochmuller H, Ascadi G, Jani A, Massie B, Karpati G. The route of administration is a major determinant of the transduction efficiency of rat tissues by adenoviral recombinants. Gene Ther 1995; 2:107–115.
7. Yang Y, Nunnes F, Berencsi K, Furth E, Gonczol E, Wilson J. Cellular immunity to viral antigens limits E1-deleted adenoviruses for gene therapy. Proc Natl Acad Sci USA 1994; 91:4407–4411.
8. Dai Y, Schwarz E, Gu D, Zhang W-W, Sarvetnick N, Verma I. Cellular and humoral immune responses to adenoviral vectors containing factor IX gene: tolerization of factor IX and vector antigens allows for long-term expression. Proc Natl Acad Sci USA 1995; 92:1401–1405.
9. Smith T, Mehaffey M, Kayda D, et al. Adenovirus mediated expression of therapeutic plasma levels of human factor IX in mice. Nature Genet 1993; 5:397–402.
10. Yei S, Mittereder N, Tang K, O'Sullivan C, Trapnell B. Adenovirus-mediated gene transfer for cystic fibrosis: quantitative evaluation of repeated in vivo vector administration to the lung. Gene Ther 1994; 1:192–200.
11. Kozarsky K, McKinley D, Austin L, Raper S, Stratford-Perricaudet L, Wilson J. In vivo correction of low density lipoprotein receptor deficiency in the Watanabe heritable hyperlipidemic rabbit with recombinant adenoviruses. J Biol Chem 1994; 269:13695–13702.
12. Yang Y, Li Q, Ertl H, Wilson J. Cellular and humoral immune responses to viral antigens create barriers to lung-directed gene therapy with recombinant adenoviruses. J Virol 1995; 69:2004–2015.
13. Yang Y, Nunes F, Berencsi K, Gonczol E, Engelhardt J, Wilson J. Inactivation of E2a in recombinant adenoviruses improves the prospect for gene therapy in cystic fibrosis. Nature Genet 1994; 7:362–369.
14. Engelhardt J, Ye X, Doranz B, Wilson J. Ablation of E2A in recombinant adenoviruses improves transgene persistence and decreases inflammatory response in mouse liver. Proc Natl Acad Sci USA 1994; 91:6196–6200.

15. Engelhardt J, Litzky L, Wilson J. Prolonged transgene expression in cotton rat lung with recombinant adenoviruses defective in E2a. Human Gene Ther 1994; 5:1217– 1229.

16. Yang Y, Trinchieri G, Wilson J. Recombinant IL–12 prevents formation of blocking IgA antibodies to recombinant adenovirus and allows repeated gene therapy to mouse lung. Nature Med 1995; 1:890–893.

17. Kolls J, Lei D, Odom G, et al. Use of transient CD4 lymphocyte depletion to prolong transgene expression of E1-deleted adenoviral vectors. Human Gene Ther 1996; 7:489–497.

18. Douglas J , Curiel D. Targeted gene therapy. Tumor Target 1995; 1:67–84.

19. Hunter E , Swanstrom R. Retrovirus envelope glycoproteins. Curr Top Microbiol Immunol 1990; 157:187–253.

20. Weiss R, Boettiger D, Murphy H. Pseudotypes of avian sarcoma viruses with the envelope properties of vesicular stomatitis virus. Virology 1977; 76:808–825.

21. Dong J, Roth M, Hunter E. A chimeric avian retrovirus containing the influenza virus hemagglutinin gene has an expanded host range. J Virol 1992; 66:7374–7382.

22. Wilson C, Reitz M, Okayama H, Eiden M. Formation of infectious hybrid virions with gibbon ape leukemia virus and human T-cell leukemia virus retroviral envelope glycoproteins and gag and pol proteins of Moloney murine leukemia virus. J Virol 1989; 63:2374–2378.

23. Emi N, Friedmann T, Yee J-K. Pseudotype formation of murine leukemia virus with the G protein of vesicular stomatitis virus. J Virol 1991; 65:1202–1207.

24. Russell S, Hawkins R, Winter G. Retroviral vectors displaying functional antibody fragments. Nucleic Acids Res 1993; 21:1081–1085.

25. Chu T-H, Martinez I, Sheay W, Dornburg R. Cell targeting with retroviral vector particles containing antibody-envelope fusion proteins. Gene Ther 1994; 1:292–299.

26. Chu T-H , Dornburg R. Retroviral vector particles displaying the antigen-binding site of an antibody enable cell-type-specific gene transfer. J Virol 1995; 69:2659–2663.

27. Somia N, Zoppe M, Verma I. Generation of targeted retroviral vectors by using single-chain variable fragment: an approach to in vivo gene delivery. Proc Natl Acad Sci USA 1995; 92:7570–7574.

28. Valsesia-Wittmann S, Drynda A, Deleage G, et al. Modifications in the binding domain of avian retrovirus envelope protein to redirect the host range of retroviral vectors. J Virol 1994; 68:4609–4619.

29. Kasahara N, Dozy A, Kan Y. Tissue-specific targeting of retroviral vectors through ligand-receptor interactions. Science 1994; 266:1373–1376.

30. Han X, Kasahara N, Kan Y. Ligand-directed retroviral targeting of human breast cancer cells. Proc Natl Acad Sci USA 1995; 92:9747–9751.

31. Cosset F-L, Morling F, Takeuchi Y, Weiss R, Collins M, Russell S. Retroviral retargeting by envelopes expressing an N-terminal binding domain. J Virol 1995; 69:6314–6322.

32. Valsesia-Wittmann S, Morling F, Nilson B, Takeuchi Y, Russell S, Cosset F-L. Improvement of retroviral retargeting by using amino acid spacers between an additional binding domain and the N terminus of Moloney murine leukemia virus SU. J Virol 1996; 70:2059–2064.

33. Neda H, Wu C, Wu C. Chemical modification of an ecotropic murine leukemia virus results in redirection of its target cell specificity. J Biol Chem 1991; 266: 14143–14146.
34. Roux P, Jeanteur P, Piechaczyk M. A versatile and potentially general approach to the targeting of specific cell types by retroviruses: application to the infection of human cells by means of major histocompatibilty complex class I and class II antigens by mouse ecotropic murine leukemia virus-derived viruses. Proc Natl Acad Sci USA 1989; 86:9079–9083.
35. Etienne-Julan M, Roux P, Carillo S, Jeanteur P, Piechaczyk M. The efficiency of cell targeting by recombinant retroviruses depends on the nature of the receptor and the composition of the artificial cell-virus linker. J Gen Virol 1992; 73:3251–3255.
36. Seth P, FitzGerald D, Willingham M, Pastan I. Pathway of adenovirus entry into cells. In: Crowell R, Lonberg-Holm K, eds. Virus Attachment and Entry Into Cells. Washington, DC: American Society for Microbiology, 1986:191–195.
37. Philipson L, Lonberg-Holm K, Pettersson U. Virus-receptor interaction in an adenovirus system. J Virol 1968; 2:1064–1075.
38. Wickham T, Mathias P, Cheresh D, Nemerow G. Integrins $\alpha v\beta 5$ and $\alpha v\beta 5$ promote adenovirus internalization but not virus attachment. Cell 1993; 73:309–319.
39. Defer C, Belin M-T, Caillet-Boudin M-L, Boulanger P. Human adenovirus-host cell interactions: comparative study with members of subgroups B and C. J Virol 1990; 64:3661–3673.
40. van Oostrum J , Burnett R. Molecular composition of the adenovirus type 2 virion. J Virol 1985; 56:439–448.
41. Ruigrok R, Barge A, Albiges-Rizo C, Dayan S. Structure of adenovirus fiber. II. Morphology of single fibres. J Mol Biol 1990; 215:589–596.
42. Boudin M-L , Boulanger P. Assembly of adenovirus penton base and fiber. Virology 1982; 116:589–604.
43. Stewart P, Burnett R, Cyrlaff M, Fuller S. Image reconstruction reveals the complex molecular organization of adenovirus. Cell 1991; 67:145–154.
44. Stewart P, Fuller S, Burnett R. Difference imaging of adenovirus: bridging the resolution gap between X-ray crystallography and electron microscopy. EMBO J 1993; 12:2589–2599.
45. Green N, Wrigley N, Russell W, Martin S, McLachlan A. Evidence for a repeating cross-β sheet structure in the adenovirus fiber. EMBO J 1983; 2:1357–1365.
46. Henry L, Xia D, Wilke M, Deisenhofer J, Gerard R. Characterization of the knob domain of the adenovirus type 5 fiber protein expressed in *E. coli*. J Virol 1994; 68:5239–5246.
47. Louis N, Fender P, Barge A, Kitts P, Chroboczek J. Cell-binding domain of adenovirus serotype 2 fiber. J Virol 1994; 68:4104–4106.
48. Chardonnet Y , Dales S. Early events in the interaction of adenoviruses with HeLa cells. Virology 1970; 40:462–477.
49. FitzGerald D, Padmanabhan R, Pastan I, Willingham M. Adenovirus-induced release of epidermal growth factor and *Pseudomonas* toxin into the cytosol of KB cells during receptor-mediated endocytosis. Cell 1983; 32:607–617.

50. Varga M, Weibull C, Everitt E. Infectious entry pathway of adenovirus type 2. J Virol 1991; 65:6061–6070.
51. Greber U, Willetts M, Webster P, Helenius A. Stepwise dismantling of adenovirus 2 during entry into cells. Cell 1993; 1993:477–486.
52. Bai M, Harfe B, Freimuth P. Mutations that alter an Arg-Gly-Asp (RGD) sequence in the adenovirus type 2 penton base protein abolish its cell-rounding activity and delay virus reproduction in flat cells. J Virol 1993; 67:5198–5205.
53. Huang S, Endo R, Nemerow G. Upregulation of integrins $\alpha v\beta 3$ and $\alpha v\beta 5$ on human monocytes and T lymphocytes facilitates adenovirus-mediated gene delivery. J Virol 1995; 69:2257–2263.
54. Ascadi G, Jani A, Massie B, et al. A differential efficiency of adenovirus-mediated in vivo gene transfer into skeletal muscle cells of different maturity. Human Mol Genet 1994; 3:579–584.
55. Michael S, Huang C, Romer M, Wagner E, Hu P, Curiel D. Binding-incompetent adenovirus facilitates molecular conjugate-mediated gene transfer by the receptor-mediated endocytosis pathway. J Biol Chem 1993; 268:6866–6869.
56. Cotten M, Wagner E, Zatloukal K, Phillips S, Curiel D, Birnstiel M. High-efficiency receptor-mediated delivery of small and large (48 kilobase) gene constructs using the endosome disruption activity of defective or chemically inactivated adenovirus particles. Proc Natl Acad Sci USA 1992; 89:6094–6098.
57. Curiel D, Agarwal S, Wagner E, Cotten M. Adenovirus enhancement of transferrin-polylysine-mediated gene delivery. Proc Natl Acad Sci USA 1991; 88:8850–8854.
58. Curiel D, Wagner E, Cotten M, et al. High-efficiency gene transfer mediated by adenovirus coupled to DNA-polylysine complexes. Human Gene Ther 1992; 3: 147–154.
59. Curiel D. High-efficiency gene transfer employing adenovirus-polylysine-DNA complexes. Nat Immun 1994; 13:141–164.
60. Michael S, Hong J, Curiel D, Engler J. Addition of a short peptide ligand to the adenovirus fiber protein. Gene Ther 1995; 2:660–669.
61. Hong J, Engler J. The amino terminus of the adenovirus fiber protein encodes the nuclear localization signal. Virology 1991; 185:758–767.
62. Novelli A, Boulanger P. Deletion analysis of functional domains in baculovirus-expressed adenovirus type 2 fiber. Virology 1991; 185:365–376.
63. Novelli A, Boulanger P. Assembly of adenovirus type 2 fiber synthesized in cell-free translation system. J Biol Chem 1991; 266:9299–9303.
64. Batley J, Wada E. Two distinct receptor subtypes for mammalian bombesin-like peptides. Trends Neurosci 1991; 14:524–528.
65. Sausville E, Lebacq-Verheyden A, Spindel E, Cuttitta F, Gazdar A, Battey J. Expression of the gastrin-releasing peptide in human small cell lung cancer. J Biol Chem 1984; 261:2451–2459.
66. Moody T, Cuttitta F. Growth factor and peptide receptors in small cell lung cancer. Life Sci 1993; 52:1161–1173.
67. Batra R, Wang-Johanning F, Wagner E, Garver R, Curiel D. Receptor-mediated gene delivery employing lectin-binding specificity. Gene Ther 1994; 1:255–260.

68. Xia D, Henry L, Gerard R, Deisenhofer J. Crystal structure of the receptor-binding domain of adenovirus type 5 fiber protein at 1.7 A resolution. Structure 1994; 2:1259–1270.
69. Ballay A, Levrero M, Buendia M, Tiollais P, Perricaudet M. In vitro and in vivo synthesis of the hepatitis B virus surface antigen and of the receptor for polymerized human serum albumin from recombinant human adenoviruses. EMBO J 1985; 4:3861–3865.
70. Berkner K, Schaffhausen B, Roberts T, Sharp P. Abundant expression of polyomavirus middle T antigen and dihydrofolate reductase in an adenovirus recombinant. J Virol 1987; 61:1213–1220.
71. Davidson D, Hassell J. Overproduction of polyomavirus middle T antigen in mammalian cells through the use of an adenovirus vector. J Virol 1987; 61:1226–1239.
72. McGrory W, Bautista D, Graham F. A simple technique for the rescue of early region I mutations into infectious human adenovirus type 5. Virology 1988; 163: 614–617.
73. Mattes M, Major P, Goldenberg D, Dion A, Hutter R, Klein K. Patterns of antigen distribution in human carcinomas. Cancer Res 1990; 50(suppl):880S–884S.
74. Leamon C, Low P. Delivery of macromolecules into living cells: a method that exploits folate receptor endocytosis. Proc Natl Acad Sci USA 1991; 88:5572–5576.
75. Gottschalk S, Cristiano R, Smith L, Woo S. Folate receptor mediated DNA delivery into tumor cells: potosomal disruption results in enhanced gene expression. Gene Ther 1994; 1:185–191.
76. Kranz D, Patrick T, Brigle K, Spinella M, Roy E. Conjugates of folate and anti-T-cell-receptor antibodies specifically target folate-receptor-positive tumor cells for lysis. Proc Natl Acad Sci USA 1995; 92:9057–9061.
77. Grubb B, Pickles R, Ye H, et al. Inefficient gene transfer by adenovirus vector to cystic fibrosis airway epithelia of mice and humans. Nature 1994; 371:802–806.
78. Dupuit F, Zahm J-M, Pierrot D, et al. Regenerating cells in human airway surface epithelium represent preferential targets for recombinant adenovirus. Human Gene Ther 1995; 6:1185–1195.
79. Goldman M, Wilson J. Expression of αvβ5 integrin is necessary for efficient adenovirus-mediated gene transfer in the human airway. J Virol 1995; 69:5951–5958.
80. Wickham T, Carrion M, Kosvedi I. Targeting of adenovirus penton base to new receptors through replacement of its RGD motif with other receptor-specific peptide motifs. Gene Ther 1995; 2:750–756.
81. Mathias P, Wickham T, Moore M, Nemerow G. Multiple adenovirus serotypes use αv integrins for infection. J Virol 1994; 68:6811–6814.

Part Four

DISEASE-DIRECTED GENE THERAPY

12

Toward Correction of the Genetic Defect in Cystic Fibrosis

LARRY G. JOHNSON and RICHARD C. BOUCHER

The University of North Carolina at Chapel Hill
Chapel Hill, North Carolina

I. Introduction

Advances in molecular cloning and genetic mouse models have led to the discovery of many genes linked to clinical disease. Gene therapy is a natural product of this research, and its application to treatment of disease is under way. Cystic fibrosis (CF) is an attractive target for gene therapy owing to the high morbidity and mortality of CF lung disease and the absence of adequate therapy. In this article, we will review pertinent issues affecting the efficiency, efficacy, and safety of gene therapy of CF lung disease derived from preclinical studies and ongoing clinical trials with recombinant adenoviruses, liposomes, and adeno-associated virus (AAV) vectors.

II. Clinical Overview

Cystic fibrosis is a common monogenic recessive disorder affecting individuals of Northern European ancestry including approximately 30,000 individuals in the United States. CF lung disease is characterized by thick, viscous airway secretions, impaired mucociliary clearance, chronic bacterial infection, bronchiectasis, and premature death (1,2). Antimicrobial therapy combined with better nutritional supplementation has led to dramatic increases in survival with an increase in median survival from less than one year of age in 1940 to ~29 years of age today (3).

CF affects a variety of epithelial tissues including airway, gastrointestinal, pancreatic, intrahepatic biliary ductal, and sweat ductal epithelia (4–6). Respiratory symptoms including cough, pneumonia, and wheezing can present early in life in CF patients and have been reported in ~50% of CF patients who are diagnosed by the age of 6 months (7). Moreover, bronchoalveolar lavage studies in infants suggest inflammation manifested as increased IL–8 levels, and neutrophilia and abnormal secretions (airway casts) may occur even in the absence of detectable airway pathogens (8,9). Over time, the cough becomes continuous and productive of viscous, purulent, and often greenish-colored sputum leading to a pattern of lung disease characterized by periods of exacerbation with partial recovery of lung function following treatment with antibiotics, and a gradual loss of function over a period of many years. The earliest lung function abnormalities are characterized by increased RV/TLC ratios, reduced maximal midexpiratory flow rates (V_{max} 25–75), and hyperinflation on chest x-ray, suggestive of small-airways disease (10–12). Evidence of luminal mucous impaction, bronchial cuffing, and bronchiectasis has also been detected, particularly on computed tomography (CT) scans of the chest (13,14). Characteristic of CF sputum microbiology is the presence of staphylococcus aureus and hemophilis influenza early (6 to 9 months of age) in the disease, which are replaced by *Pseudomonas aeruginosa* as the disease progresses. In the end stages of disease, patients typically have multiply resistant *Pseudomonas aeruginosa*, frequently in a mucoid form; and *Burkholderia (Pseudomonas) cepacia* may also be a late finding.

Gastrointestinal tract manifestations are among the earliest manifestations of the disease, with approximately 10% of patients presenting in the first 24 hours of life with gastrointestinal obstruction termed meconium ileus. Meconium ileus equivalent (distal intestinal obstruction) occurs in children and young adults. This entity is characterized by right lower quadrant pain, loss of appetite, emesis, and often a palpable mass, and can be confused with appendicitis. Intestinal abnormalities are frequently complicated by pancreatic insufficiency in many CF patients due to insufficient pancreatic enzyme secretion with consequent malabsorption of fat and fat-soluble vitamins. Because pancreatic islet cells are typically spared, hyperglycemia and a requirement for insulin present late in the disease (6).

Hepatobiliary disease is the second leading cause of death in CF. Expression of hepatobiliary disease is variable and includes cholestasis, focal biliary and multilobular cirrhosis, portal hypertension, and hepatic failure (5,15,16). While some researchers speculate that the frequency of overt liver disease may increase as survival increases in CF, other studies suggest that presentation of clinically overt disease peaks in adolescence (15,16).

In the sweat gland, normal volumes of sweat are secreted. Because CF patients are unable to absorb sodium chloride from sweat as it moves through the sweat duct, CF sweat is characterized by raised sweat chloride (Cl^-) and sodium (Na^+) concentrations (4,17). This observation has long been the foundation for the sweat chloride concentration as a diagnostic test for the disease.

Late-onset puberty is common to males and females with CF, presumably secondary to the effects of chronic lung disease and inadequate nutrition (2,18). More than 95% of male patients with CF are azospermic, reflecting obliteration of the vas deferens, and 20% of CF women are infertile due to abnormalities of cervical mucus. However, 90% of completed pregnancies produce viable infants, and CF women are generally able to breastfeed infants normally.

III. Molecular Pathophysiology

A. The Gene

More than 500 mutations have been reported since the cloning of the CF gene, the cystic fibrosis transmembrane conductance regulator (*CFTR*) gene, in 1989 (19,20). The most common mutation in *CFTR* is a 3-bp deletion leading to a deletion of phenylalanine (F) at position 508 (ΔF508) of the protein product which is present on ~70% of CF chromosomes (1,21,22). This mutation leads to abnormal intracellular processing and trafficking of mutant CFTRs such that CFTR is degraded prior to reaching the cell membrane. Failure of mutant CFTR to reach the membrane causes defective apical membrane chloride (Cl^-) conductance in epithelial cells affected by this disorder (23). Alleles G551D, G542X, 621+1G→T, R117H, W1282X, N1303K, and Δ3849+10 kb C→T account for another 10% to 20% of mutations (1,21,22). Four different classes of CFTR mutations have been proposed (21), including defective protein production (class I), defective processing (class II), defective regulation (class III), and defective conduction (class IV). ΔF508 falls within class II, G551D into class III, R117H into class IV, and G542X into class I. Interestingly, class IV mutations tend to be associated with pancreatic sufficiency, a milder disease course, and, frequently, congenital bilateral absence of the vas deferens (CBAVD), especially R117H (21,22).

B. CFTR Localization

CFTR is expressed in most human epithelial tissues including airway epithelia, intestinal crypts, kidney tubules, intrahepatic biliary ducts, and pancreatic ducts

(24–29). CFTR expression in the lung in utero is high in airways and absent in airway submucosal glands (30,31). In contrast, CFTR expression in the adult lung is highest in the serous cells of submucosal glands and in submucosal gland duct cells in the cartilaginous airways (29). Expression in the superficial airways is generally low (1 to 2 mRNA transcripts per cell) except for the ~1% to 10% of airway cells in the distal airways that express high levels of CFTR (25,29,32). Alveolar epithelia in adult lung do not express appreciable amounts of CFTR.

C. CFTR Function

Current evidence suggests that CFTR is a cAMP-mediated Cl^- channel (33–36). However, CFTR also performs several regulatory functions including (1) regulation of the outwardly rectifying Cl^- channel [ORCC (37,38)] and the alternative Cl^- conductance Cl_a (39), the putative channel stimulated by calcium-mediated agonists, and (2) regulation of the epithelial Na^+ channel [ENaC (40)]. Regulation of ENaC has special relevance since raised Na^+ absorption, a feature of CF airway epithelia, leads to the raised airway transepithelial potential difference measured in CF patients (41–44). The expression of the ENaC subunits (α, β) in both serous and mucous gland cells suggest the possible presence of Na^+ hyperabsorption in submucosal glands as well (45).

Patch clamp studies suggest that the mechanisms responsible for Na^+ hyperabsorption in CF airway epithelia reflect, in part, an increase in apical cell membrane Na^+ channel activity (46). Because relative levels of ENaC subunit (α, β, γ) mRNAs in superficial airway epithelial cells are similar in CF and normal airway epithelia (45), the increase in Na^+ channel activity does not appear to arise from an increase in channel number. Rather, wild-type CFTR acts as a cAMP-regulated inhibitor of Na^+ channel activity permitting an increased rate of transepithelial Na^+ transport when CFTR, the inhibitor, is absent or mutated (40).

D. Mechanisms of Disease Pathogenesis

The mechanisms by which these defects in Cl^- and Na^+ transport in the superficial airways combine with lack of CFTR function in submucosal glands to predispose CF airways to chronic infection by *Staphylococcus aureus* and *Pseudomonas* have not been elucidated. Prince and colleagues have suggested that increased binding of *Pseudomonas* to the increased number of asialylated receptors in CF epithelia stimulates the epithelial cells to secrete IL–8, which initiates the cycle of inflammation and infection in CF lungs (47,48). This theory does not take into consideration the relationship to the demonstrated ion transport defects in CF, nor does it account for the role of other bacterial organisms that precede *Pseudomonas* colonization in CF airways.

Another theory suggests that excessive Na$^+$ absorption and defective Cl$^-$ secretion in CF lead to early obstruction of bronchioles and submucosal gland ducts. Defective secretion of a variety of antibacterial and anti-inflammatory compounds from serous cells of submucosal glands and distal airways, including lysozyme, secretory IGA, peroxidase, and antileukoproteases, may subsequently promote proliferation of bacterial pathogens such as *Pseudomonas* within the airway lumen (49).

Recent studies have suggested alternative mechanisms of pathogenesis of CF lung disease. In a study by Mizgerd et al. (50), neutrophil bactericidal activity and phagocytosis was inhibited by low extracellular Na$^+$ concentrations. Extracellular Na$^+$ concentrations may also be low in CF airways causing defective neutrophil function with subsequent proliferation of bacteria such as *Haemophilus influenzae*, *Staphylococcus aureus*, and *Pseudomonas aeruginosa* in CF airways. Inhibition of Na$^+$ transport rates by wild-type CFTR, such as occurs in non-CF airways, would be expected to maintain airway surface fluid Na$^+$ concentrations at levels permitting normal function of airway neutrophils and, hence, better bacterial killing and clearance.

In contrast, a study by Smith et al. suggests that high extracellular Cl$^-$ concentrations resulting from defective Cl$^-$ secretion combined with Na$^+$ hyperabsorption within the surface fluid lining CF airways inhibits bactericidal activity of a heat-stable, low-molecular-weight, defensinlike molecule secreted onto the airway surfaces (51). In the absence of defective Cl$^-$ secretion and Na$^+$ hyperabsorption, as occurs in non-CF airways (52,53), the Cl$^-$ concentration is significantly lower than in CF, permitting efficient killing of low colony counts of inhaled and/or aspirated bacterial organisms such as *Pseudomonas aeruginosa* and *Staphylococcus aureus*. Importantly, adenovirus-mediated expression of CFTR in a well-differentiated CF primary airway epithelial cell cultures restored bactericidal activity to the surface fluid with subsequent efficient killing of *Pseudomonas* applied to the apical surface of the transduced CF cells.

In a third study, defective internalization of *Pseudomonas* by CF airway epithelial cells relative to CF cells transduced with wild-type CFTR has been proposed as a method limiting clearance of *Pseudomonas aeruginosa* from the airways (54). In this study, transformed human CF airway epithelial cells internalized significantly fewer bacteria than CF airway epithelial cells that had been stably transduced with wild-type CFTR. This decreased uptake of bacteria into CF airway cells was specific for *Pseudomonas aeruginosa* and could be ameliorated by growing the cells at 26°C, which allows the mutant CFTR to reach the membrane. Furthermore, *Pseudomonas aeruginosa* variants with an incomplete LPS structure, such as might arise during chronic infection with these organisms, appeared to further impair internalization of *Pseudomonas* in CF airway cells. This study suggests that airway epithelial cells may play a role in the phagocytic uptake and clearance of *Pseudomonas* from human airways that is defective when CFTR is not expressed in the membrane. However, phagocytosis of bacteria, a feature previously attributed to poorly differentiated or wound-repairing cells, may not be a feature of well-differentiated, pseudostratified columnar epithelia (55) such that the benefits of bacterial internalization on airway clearance of these organisms would not occur. Furthermore, the specificity of this mechanism for *Pseudomonas* does not incorporate the roles of bacterial pathogens that precede *Pseudomonas* infection, nor does it account for the role of ion transport defects in the disease.

IV. Therapeutic Approaches to Gene Therapy

A. Rationale and Targets for CF Gene Therapy

Because CF is an autosomal-recessive disorder in which heterozygotes exhibit a normal phenotype, introduction of a single wild-type (normal) copy of the gene into the defective CF cell should restore the normal phenotype. Introduction of a wild-type (normal) copy of *CFTR* into CF airway epithelial cells using retrovirus, vaccinia virus, and adenoviral vectors has been shown to restore CFTR-mediated Cl- transport function to these cells, consistent with this concept (39,56,57). Furthermore, Olsen and colleagues have demonstrated persistence of CFTR-mediated Cl- transport function in cultured CF epithelial cells for up to 6 months following retrovirus-mediated gene transfer (58). These studies demonstrating restoration of normal Cl⁻ transport following in vitro gene transfer of wild-type (i.e., normal) CFTR have established the feasibility of gene therapy for CF.

Yet, the in vivo cellular targets for CF gene therapy in humans have not been clearly elucidated. Whether CF lung disease starts in the superficial ciliated columnar epithelial cells lining the lumen of the small airways or the serous cells of submucosal glands remains controversial. The site of disease initiation is relevant since strategies for gene therapy of CF that target the superficial columnar airway epithelium by luminal delivery may not effectively target the glands, while strategies that target the basolateral membrane of the superficial epithelia (basal cells) and the submucosal glands via intravenous delivery may present more of a challenge for airway delivery. Current strategies primarily target the airways via the lumen.

B. Preclinical Studies

Current approaches to gene therapy of cystic fibrosis have primarily utilized adenoviruses, cationic liposomes, and AAV to achieve gene transfer. Although other vectors are under development, we shall focus in this review on those vectors that have undergone sufficient investigation to permit use and/or con-

Table 1 Clinical Gene Transfer Safety and Efficacy Trials in Cystic Fibrosis

Principal investigator(s)	Vector	Description of trial
Crystal	Adenovirus	A phase I study, in CF patients, of the safety, toxicity, and biological efficacy of a single administration of a replication-deficient recombinant adenovirus carrying the cDNA of the normal human cystic fibrosis transmembrane conductance regulator gene in the lung
Crystal	Adenovirus	Evaluation of repeat administration of a replication-deficient recombinant adenovirus containing the normal CF transmembrane conductance regulator cDNA to the airways of individuals with CF
Welsh, Smith	Adenovirus	Cystic fibrosis gene therapy using an adenovirus vector: in vivo safety and efficacy in nasal epithelium
Welsh	Adenovirus	Adenovirus-mediated gene transfer of CFTR to the nasal epithelium and maxillary sinus of patients with CF
Wilson	Adenovirus	Gene therapy of cystic fibrosis lung disease using E1-deleted adenoviruses: a phase I trial
Boucher, Knowles	Adenovirus	Gene therapy for CF using E1-deleted adenovirus: a phase I trial in the nasal cavity
Wilmott, Whitsett	Adenovirus	A phase I study of gene therapy of CF utilizing a replication-deficient recombinant adenovirus vector to deliver the human CF transmembrane conductance regulator cDNA to the airways
Dorkin	Adenovirus	Adenovirus-mediated gene transfer for CF: safety of single administration in the lung (lobar instillation)
Dorkin	Adenovirus	Adenovirus-mediated gene transfer for CF: safety of single administration in the lung (aerosol administration)
Bellon	Adenovirus	Respiratory epithelium, in vivo
Alton, Geddes	Cationic liposomes	Liposome-mediated CFTR gene transfer to the nasal epithelium of patients with CF
Welsh, Zabner	Cationic liposomes	Cationic lipid-mediated gene transfer of CFTR: safety of a single administration to the nasal epithelia

(continued)

Table 1 Continued

Principal investigator(s)	Vector	Description of trial
Sorscher, Logan	Cationic liposomes	Gene therapy for CF using cationic liposome-mediated gene transfer: a phase I trial of safety and efficacy in the nasal airway
Porteus, Innes	Cationic liposomes	DOTAP liposome delivery of gene therapy for CF—a phase I trial in the human nose
Flotte	Adeno-associated virus	A phase I study of an adeno-associated virus-CFTR gene vector in adult CF patients with mild lung disease
Gardner	Adeno-associated virus	A phase I/II study of tgAAV-CF for the treatment of chronic sinusitis in patients with CF

sideration for use in Phase I or II human clinical trials (Table 1). The general features of these vectors have been extensively reviewed elsewhere (59,60) and in preceding chapters of this volume. The relevant features of these vectors can generally be grouped with regard to the following: efficiency, efficacy, and safety.

Efficiency

Efficiency of gene transfer has been defined as the percent or fraction of cells transduced. An in vitro study by Johnson and colleagues suggested that as few as 6% to 10% of CF airway epithelial cells must be corrected in order to restore maximal CFTR-mediated Cl^- transport function to an entire epithelial sheet (61). In contrast, correction of all the cells in an epithelial sheet may be necessary to normalize Na^+ hyperabsorption (39). Because it is not known whether correction of the Cl^- or Na^+ transport defects will correct or prevent CF lung disease, transduction of all the columnar cells in the epithelium with a low level of endogenous CFTR, mimicking endogenous expression of CFTR, may be the best strategy.

Ad Vectors

Adenovirus (Ad) vectors have dominated gene therapy efforts for cystic fibrosis in the United States. Efficient Ad-mediated gene transfer to primary human airway epithelia in vitro and to cotton rat airway epithelia in vivo has routinely been reported (62–66). However, in vivo Ad-mediated gene transfer to the airway epithelia of nonhuman primates tends to be inefficient and patchy (67). These results can perhaps be explained by a study by Grubb et al. (68), which suggests that Ad vectors transduce different cell types in vivo with different efficiencies. Using a model of mechanical injury, an Ad-*lacZ* vector efficiently transduced basal cells which were the predominant cell type at the site of mechanical injury in human and mouse tracheal explants, whereas columnar cells in undamaged areas were resistant to gene transfer. A recent study by Pickles et al. (69) has confirmed this observation in model systems of well-differentiated rat and human airway epithelia and extended it to human intrapulmonary (bronchial) airways. Adenoviruses bind to an unidentified high affinity cell surface receptor with cellular uptake mediated by $\alpha_v\beta_5$ integrins (70) which may not be expressed in well differentiated columnar cells, limiting cellular entry and hence efficient gene transfer (71).

These data are consistent with the observations of Dupuit et al. (72) in excised human airway specimens demonstrating preferential transduction of undifferentiated regenerating or wound repairing cells by Ad vectors, but not well-differentiated pseudostratified columnar epithelia. However, these studies contradict an earlier study by Mastrangeli et al. (73), which reported in vivo transduction of columnar cells in cotton rats, but this discrepancy may have reflected either species differences or transduction of basal cells that subsequently differentiated into columnar cells. Studies in in vitro and in vivo model systems of differentiated airway epithelia suggest that increasing the duration of Ad vector incubation with differentiated epithelia may increase the efficiency of gene transfer presumably through nonspecific mechanisms (74,75).

The desire to prevent development of lung disease in infants and children has led several investigators to examine the feasibility of gene transfer to the fetal airways. In a study by Ballard et al. (76), instillation of an Ad-*lacZ* vector to major bronchi of 20–24 week gestation human fetal lungs permitted efficient gene transfer to epithelial cells of airways and terminal saccules within 24 hours. Interestingly, infection of airway explant cultures from these fetal lungs demonstrated transgene expression primarily in the peripheral or wound repairing type cells consistent with the findings of Dupuit et al. (72). McCray and colleagues have reported relatively efficient (transduction of ~18% of cells in the trachea) Ad-mediated gene transfer to fetal lamb airways following intratracheal administration in utero (77). Unfortunately, expression was limited by acute morphologic and inflammatory responses (see safety section below). Efficient Ad-mediated gene transfer to fetal rat airways following delivery into the amniotic fluid has also been reported (78).

Because adenovirus is a transient expression system, efficiency of gene transfer with repetitive administration of vector is a key issue. The generation of neutralizing antibody following initial airway infection with adenovirus has been shown to inhibit subsequent infection of susceptible cells with this vector-limiting gene transfer (79–82). Recently, alternate dosing with Ad vectors from different subgroups has been proposed as a strategy to circumvent the anti-Ad humoral response limiting gene transfer with repetitive administration (83). In these experiments, intratracheal administration of wild-type Ad5 (subgroup C), but not wild-type Ad4 (subgroup E) or wild-type Ad30 (subgroup D), prevented subsequent gene transfer mediated by intratracheal administration of an Ad5-*CAT* or Ad5-*lacZ* vector (subgroup C) 7 days later. An alternative strategy to enable consistent gene transfer with repetitive administration of Ad vectors is the inhibition of neutralizing antibody production by immunosuppressive therapy. Yang et al. (84) have reported that intratracheal and/or intraperitoneal delivery of IL–12 and interferon gamma (IFNγ) markedly reduced (60-fold) production of vector specific neutralizing and IgA antibodies in BAL fluid of mice infected with an Ad5 vector. Furthermore, co-administration of IL-12 and IFNγ with an Ad-*lacZ* vector permitted successful gene transfer 28 days after previous intratracheal dosing with an Ad-*ALP* (alkaline phosphatase) vector.

Cationic Liposomes

Efficient cationic liposome-mediated gene transfer has been reported in vitro in a variety of cell types (85–89) and in the alveolar region of the lung in vivo (90). However, the efficiency of cationic liposome-mediated gene transfer to well differentiated airway epithelia in vivo is low and patchy. Studies designed to identify rate limiting factors for liposome-mediated gene transfer in undifferentiated cell lines in vitro suggest that nuclear (but not cellular) entry is rate-limiting in nontransfectable undifferentiated cells (91). In contrast, gene transfer to well-differentiated airway epithelial cells is limited by failure of DNA-liposome complexes to enter the cell, preventing efficient gene transfer (55). This inability to enter differentiated airway epithelial cells arises from a loss of phagocytic entry mechanisms in differentiated columnar airway epithelial cells (55). The issue of repetitive dosing with this transient expression vector system has not been addressed in detail.

Adeno-Associated Virus Vectors

Efficient dose-dependent gene transfer to transformed airway epithelial cells has been reported with this vector (92–93). However, AAV vectors have been shown to transduce primary airway epithelia much less efficiently than immortalized airway cells (94). The decreased transduction efficiency was not due to differences in vector entry since similar quantities of vector were detectable in primary and immortalized airway cells by PCR 2 days postinfection. Rather, vector genomes in primary airway epithelial cells remained episomal, unintegrated, and single-stranded. The inefficient conversion of single-stranded episomes to double-stranded DNA, a requirement for transgene expression, in this study has raised questions regarding the proper interpretation of the immunohistochemical detection of CFTR in rabbit airways transduced with AAV-*CFTR* in vivo (95). The lack of highly specific anti-CFTR antibodies contributes to the controversy.

Efficacy

Functional correction of the CFTR-mediated Cl⁻ permeability defect has been used as a working definition of efficacy following transduction of cells by wild-type human CFTR. Detection of CFTR protein by immunohistochemistry and transduced mRNA by in situ hybridization and RT-PCR are important confirmatory measures of gene expression, especially when gene transfer is below the level of detection by functional assays.

Adenoviruses

Ad-mediated correction of the CF Cl⁻ permeability defect in a CF pancreatic cell line in vitro has been reported (66). Ad-mediated correction of the Cl⁻ permeability defect in primary human CF airway epithelia at a multiplicity of infection (MOI; number of infectious particles per cell) as low as 0.1 to 1 has also been reported (64). In contrast, correction or normalization of raised Na^+ transport in primary CF airway epithelia by Ad vectors required a high MOI and transduction of the virtually all of the CF cells within the epithelial sheet (39). These data are consistent with a report of Ad-mediated *CFTR* gene transfer in a mouse model of CF (68). In this model, partial correction (~50%) of the Cl⁻ transport defect and no correction of Na^+ transport was detected in the nasal epithelium following in vivo Ad-mediated *CFTR* gene transfer despite a high MOI and repetitive daily administration (four consecutive days) of vector.

Cationic Liposomes

Recently, novel cationic lipids have been reported in preliminary studies to correct the CFTR-mediated Cl⁻ transport defect in CF airway epithelia in vitro (96,97). The level of correction detected corresponds to an MOI of 10 in cells infected with an Ad-*CFTR* vector. Two reports also suggest that cationic liposome-mediated gene transfer can restore CFTR-mediated Cl⁻ transport to airways in transgenic (knockout) mouse models of cystic fibrosis (98,99). In a study performed in rats, cationic liposome-mediated transduction of the *lacZ* cDNA has produced high-efficiency transduction of rat airways in vivo following direct airway instillation (100). Instillation of CFTR plasmid DNA-liposome complexes in this same study significantly enhanced forskolin (cAMP)-stimulated currents over baseline currents measured in untransfected control and lacZ-treated tracheas.

AAV

Despite a limitation in the size of the cDNA insert (~4.5 kb), in vitro correction of the CF Cl⁻ permeability defect has been reported in CF airway epithelial cells using AAV vectors encoding wild-type and/or truncated CFTR cDNAs (37,101). Flotte and co-workers (95) have also reported expression of CFTR from an AAV vector for up to 6 months by immunohistochemistry and by RT-PCR following bronchoscopic delivery to rabbit lung.

Safety

Adenoviruses

Several safety concerns have been raised with Ad vectors, including an immune/inflammatory responses, ectopic expression, and integration. Ad vectors have been shown to induce a dose-related lymphocyte predominant inflammatory response in murine, cotton rat, and baboon lungs (102–107). The inflammatory response is mediated by cytotoxic (CD8) T-cells and may be induced, in part, by production of viral gene products. Despite the E1 deletion, first-generation vectors have been shown to express late gene products, e.g., hexon. Second-generation, E2a-defective vectors reduce, but do not completely eliminate, late gene expression (105–107). These E2a defective vectors have been associated with a modest reduction in the inflammatory response, but a longer duration of the inflammation. By reducing the inflammatory response, E2a defective vectors have been associated with more prolonged transgene expression in nonhuman primates and in rodents (105–107).

Second-generation E1-deleted Ad2 vectors have also been developed in which most of the E4 region has been deleted, except for open reading frame 6 (ORF6), in an attempt to block all late gene expression. Recent studies in nonhuman primates examining toxicity with repetitive administration (every 3 weeks for up to 11 doses) of this vector suggest that at doses of up to 3×10^9 infectious units (IU) delivered to a single lobe of lung, histopathologic changes of inflammation are minimal (108). However, doses of vector $> 3 \times 10^9$ IU generated the expected histologic changes of inflammation. Recently, an adenoviral packaging cell line that deletes all of E1 and E4, including ORF6, has been reported (109), and efforts are under way to develop vectors that have all of the early and late region genes deleted.

Delivery of Ad vectors to fetal animals is also associated with an inflammatory response. McCray et al. (77) have demonstrated that intratracheal delivery of Ad2 vectors encoding *lacZ* and *CFTR* transgenes in utero was associated with reactive hyperplasia and squamous metaplasia 3 days post vector delivery, a mononuclear cell inflammatory infiltrate 7 days post vector delivery, and persistence of vector in amniotic fluid for up to 7 days post vector delivery, suggestive of replication. The inflammatory cell infiltrate was surprising since fetal animals have generally been presumed to be immune-tolerant hosts.

Neurogenic inflammation has recently been described with intra-airway administration of Ad vectors (110–112). A 2.5-fold greater increase in airway vascular permeability post capsaicin stimulation of airway sensory nerves has been reported in rat airways infected in vivo with an E1, E3-deleted Ad vector (Ad5-CMV*lacZ*) compared to rat airways sham-infected with vehicle (110). This effect was dose-dependent and could be reduced, but not abolished, by UV-psoralen inactivation of the Ad5-CMV*lacZ* vector (111), consistent with inhibition of viral gene expression. Inhibition of the potentiation in capsaicin-stimulated airway vascular permeability in rats dosed with vector by a selective Substance P (NK1) receptor antagonist (112) confirmed that this phenomenon was mediated by neurogenic inflammation. It also suggested that NK receptor antagonists may play a role in modulating the acute inflammatory response induced by in vivo administration of Ad vectors in the lung.

Neutralizing antibodies to group C adenovirus (the serogroup for Ad5 and Ad2) are common in human serum. Furthermore, circulating neutralizing antibodies and mucosal IgA antibodies have been identified following intra-tracheal and intranasal administration of Ad vectors to rodents (79–82). These antibodies bind Ad vector upon subsequent dosing, inhibiting effective gene transfer with repetitive administration. As discussed above, IL–12 and IFNγ coadministration with vector can significantly decrease production of neutralizing antibodies (84). Interestingly, anti-β-gal antibodies to transduced lacZ, a foreign protein in mammalian tissues, have been identified in rodents following administration of Ad-*lacZ* vectors (79). Whether transduced CFTR in CF individuals functions as an antigen for the induction of anti-CFTR antibodies has not been determined.

The ability of Ad vectors to infect many different cell types raises concern with regard to ectopic expression of CFTR. A study in fibroblasts, a cell type that does not express CFTR endogenously, demonstrated a slower growth rate and different electrophysiological properties, i.e., depolarized membrane potentials, following expression of CFTR at moderate to high levels when compared to fibroblasts that expressed CFTR at low levels or not at all (113). In contrast, transgenic mice that overexpress CFTR in pulmonary epithelia do not differ from control mice in their lung development, somatic growth characteristics, or reproductive function (114). These data suggest that the adverse effects from overexpression of CFTR may be manifested primarily in cells of nonepithelial origin that lack distinct apical and basolateral domains.

Adenoviruses have been used to efficiently transform human and rodent cell lines at frequencies up to 100-fold greater than obtained by transfection with calcium phosphate (115). Transformation has been mediated through the E1 region genes, although substitution of simian virus 40 (SV40) sequences for the E1 genes will also enable stable transformation. The presence of E1a sequences in low copy numbers (<1/cell) in 13% of airway samples from CF patients and 21% of airway samples from non-CF individuals raises concerns that integration of Ad sequences in airway epithelia could potentially complement E1-deleted vectors (116). However, since E4 sequences are only rarely detectable and L2 sequences are never detectable, it has been hypothesized that the E1a sequences detected represent fragments and may not be sufficient to support replication of E1 deleted Ad vectors. In vitro studies by Olsen et al. (117) suggest that integration of full-length vectors may occur in airway cells following Ad-mediated gene transfer at frequencies of up to 1 in 2000. Given the requirement for repetitive dosing with in vivo Ad-mediated gene transfer of CFTR, the risk of insertional mutagenesis may be higher than previously believed.

Infection of cultured primary human airway epithelial cells with an E1, E3-deleted first-generation Ad5 vector encoding the lacZ cDNA (Ad5-CMV*lacZ*) has been shown to induce apoptosis (programmed cell death) and a decrease in recruitment of cells into S phase of the cell cycle compared to control cells sham-infected with vehicle (118). The induction of apoptosis and cell cycle alterations detected were dose-dependent and present even at an MOI as low as 1. These phenomena were unique since these effects have previously been attributed to interactions of E1a and E1b regions with p53 (119–122) which have been removed from the Ad5-CMV*lacZ* vector. Furthermore, UV inactivation of Ad5-CMV*lacZ* abolished vector induced apoptosis and cell cycle alteration in cultured primary human airway epithelial cells, suggesting that adenoviral genes other than E1a and E1b may be responsible for these effects. Because the chronically inflamed airway epithelium in CF patients has a greater percentage of proliferating cells than that observed in normal airways (123), any inhibition of the cell proliferative response by Ad vectors may be harmful to reparative responses in CF airways. Increased amounts of apoptosis occurring in Ad-infected cells may also contribute to the limited duration of transgene expression that occurs with first-generation Ad vectors in vivo.

Cationic Liposomes

Liposomes have generally been associated with a good safety profile in vivo. Canonico et al. (124) have examined the toxicity profile of cationic liposome-mediated expression of the human α_1-AT gene in rabbit lung in vivo following both aerosol and intravenous delivery using the cationic liposome DOTMA/DOPE (Lipofectin) complexed to 500 μg of α_1-AT plasmid DNA in a 1:5 (w/w) DNA:lipid ratio. Expression, detected by Northern blot and immunohistochemistry, was observed in endothelium, smooth muscle, ciliated airway epithelial cells, and in alveolar macrophages following IV delivery. Expression in alveolar and airway epithelium was greater following aerosol dosing than intravenous dosing. Importantly, no adverse effects of weekly injections of 500 μg of a pCMV human α_1-AT plasmid complexed to 2500 μg of Lipofectin were detectable by lung histology, lung compliance and resistance measurements, or measurements of gas exchange over 4 weeks. This study confirms the notion that conventional cationic lipids used in human clinical studies are safe. However, the low-toxicity profile (125,126), which has generally made liposome-mediated gene transfer attractive, may not be a feature of the newer lipids which in preliminary studies, induce a dose dependent acute neutrophil predominant inflammatory response following intratracheal administration to murine lung (127). Should these newer preparations supplant the more conventional liposomes, the relative toxicity of these lipids following in vivo delivery to the lung by intracheal administration versus aerosol delivery, which may confine vector solely to the airways, will need to be addressed.

AAV

In vitro studies suggest that AAV vectors, unlike wild-type AAV which integrates site specifically (128,129), integrate into multiple random sites (130) so that a finite risk of insertional mutagenesis persists with this vector. In other in vitro systems, persistent expression has been linked to episomal expression. To date, immunogenic and inflammatory responses have not been reported in preclinical studies in the lungs of animals dosed with AAV vectors (131).

C. Clinical Trials

Clinical gene transfer safety and efficacy trials in airway epithelia of CF patients have been initiated in the United States and in Europe. The majority of these trials utilize Ad vectors to express CFTR, although the number of trials of liposome and AAV-mediated gene transfer to have recently been increasing (Table 1).

Adenovirus Vectors

Of the adenovirus vector trials (Table 1), the initial five studies at five different centers were Phase I trials evaluating single dosing of Ad-*CFTR* vectors in the nasal and/or lower airway epithelia of CF patients (132–137). Two trials have been initiated which will evaluate the feasibility of repetitive dosing of Ad vectors (138–139). Trials investigating the safety and efficacy of aerosolized adenovirus vector administration have also been initiated (Table 1).

Data have now been published from three different trials. In an uncontrolled, unblinded study of three patients, Zabner et al. (132) reported correction of the Cl⁻ transport defect in the nasal epithelium of CF subjects at estimated MOIs of 1, 3, and 25. They also reported expression of CFTR in these patients by RNA-specific (RS)-PCR in two of the three subjects studied at MOIs of 3 and 25. The major criticism of this study has been the nasal potential difference technique used to measure Cl⁻ secretion does not rigorously discriminate CF from normal airways in vivo (140).

Crystal and colleagues (133) have reported mRNA expression by RT-PCR and immunohistochemical detection of CFTR from bronchial brushings in one out of four (not the same patient) patients in their study of Ad-mediated gene transfer to nasal airway epithelia followed 24 hours later by delivery to the bronchial epithelium. Of note, one patient at a dose of 2×10^9 pfu delivered to the right bronchus developed a systemic and local inflammatory syndrome starting ~12 to 24 hours after CFTR administration characterized by headache, fatigue, fever, tachycardia, hypotension, pulmonary infiltrates, and a decrease in lung function. Associated with this clinical syndrome was a greater increase in IL–6 levels in this individual relative to the increase in serum IL–6 levels in the other three study subjects post Ad-*CFTR* vector administration. This particular subject was treated with broad-spectrum antibiotics, antipyretics, nasal oxygen, and IV fluids with resolution of clinical signs and symptoms by 14 days, although the chest radiographic abnormalities persisted for up to 25 days and lung function did not return to baseline for 30 days. More recently, these same investigators have reported their functional data from administration of Ad-*CFTR* to the nasal epithelium of CF patients in an unblinded protocol at doses ranging from 2×10^5 to $2 \times 10^{8.5}$ pfu (141). Partial correction of both Na⁺ hyperabsorption and Cl⁻ secretion was detected by the nasal PD technique when averaged over 14 days. Surprisingly, a dose-dependent relationship was not apparent in this particular study.

In a double-blinded, vehicle controlled, dose escalation study by Knowles and colleagues (142), transduced CFTR-mRNA was detected by RT-PCR and/or in situ hybridization in five of the six patients at the highest doses (MOI 100 and 1000) and in one of six patients at lower MOIs (MOI ≤ 10). No functional restoration of CFTR-mediated Cl⁻ transport or normalization of Na⁺ transport was detected using the nasal potential difference technique. The functional data suggested that the proportion of cells transduced was less than 1%, consistent with morphometric analysis of in situ hybridization studies of mucosal biopsies from these subjects. The advent of mucosal inflammation in the Ad5-CB*CFTR* dosed nostril in two of three patients at MOI=1000 precluded further increases in the dose of vector instilled. In vivo studies in rats suggest that the mucosal inflammation noted above may have reflected vector-induced neurogenic inflammation (110–112). A 15-fold increase in neutralizing antibody titer was detected in one of the high-dose patients.

The data from published reports would suggest that the efficiency of Ad-mediated transduction of CFTR in the nasal epithelium of CF patients is low. Moreover, safety concerns have now been raised in humans with this vector system. Further improvements in Ad-mediated gene transfer efficiency will likely be required to safely achieve sufficient efficacy.

Cationic Liposomes

Four trials of cationic liposome-mediated CFTR gene transfer to the nasal epithelium of CF patients have been initiated or proposed (Table 1). Only one study has been published to date. Caplen et al. (143) reported a double-blinded, placebo-controlled trial in which nine CF subjects received CFTR plasmid DNA (pSV-CFTR) complexed with DC-Chol/DOPE liposomes in a 1:5 (w/w) DNA to lipid ratio (143). Six CF subjects received only DC-Chol/DOPE liposomes to the nasal epithelia. The doses of DNA utilized in the study were 10, 100, and 300 μg plasmid DNA delivered by nasal spray. The highest dose was delivered in 200-μl aliquots to each nostril every 10 min requiring a total time of 7.5 hours. Nasal potential difference measurements revealed a mean hyperpolarization of the nasal potential difference (more negative potential difference) following low Cl⁻ perfusion in patients receiving the CFTR plasmid DNA-liposome complex that was ~20% of that measured in normals. No differences in nasal PDs between CF controls dosed with liposomes only and CF patients dosed with DNA-liposome complexes were detected following treatment with isoprenaline, a cAMP-mediated agonist. RT-PCR detected vector-derived CFTR mRNA in nasal biopsies of five of eight patients who received the DNA liposome complex, but was also positive in one of five patients who received placebo (liposomes only). Importantly, no toxicity was observed. In a separate publication, these investigators have reported that DC-Chol/DOPE liposomes without plasmid DNA, when delivered to the nasal epithelium in six normal and three CF subjects, did not alter nasal ion transport parameters or lung function, or alter antibiotic sensitivities of CF sputum bacterial isolates (144).

Although no results have been reported to date, several other clinical trials are well under way. Based on the limited published data from the liposome trials thus far, improvements in gene transfer efficiency will likely be required if they are to be of therapeutic benefit.

AAV

Two clinical safety and efficacy trials of AAV vectors for CF gene therapy have been developed (Table 1). No data have been reported from these studies.

V. Conclusions

Advances in understanding the molecular basis of acquired and genetic diseases have raised hopes for innovative treatments such as gene therapy. Despite an increasing number of clinical gene transfer protocols for CF (Table 1), the pathophysiologic mechanisms by which CFTR dysfunction causes disease and how best to reverse this process remains undefined. A precise definition of the appropriate site (airways versus glands) for targeting gene transfer is a first requirement. Following the assumption that the superficial epithelium is the target, the initial human trials indicate that transduction efficiency of current gene transfer vectors in vivo is low. Thus, a major focus of future research will be to define the barriers for inefficient gene transfer and to incorporate these concepts into new vector design. In this process, each patient studied offers insight into factors affecting gene transfer efficiency and efficacy while illustrating the challenges that must be overcome for gene therapy of CF to become a clinical reality.

Acknowledgments

The authors wish to thank Ms. Julia Morris, Ms. Susan Boyles, and Ms. Elizabeth Godwin for their assistance in the preparation of this manuscript.

References

1. Tsui LC, Buchwald M. Adv Human Genet 1991; 20:153–166.
2. Boat TF, Welsh MJ, Beaudet AL. Cystic fibrosis. In: Scriver CR, Beaudet AL, Sly WS, Valle D, eds. The Metabolic Basis of Inherited Disease. New York: McGraw-Hill, 1989:2649–2680.
3. Fiel SB, FitzSimmons S, Schidlow D. Evolving demographics of cystic fibrosis. Sem Respir Crit Care Med 1994; 15(5):349–355.
4. Quinton P. Cystic fibrosis: a disease in electrolyte transport. Faseb J 1990; 4:2709–2717.
5. di Sant' Agnese PA, Hubbard VS. The hepatobiliary system. In: Taussig LM, ed. Cystic Fibrosis. New York: Thieme-Stratton, 1984:296–322.
6. Handwerger S, Roth J, Gorden P, di Sant' Agnese PA, Carpenter DF, Peter G. Glucose intolerance in cystic fibrosis. N Engl J Med 1969; 281:451–461.
7. Cystic Fibrosis Foundation National CF Patient Registry. 1994 Annual Data Report. Bethesda, Md: Cystic Fibrosis Foundation, 1995.
8. Khan TZ, Wagener JS, Bost T, Martinez J, Accurso F.J, Riches DW. Early pulmonary inflammation in infants with cystic fibrosis. Am J Respir Crit Care Med 1995; 151:1075–1082.
9. Armstrong DS, Grimwood K, Carzino R, Carlin JB, Olinsky A, Phelan PD. Lower respiratory infection and inflammation in infants with newly diagnosed cystic fibrosis. Br Med J 1995; 310:1571–1572.
10. Levison H, Godfrey S. Pulmonary aspects of cystic fibrosis. In: Mangos JA, Talamo RE, eds. Cystic Fibrosis: Projections Into the Future. New York: Stratton Intercontinental Medical Book Corp, 1976:3–24.
11. Brasfield D, Hicks G, Soong SJ, Peters J, Tiller RE. Evaluation of a scoring system of the chest radiograph in cytic fibrosis: a collaborative study. Am J Roentgenol 1980; 134:1195–1198.

12. Tepper RS, Montgomery GL, Ackerman V, Eigen H. Longitudinal evaluation of pulmonary function in infants and very young children with cystic fibrosis. Pediatr Pulmonol 1993; 16:96–100.

13. Stiglbauer R, Schurawitzki H, Eichler I, Vergesslich KA, Gotz M. High resolution CF in children with cystic fibrosis. Acta Radiol 1992; 33:548–553.

14. Lynch DA, Brasch RC, Hardy KA, Webb WR. Pediatric pulmonary disease: assessment with high resolution ultrafast CT. Radiology 1990; 176:243–248.

15. Scott-Jupp R, Lama M, Tanner MS. Prevalence of liver disease in cystic fibrosis. Arch Dis Child 1991; 66:698–701.

16. Feigelson J, Anagnostopoulos C, Poquet M, Pecau Y, Munck A, Navarro J. Liver cirrhosis in cystic fibrosis—therapeutic implications and long term follow up. Arch Dis Child 1993; 68:653–657.

17. Quinton PM. Defective epithelial ion transport in cystic fibrosis. Clin Chem 1989; 35: 726–730.

18. Kaplan E, Schwachman H, Perlmutter AD, Rule A, Knaw KT, Holsclaw DS. Reproductive failure in males with cystic fibrosis. N Engl J Med 1968; 279:65–69.

19. Rommens JM, Ianuzzi MC , Kerem B-S, et al. Identification of the cystic fibrosis gene: chromosome walking and jumping. Science 1989; 245:1059–1065.

20. Riordan JR, Rommens JM, Kerem, B-S, et al. Identification of the cystic fibrosis gene: cloning and characterization of complementary DNA. Science 1989; 245: 1066–1073.

21. Welsh MJ, Smith AE. Molecular mechanisms of CFTR chloride channel dysfunction in cystic fibrosis. Cell 1993; 73:1251–1254.

22. Novelli G, Sangiuolo F, Maceratesi P, Dallapiccola B. The up-to-date molecular genetics of cystic fibrosis. Biomed Pharmacother 1994; 48:455–463.

23. Cheng SH, Gregory RJ, Marshall J, et al. Defective intracellular transport and processing of CFTR is the molecular basis of most cystic fibrosis. Cell 1990; 63:827–834.

24. Cohn JA, Melhus O, Page LJ, Dittrich KL, Vigna SL. CFTR: development of high-affinity antibodies and localization in sweat gland. Biochem Biophys Res Commun 1991; 181:36–43.

25. Trapnell BC, Chu CS, Paakko PK, et al. Expression of the cystic fibrosis transmembrane conductance regulator gene in the respiratory tract of normal individuals and individuals with cystic fibrosis. Proc Natl Acad Sci USA 1991; 88:6565–6569.

26. Crawford I, Maloney PC, Zeitlin PL, et al. Immunocytochemical localization of the cystic fibrosis gene product CFTR. Proc Natl Acad Sci USA 1991; 88:9262–9266.

27. Marino CR, Matovcik LM, Gorelick FS, Cohn JA. Localization of the cystic fibrosis transmembrane conductance regulator in pancreas. J Clin Invest 1991; 88:712–716.

28. Denning GM, Ostedgaard LS, Cheng SH, Smith AE, Welsh MJ. Localization of cystic fibrosis transmembrane regulator in chloride secretory epithelia. J Clin Invest 1992; 89:339–349.

29. Englehardt JF, Yankaskas JR, Ernst SA, et al. Submucosal glands are the predominant site of CFTR expression in human bronchus. Nature Genet 1992; 2:240–247.

30. Trezise AEO, Chambers JA, Wardle CJ, Gould S, Harris A. Expression of the cystic fibrosis gene in human foetal tissues. Human Mol Genet 1993; 2:213–218.

31. Tizzano EF, O'Brodovich H, Chitayat D, Benichou J-C, Buchwald M. Regional expression of CFTR in developing human respiratory tissues. Am J Respir Cell Mol Biol 1994; 10:355–362.
32. Jacquot J, Puchelle E, Hinnrasky C, et al. Localization of the cystic fibrosis transmembrane conductance regulator in airway secretory glands. Eur Respir J 1993; 6:169–176.
33. Anderson MP, Gregory RJ, Thompson S, et al. Demonstration that CFTR is a chloride channel by alteration of its anion selectivity. Science 1991; 253:202–205.
34. Bear CE, Duguay F, Naismith AL, Kartner N, Hanrahan JW, Riordan JR. Cl⁻ channel activity in *Xenopus* oocytes expressing the cystic fibrosis gene. J Biol Chem 1991; 266:19142–19145.
35. Kartner N, Hanrahan JW, Jensen TJ, et al. Expression of the cystic fibrosis gene in non-epithelial invertebrate cells produces a regulated anion conductance. Cell 1991; 64:681–691.
36. Bear CE, Li CH, Kartner N, et al. Purification and functional reconstitution of the cystic fibrosis transmembrane conductance regulator (CFTR). Cell 1992; 68:809–818.
37. Egan M, Flotte T, Afione S, et al. Defective regulation of outwardly rectifying Cl⁻ channels by protein kinase A corrected by insertion of CFTR. Nature 1992; 358: 581–584
38. Gabriel SE, Clarke LL, Boucher RC, Stutts MJ. CFTR and outwardly rectifying chloride channels are distinct proteins with a regulatory relationship. Nature 1993; 363:263–268.
39. Johnson LG, Boyles SE, Wilson J, Boucher RC. Normalization of raised sodium absorption and raised calcium-mediated chloride secretion by Ad-mediated expresssion of cystic fibrosis transmembrane conductance regulator in primary human cystic fibrosis airway epithelial cells J Clin Invest 1995; 95:1377–1382.
40. Stutts MJ, Canessa CM, Olsen JC, et al. CFTR as a cAMP-dependent regulator of sodium channels. Science 1995; 269:847–850.
41. Boucher RC, Stutts MJ, Knowles MR, Cantley L, Gatzy JT. Na⁺ transport in cystic fibrosis respiratory epithelia: abnormal basal rate and response to adenylate cyclase activation. J Clin Invest 1986; 78:1245–1252.
42. Willumsen NJ, Boucher RC. Transcellular sodium transport in cultured cystic fibrosis human nasal epithelium. Am J Physiol 1991; 261 (Cell Physiol 30):C332–C341.
43. Knowles MR, Stutts MJ, Spock A, Fischer N, Gatzy JT, Boucher RC. Abnormal ion permeation through cystic fibrosis respiratory epithelium. Science 1983; 221: 1067–1070.
44. Knowles MR, Gatzy J, Boucher R. Increased bioelectric potential difference across respiratory epithelia in cystic fibrosis. N Engl J Med 1981; 305:1489–1495.
45. Burch LH, Talbot CR, Knowles MR, Canessa CM, Rossier BC, Boucher RC. Relative expression of the human epithelial Na⁺ channel subunits in normal and cystic fibrosis airways. Am J Physiol 1995; 269 (Cell Physiol 38):C511–C518.
46. Chinet TC, Fullton JM, Yankaskas JR, Boucher RC, Stutts MJ. Mechanism of sodium hyperabsorption in cultured cystic fibrosis nasal epithelium: a patch-clamp study. Am J Physiol 1994; 266 (Cell Physiol 35):C1061–C1068.

47. Prince A. Adhesins and receptors of *Pseudomonas aeruginosa* associated with infection of the respiratory tract. Microb Pathog 1992; 13:251–260.
48. Saiman L, Prince A. *Pseudomonas aeruginosa* pili bind to asialo GM1 which is increased on the surface of cystic fibrosis epithelial cells. J Clin Invest 1993; 92: 1875–1880.
49. Widdicombe JH. Role of serous cells in pathogenesis of CF. Pediatr Pulmonol 1995; 12(suppl):69–70.
50. Mizgerd JP, Kobzik L, Warner AE, Brain JD. Effects of sodium concentration on human neutrophil bactericidal functions. Am J Physiol 1995; 269:L388–L393.
51. Smith JJ, Travis SM, Greenberg EP, Welsh MJ. Cystic fibrosis airway epithelia fail to kill bacteria because of abnormal airway surface fluid. Cell 1996; 85:229–236.
52. Joris L, Dab I, Quinton PM. Elemental composition of human airway surface fluid in healthy and diseased Airways. Am Rev Respir Dis 1993; 148:1633–1637.
53. Gilljam H, Ellin A, Strandvik B. Increased bronchial chloride concentration in cystic fibrosis. Scand J Clin Lab Invest 1989; 49:121–124.
54. Pier GB, Grout M, Zaidi TS, et al. Role of mutant CFTR in hypersusceptibility of cystic fibrosis patients to lung infections. Science 1996; 271:64–67
55. Matsui H, Johnson LG, Randell SH, Boucher RC. Loss of binding and entry of liposome-DNA complexes decreases transfection efficiency in differentiated rat tracheal epithelial cells. J Biol Chem 1997; 272:1117–1126.
56. Drumm ML, Pope HA, Cliff WH, et al. Correction of the cystic fibrosis defect in vitro by retrovirus-mediated gene transfer. Cell 1990; 62:1227–1233.
57. Rich DP, Anderson MP, Gregory RJ, et al. Expression of the cystic fibrosis transmembrane conductance regulator corrects defective chloride channel regulation in cystic fibrosis airway epithelial cells. Nature 1990; 347:358–363.
58. Olsen JC, Johnson LG, Stutts MJ, et al. Correction of the apical membrane chloride permeability defect in polarized cystic fibrosis airway epithelia following retroviral-mediated gene transfer. Human Gene Ther 1992; 3:253–266.
59. Johnson LG. Gene therapy for cystic fibrosis. Chest 1995; 107(suppl):77S–83S.
60. Colledge WH. Cystic fibrosis gene therapy. Curr Opinion Genet Dev 1994; 4:466–471.
61. Johnson LG, Olsen JC, Sarkadi B, Moore KL, Swanstrom R, Boucher RC. Efficiency of gene transfer for restoration of normal airway epithelial function in cystic fibrosis. Nature Genet 1992; 2:21–25.
62. Engelhardt JF, Yang Y, Stratford-Perricaudet LD, et al. Direct gene transfer of human CFTR into human bronchial epithelia of xenografts with E1-deleted adenoviruses. Nature Genet 1993; 4:27–34.
63. Rich DP, Couture LA, Cardoza LM, et al. Development and analysis of recombinant adenoviruses for gene therapy of cystic fibrosis. Human Gene Ther 1993; 4:461–476.
64. Zabner J, Couture LA, Smith AE, Welsh MJ. Correction of cAMP-stimulated fluid secretion in cystic fibrosis airway epithelia: efficiency of Ad-mediated gene transfer in vitro. Human Gene Ther 1994; 5:585–593.

65. Rosenfeld MA, Siegfried W, Yoshimura K, et al. Ad-mediated transfer of a recombinant α_1-antitrypsin gene to the lung epithelium in vivo. Science 1991, 252: 431–434.
66. Rosenfeld MA, Yoshimura K, Trapnell BC, et al. In vivo transfer of the human cystic fibrosis transmembrane conductance regulator gene to the airway epithelium. Cell 1992; 68:143–155.
67. Engelhardt JF, Simon RH, Yang Y, et al. Ad-mediated transfer of the CFTR gene to lung of non-human primates: biological efficacy study. Human Gene Ther 1993; 4:759–769.
68. Grubb BR, Pickles RJ, Ye H, et al. Inefficient gene transfer by adenovirus vector to cystic fibrosis airway epithelia of mice and humans. Nature 1994; 371: 802–806.
69. Pickles RJ, Barker PM, Ye H, Boucher RC. Efficient Ad-mediated gene transfer to basal but not columnar cells of cartilaginous airway epithelia. Human Gene Ther 1996; 7:921–931.
70. Wickham TJ, Mathias P, Cheresh DA, Nemerow GR. Integrins $\alpha_v\beta_3$ and $\alpha_v\beta_5$ promote adenovirus internalization but not virus attachment. Cell 1993; 73:309–319.
71. Goldman MJ, wilson JM. Expression of $\alpha_v\beta_5$ integrin is necessary for efficient Ad-mediated gene transfer in the human airway. J. Virol. 1995; 69:5951–5958.
72. Dupuit F, Zahm JH.-M, Pierrot D, et al. Regenerating cells in human airway surface epithelium represent preferential targets for recombinant adenovirus. Human Gene Ther 1995; 6:1185–1193.
73. Mastrangeli A, Danel C, Rosenfeld MA, et al. Diversity of airway epithelial cell targets for in vivo recombinant Ad-mediated gene transfer. J Clin Invest 1993; 91:225–234.
74. Pilewski JM, Engelhardt JF, Bavaria JE, Kaiser LR, Wilson JM, Albelda SM. Ad-mediated gene transfer to human bronchial submucosal glands using xenografts. Am J Physiol 1995; 268 (Lung Cell Mol Physiol 12):L657–L665.
75. Zabner J, Moninger T, Zeiher B, Wadsworth S, Smith A, Welsh MJ. Gene transfer to ciliated airway epithelia by recombinant adenovirus. Pediatr Pulmonol 1995; 12(suppl):224–225.
76. Ballard PL, Zepeda ML, Schwartz M, Lopez N, Wilson JM. Ad-mediated gene transfer to human fetal lung ex vivo. Am J Physiol 1995; 268:1839-1845.
77. McCray PB Jr, Armstrong K, Zabner J, et al. Adenoviral-mediated gene transfer to fetal pulmonary epithelia in vitro and in vivo. J Clin Invest 1995; 95:2620–2632.
78. Sekhon HS, Larson JE. In utero gene transfer into the pulmonary epithelium. Nature Med 1995; 1:1201–1203.
79. Van Ginkel FW, Liu C-G, Simecka JW, et al. Intratracheal gene delivery with adenoviral vector induces elevated systemic IgG and mucosal IgA antibodies to adenovirus and b-galactosidase. Human Gene Ther 1995; 6:895–903.
80. Kaplan JM, St. George JA, Pennington SE, et al. Humoral and cellular immune responses of nonhuman primates to long-term repeated lung exposure to Ad2/ CFTR2. Gene Ther 1996; 3:117–127.
81. Dong J-Y, Wang D, Van Ginkel FW, Pascual DW, Frizzell RA. Systematic analysis of repeated gene delivery into animal lungs with a recombinant adenovirus vector. Human Gene Ther 1996; 7:319–331.

82. Yei S, Mittereder N, Tang K, O'Sullivan C, Trapnell BC. Ad-mediated gene transfer for cystic fibrosis: quantitative evaluation of repeated in vivo vector administration to the lung. Gene Ther 1994; 1:192–200.
83. Mastrangeli A, Harvey B-G, Yao J, et al. Sero-switch Ad-mediated in vivo gene transfer: circumvention of anti-adenovirus humoral immune defenses against repeat adenovirus vector administration by changing the adenovirus serotype. Human Gene Ther 1996; 7:79–87.
84. Yang Y, Trinchieri G, Wilson JR. Recombinant IL-12 prevents formation of blocking IgA antibodies to recombinant adenovirus and allows repeated gene therapy to mouse lung. Nature Med 1995; 1:890893.
85. Zhou X, Huang L. DNA transfection mediated by cationic liposomes containing lipopolylysine: characterization and mechanism of action. Biochim Biophys Acta 1994; 1189:195–203.
86. Debs R, Pian M, Gaensler K, Clements J, Friend DS, Dobbs L. Prolonged transgene expression in rodent lung cells. Am J Respir Cell Mol Biol 1992; 7:406–413.
87. Jarnagin WR, Debs RJ, Wang S-S, Bissell DM. Cationic lipid-mediated transfection of liver cells in primary culture. Nucleic Acids Res 1992; 20:4205–4211.
88. Felgner PL, Gadek TR, Holm M, et al. Lipofection: a highly efficient lipid-mediated DNA transfection procedure. Proc Natl Acad Sci USA 1987; 84:7413–7417.
89. Lu L, Zeitlin PL, Guggino WB, Craig RW. Gene transfer by lipofection in rabbit and human secretory epithelial cells. Pflugers Arch 1989; 415:198–203.
90. Stribling R, Brunette E, Liggitt D, Gaensler K, Debs R. Aerosol gene delivery in vivo. Proc Natl Acad Sci USA 1992; 89:11277–11281.
91. Zabner J, Fasbender AJ, Moninger T, Poellinger KA, Welsh MJ. Cellular and molecular barriers to a gene transfer by a cationic lipid. J Biol Chem 1995; 270: 18997–19007.
92. Flotte TR, Solow R, Owens RA, Afione S, Zeitlin PL, Carter BJ. Gene expression from adeno-associated virus vectors in airway epithelial cells. Am J Respir Cell Mol Biol 1992; 7:349–356.
93. Flotte TR, Afione SA, Zeitlin PL. Adeno-associated virus vector gene expression occurs in nondividing cells in the absence of vector DNA integration. Am J Respir Cell Mol Biol 1994; 11:517–521.
94. Halbert CL, Alexander IE, Wolgamot GM, Miller AD. Adeno-associated virus vectors transduce primary cells much less efficiently than immortalized cells. J Virol 1995; 69:1473–1479.
95. Flotte TR, Afione SA, Conrad C, et al. Stable in vivo expression of the cystic fibrosis transmembrane conductance regulator with an adeno-associated virus vector. Proc Natl Acad Sci USA 1993; 90:10613–10617.
96. Jiang C, Marshall J, Siegel C, et al. Correction of chloride transport defect in cystic fibrosis airway cells by cationic lipid-mediated gene transfer. Am J Respir Crit Care Med 1995; 151(4):A124. Abstract.
97. Marshall J, Lee E, Siegel C, et al. Optimized gene transfer into airway cells in vitro and in vivo using novel and improved cationic lipids. Pediatr Pulmonol 1994; 10 (suppl):223. Abstract.

98. Hyde SC, Gill DR, Higgins CF, et al. Correction of the ion transport defect in cystic fibrosis transgenic mice by gene therapy. Nature 1993; 362: 250–255.

99. Alton EWFW, Middleton PG, Caplen NJ, et al. Non-invasive liposome-mediated gene delivery can correct the ion transport defect in the cystic fibrosis mutant mice. Nature Genet 1993; 5:135–142.

100. Logan JJ, Bebok Z, Walker LC, et al. Cationic lipids for reporter gene and CFTR transfer to rat pulmonary epithelium. Gene Ther 1995; 2:38–49.

101. Flotte TR, Afione SA, Solow R, et al. Expression of the cystic fibrosis transmembrane conductance regulator from a novel adeno-associated virus promoter. J Biol Chem 1993; 268:3781–3790.

102. Yang Y, Li Q, Ertl HCJ, Wilson JM. Cellular and humoral immune responses to viral antigens create barriers to lung-directed gene therapy with recombinant adenoviruses. J Virol 1995; 69(4):2004–2015.

103. Simon RH, Engelhardt JF, Yang Y, et al. Ad-mediated transfer of the CFTR gene to lung of non-human primates: toxicity study. Human Gene Ther 1993; 4:771–780.

104. Yei S, Mittereder N, Wert S, Whitsett JA, Wilmott RA, Trapnell BC. In vivo evaluation of the safety of Ad-mediated transfer of the human cystic fibrosis transmembrane conductance regulator cDNA to the lung. Human Gene Ther 1994; 5:731–744.

105. Goldman MJ, Litzky LA, Engelhardt JF, Wilson JM. Transfer of the CFTR gene to the lung of nonhuman primates with E1-deleted E2a-defective recombinant adenoviruses: A preclinical toxicology study. Human Gene Ther 1995; 6: 839–851.

106. Yang Y, Nunes FA, Berencsi K, Gonczol E, Engelhardt JR, Wilson JM. Inactivation of E2a in recombinant adenoviruses limits cellular immunity and improves the prospect for gene therapy of cystic fibrosis. Nature Genet 1994; 7:362–369.

107. Engelhardt JF, Litzky L, Wilson JM. Prolonged transgene expression in cotton rat lung with recombinant adenoviruses defective in E2a. Human Gene Ther 1994; 5:1217–1229.

108. St George JA, Pennington SE, Kaplan JM, et al. Biological response of nonhuman primates to long-term repeated lung exposure to Ad2/CFTR2. Gene Ther 1996; 3:103–116.

109. Wang QC, Jia X-C, Finer MH. A packaging cell line for propagation of recombinant adenovirus vectors containing two lethal gene-region deletions. Gene Ther 1995; 2:775–783.

110. Piedimonte G, Pickles RJ, Lehmann JR, Costa DL, Boucher RC. Replication-deficient adenoviral adenoviral vector for gene therapy exhibits a strong potentiating effect on airway neurogenic inflammation (abstr). Pediatr Pulmonol 1995; (suppl 12): 224.

111. Piedimonte G, Pickles RJ, Lehmann JR, Costa DL, Boucher RC. Adenoviral vector-induced inflammation is reduced but not abolished by UV-psoralen inactivation (abstr). Pediatr Pulmonol 1995; (suppl 12): 224.

112. Piedimonte G, Lehmann JR, Pickles RJ, Costa DL, Boucher RC. Adenoviral vector-induced neurogenic inflammation in rat airways can be prevented by a

selective substance P (NK1) receptor antagonist. Ninth Annual North American Cystic Fibrosis Conference, Dallas, Texas, October 12–15, 1995. Abstract.

113. Stutts MJ, Gabriel SE, Olsen JC, et al. Functional consequences of heterologous expression of the cystic fibrosis transmembrane conductance regulator in fibroblasts. J Biol Chem 1993; 268:20653–20658.

114. Whitsett JA, Dey CA, Stripp BR, et al. Human cystic fibrosis transmembrane conductance regulator directed to respiratory epithelial cells of transgenic mice. Nature Genet 1992; 2:13–20.

115. Berkner KL. Development of adenovirus vectors for the expression of heterologous genes. Biotechniques 1988; 6:616–629.

116. Eissa NT, Chu C-S, Danel C, Crystal RG. Evaluation of the respiratory epithelium of normals and individuals with cystic fibrosis for the presence of adenovirus E1a sequences relevant to the use of E1a⁻ adenovirus vectors for gene therapy for the respiratory manisfestations of cystic fibrosis. Human Gene Ther 1994; 5:1105–1114.

117. Olsen JC, Huang W-H, Johnson LG, Boucher RC. Persistence of adenoviral vector gene expression in CF airway cells is due to integration of vector sequences into chromosomal DNA. Pediatr Pulmonol 1994; 10(suppl):230. Abstract.

118. Teramoto S, Johnson LG, Huang W, Leigh MW, Boucher RC. Effect of adenoviral vector infection on cell proliferation in cultured primary airway epithelial cells. Human Gene Ther 1995; 6:1045–1053.

119. Debbas M, White E. Wild-type p53 mediates apoptosis which is inhibited by E1b. Genes Dev 1993; 7:546–554.

120. Lowe SW, Ruley EH. Stabilization of the p53 tumor suppressor is induced by adenovirus 5 E1a and accompanies apoptosis. Genes Dev 1993, 7:535–545.

121. Rao L, Debbas M, Sabbatini P, Hockenberry D, Korsmyer S, White E. The adenovirus E1a proteins induce apoptosis which is inhibited by the E1b 19-kda and Bcl-2 proteins. Proc Natl Acad Sci USA 1992; 89:7742–7746.

122. White E, Sabbatini P, Debbas M, Wold WSM, Kusher DI, Gooding L. The 19-kilodalton adenovirus E1b transforming protein inhibits progammed cell death and prevents cytolysis by tumor necrosis factor α. Mol Cell Biol 1992; 12:2570–2580.

123. Leigh MW, Kylander JE, Yankaskas JR, Boucher RC. Cell proliferation in bronchial epithelium and submucosal glands of cystic fibrosis patients. Am J Resp Cell Mol Biol 1995; 12:605–612.

124. Canonico AE, Plitman JD, Conary JT, Meyrick BO, Brigham KL. No lung toxicity after repeated aerosol or intravenous delivery of plasmid-cationic liposome complexes. J Appl Physiol 1994; 77(1):415–419.

125. Thomas DA, Myers MA, Wichert B, Schreier H, Gonzalez-Rothi R. Acute effects of liposome aerosol inhalation on pulmonary function in healthy human volunters. Chest 1991; 99:1268–1270.

126. Stewart MJ, Plautz GE, Del Buono L, et al. Gene transfer in vivo with DNA-liposome complexes: safety and acute toxicity in mice. Human Gene Ther 1992; 3:267–275.

127. Garlick DS, Nichols M, Vaccaro C, et al. Pulmonary toxicity assoicated with intranasal instillation of cationic lipid:DNA complexes in mice. Pediatr Pulmonol 1995; 12(suppl):221.
128. Samulski RJ, Zhu X, Xiao X, et al. Targeted integration of adeno-associated virus (AAV) into human chromosome 19. EMBO J 1991; 10:3941–3950.
129. Kotin RM, Menninger JC, Ward DC, Berns KI. Mapping and direct visualization of a region-specific viral integration site on chromosome 19q–13qter. Genomics 1991; 10:831–834.
130. Kearns W, Afione S, Cutting G, Carter B, Pearson P, Flotte T. AAV-CFTR vectors integrate at multiple chromosomal sites in a CF bronchial epithelial cell line. Pediatr Pulmonol 1993; 9(suppl):242.
131. Flotte TR, Carter BJ. Adeno-associated virus vectors for gene therapy. Gene Ther 1995; 2:357–362.
132. Zabner J, Couture LA, Gregory RJ, Graham SM, Smith AE, Welsh MJ. Ad-mediated gene transfer transiently corrects the chloride transport defect in nasal epithelia of patients with cystic fibrosis. Cell 1993; 68:207–216.
133. Crystal RG, McElvaney NG, Rosenfeld MA, et al. Administration of an adenovirus containing the human CFTR cDNA to the respiratory tract of individuals with cystic fibrosis. Nature Genet 1994; 8:42–51.
134. Crystal RG, Jaffe A, Brody S, et al. A phase I study in cystic patients of the safety toxicity and biological efficacy of a single administration of a replication deficient recombinant adenovirus carrying the cDNA of the normal cystic fibrosis transmembrane conductance regulator gene in the lung. Human Gene Ther 1995; 6:643–666.
135. Wilson JM, Engelhardt JF, Grossman M, Simon RH, Yang Y. Gene therapy of cystic fibrosis lung disease using E1-deleted adenoviruses: a phase I trial. Human Gene Ther 1994; 5:501–519.
136. Boucher RC, Knowles MR, Johnson LG, et al. Gene therapy for cystic fibrosis using E1-deleted adenovirus: A phase I trial in the nasal cavity. Human Gene Ther 1994; 5:615–639.
137. Wilmott RW, Whitsett JA, Trapnell B, et al. Gene therapy for cystic fibrosis utilizing a replication deficient adenovirus vector to deliver the human cystic fibrosis transmembrane conductance regulator cDNA to the airways. A phase I study. Human Gene Ther 1994; 5:1019–1057.
138. Crystal RG, Mastrangeli A, Sanders A, Cet al. Evaluation of repeat administration of a replication deficient recombinant adenovirus containing the normal cystic fibrosis transmembrtane conductance regulator cDNA to the airways of individuals with cystic fibrosis. Human Gene Ther 1995; 6:667–703.
139. Welsh MJ, Zabner J, Graham SM. Adenovirus-mediated gene transfer for cystic fibrosis. Part A: Safety of dose and repeat administration in the nasal epithelium. Part B: Clinical efficacy in the maxillary sinus. Human Gene Ther 1995; 6:205-218.
140. Knowles MR, Paradiso AM, Boucher RC. In vivo nasal potential difference: Techniques and protocols for assessing efficacy of gene transfer in cystic fibrosis. Human Gene Ther 1995; 6:445–455.

141. Hay JG, McElvaney NG, Herena J, Crystal RG. Modification of nasal epithelial potential differences of individuals with cystic fibrosis consequent to local administration of a normal CFTR cDNA adenovirus gene transfer vector. Human Gene Ther 1995; 6:1487–1496.
142. Knowles MR, Hohneker K, Zhou Z-Q, et al. A controlled study of adenoviral vector mediated gene transfer in the nasal epithelium of patients with cystic fibrosis. N Engl J Med 1995; 333:823–831.
143. Caplen NJ, Alton EWFW, Middleton PG, et al. Liposome-mediated CFTR gene transfer to the nasal epithelium of patients with cystic fibrosis. Nature Med 1995; 1:39–46.
144. Middleton PG, Caplen NJ, Gao X, et al. Nasal application of the cationic liposome DC-Chol:DOPE does not alter ion transport lung function or bacterial growth. Eur Respir J 1994; 7:442–445.

13

Correction of the Genetic Defect in Alpha-1 Antitrypsin Deficiency by Somatic Gene Therapy

RANDY C. EISENSMITH and SAVIO L. C. WOO

Mount Sinai School of Medicine
New York, New York

I. Physiological Function of hAAT and Pathogenesis of Pulmonary Emphysema

Human α_1-antitrypsin (hAAT) is a major serine protease inhibitor. This serum protein is synthesized as a single 394-amino acid polypeptide with a molecular weight of approximately 52,000 daltons (1–3), has a carbohydrate content of about 12%, and contains 6 to 8 sialic acid residues (4,5). Although synthesized and secreted primarily from the liver (6), limited amounts of hAAT are also produced in macrophages (7,8) and monocytes (8,9) as well as parenchymal cells of the lung (10), kidney, and intestine (11,12). hAAT inhibits a variety of serine proteases such as trypsin, chymotrypsin, elastase, renin, urokinase, collagenase, and proteolytic enzymes released from leukocytes and bacteria (13). However, its most important action is as an inhibitor of neutrophil elastase (5,14). The clinical importance of hAAT was recognized following the observation that individuals with markedly deficient hAAT levels in serum develop chronic obstructive pulmonary emphysema (15) and/or infantile liver cirrhosis (16).

Although hAAT circulates in the serum at a concentration of approximately 1 to 2 mg/ml (17), its major site of physiologic action is not in the blood but rather in the lower respiratory tract of the lung. Passive diffusion of hAAT into the alveoli of the lung protects these structures against neutrophil elastase-mediated proteolysis (18,19). hAAT inactivates elastase by stoichiometric formation of a covalent complex, thereby maintaining a protease-antiprotease balance in the lung. If physiologic or pathologic conditions perturb this balance in favor of the elastase, the elastin polymers in the elastic fibers will be hydrolyzed. The resulting tissue damage to the major components of the connective tissue matrix of the alveolar structure of the lung will be permanent (20).

Elastic fibers play a critical role in normal lung function, particularly in modulating elastic recoil (21,22). Collapse of this polymeric structure accumulated over a period of years creates a chronic condition of permanent enlargement of the air spaces distal to the terminal bronchioles, accompanied by destruction of the alveolar wall. This condition is clinically defined as emphysema (20). The balance of elastase-antielastase activities in the lung can be perturbed genetically in favor of elastase if the individual has an inborn deficiency in hAAT. The imbalance is worsened by conditions that lead to an inflammatory response in the lower respiratory tract by increasing the neutrophil elastase locally.

Another major contributing factor that accelerates the development of pulmonary emphysema in deficient individuals is cigarette smoking. Not only will the particulate matter in cigarette smoke elicit a macrophage response and neutrophil migration to the lung, but the oxidants in the smoke can also inactivate hAAT by oxidation of the methionine residue at the active site (23). Patients with hAAT deficiency will generally develop emphysema in their 40s, and this development can be accelerated by 10 to 20 years in patients who smoke. Once emphysema occurs, deterioration of lung tissue will continue for several years until death, and there is no effective conventional therapy for this chronic condition.

II. Molecular Genetics of hAAT Deficiency

The synthesis of hAAT is controlled by an autosomal and allelic system (5,24). The hAAT gene, located at 14q32.1 (25,26), is approximately 12.2 kb in length and contains 5′ noncoding (IA-C) and four coding (II-V) exons (27). Trancription of the gene in liver begins at approximately the middle of exon IC (27); transcripts in macrophages and monocytes can start from the beginning of exon IA or the middle of exon IC (10). Over 75 different electrophoretic phenotypes of the protein have been classified (14,23). The hAAT locus has been assigned the term Pi (P for protease and i for inhibitor), and each phenotype has been assigned a capital letter. Most individuals are homozygotes for PiM, which has a gene frequency of 0.95. The most common electrophoretic variant in individuals with hAAT deficiency is PiZ, which is the result of a $Glu^{342} \rightarrow Lys^{342}$ missense mutation (28–30). Individuals homozygous for the PiZ phenotype have serum hAAT levels of only 10% to 15% of that of the PiM individuals. The level of hAAT in heterozygous individuals of the MZ phenotypes is about 50% to 60% of normal (24), suggesting that the defect resides with the hAAT gene itself and not through a trans-regulatory mechanism. A second common variant with reduced plasma concentration is Pi S, which is caused by a $Glu^{264} \rightarrow Val^{264}$ missense mutation (27,31,32).

hAAT deficiency is inherited as an autosomal codominant trait, and the gene frequency of the PiZ phenotype in Caucasians of Northern European ancestry is 0.02 to 0.03 (5, 32). PiZ is more common in the United States, with a gene frequency of 0.03 to 0.04 (33). It was observed clinically that reduction in serum hAAT levels in ZZ homozygotes (12% of normal) and SZ heterozygotes (35% of normal) predisposed these individuals to the development of pulmonary emphysema, such that the ZZ homozygotes have a risk 30 times greater than that of the general population (5). It has been estimated that about 1/4000 and 1/800 among the Caucasian population in the United States are ZZ homozygotes and SZ heterozygotes, respectively. With a population of over 250 million, the statistics would suggest that there are 60,000 ZZ homozygotes and 300,000 SZ heterozygotes in the United States. Although there is a broad range in the age of onset and the severity of the disease state, 80% to 90% of ZZ homozygotes will develop panacinar emphysema, most prominent in the lower lung zone (5).

III. Enzyme Replacement Therapy for hAAT Deficiency

The traditional approach to the treatment of hAAT-deficient patients is the intravenous infusion of hAAT. Crystal and colleagues at the National Institutes of Health demonstrated that therapeutic levels of hAAT can be reached in the lung of patients weighing 100 kg by weekly infusions of 6 g hAAT purified from human plasma (34). Several prospective (34,35) and retrospective (36) Phase I and Phase II studies of the safety and efficacy of long-term augmentation therapy in patients receiving intravenous infusions of hAAT at 60 mg/kg weekly or higher doses demonstrated a protective level of antineutrophil elastase activity in the lung lavage fluid and a stabilization in lung function. However, no conventional Phase III clinical trial has yet been performed to test the efficacy of hAAT augmentation therapy due to a lack of adequate hAAT supply and the high costs of long-term treatment (14).

An alternative to systemic enzyme infusion is the direct delivery of hAAT to the lung alveoli by means of aerosol preparations. Preliminary studies in laboratory animals demonstrated that hAAT could pass through epithelial layer of the lower respiratory tract and gain access to the interstitium of the alveolar walls (37). The aerosol administration of hAAT at a dose of 100 mg twice daily raised plasma hAAT levels and lung antineutrophil elastase activity well beyond the protective threshold level in the lavage fluids from the lung, with apparently no adverse clinical reactions (38). Whether this route of administration can deliver hAAT into the interstitial space of the aveolar structure at protective levels in patients has not been demonstrated.

The apparent success of enzyme replacement strategies and the desire to eliminate the potential risk of opportunistic infections associated with the use of products derived from human plasma has stimulated a number of biotechnology firms to produce the human protein through recombinant means. Bacterial and yeast systems have been used to produce massive quantities of hAAT without the carbohydrate side chains. Although the protein is perfectly functional to inactivate neutrophil elastase, its half-life in blood is reduced dramatically (from 6 days to several hours), as it is rapidly cleared by the kidney (39). Systems using mammalian cells produced the properly glycosylated protein, but the levels are too low to be practical. The most innovative approach has been the attempt to produce this protein in the milk of transgenic farm animals, and it has been reported that a level of 60 g/L of hAAT in the milk of transgenic sheep has been achieved (40).

At the present time, the only reliable source of hAAT is still spent human plasma. This biological product has been approved by the Food and Drug Administration as an "orphan drug" for treatment of hAAT deficiency but the supply is obviously limited. While this problem might be resolved in the future by the transgenic farm animal technology, whether "enzyme replacement" can be the effective treatment of chronic disorders by repeated infusion of massive quantities of recombinantly produced protein for the lifetime of the patients without side effects remains an open question.

IV. Somatic Gene Therapy for hAAT Deficiency

Theoretically, introduction of the functional hAAT gene into somatic cells of deficient individuals could provide a population of cells capable of synthesizing and secreting hAAT into the circulatory system. The protein would then diffuse into the alveolar spaces of the lung and provide the necessary antiprotease function to prevent tissue destruction by polymorphonuclear neutrophil elastase. It is important to note that this form of genetic therapy can only be effective in ameliorating the pulmonary manifestations of hAAT deficiency. This treatment is not expected to have any efficacy toward the hepatic disease that occurs in 10% to 15% of the patients, which might be caused by excessive accumulation of the mutant protein in the liver (41). Because of the high level of hAAT gene expression necessary for the production of therapeutic levels of hAAT in plasma (normal level = 1 to 2 mg/ml), somatic gene therapy will require highly effective means for transfer of genes into target cells in vivo.

A. Retrovirus-Mediated Gene Therapy for hAAT Deficiency

Initial studies in this area examined the utility of vectors derived from recombinant retroviruses. The Moloney murine leukemia virus (MMLV) represents an excellent vector system in that it has the properties of being able to efficiently transduce a wide variety of cells leading to stable integration of the proviral genome into the chromosome of the host cell. The initial study using retroviral vectors involved the ex vivo approach (42). In this study, mouse fibroblasts were transduced with a retroviral vector expressing hAAT and a clonal isolate was transplanted into the peritoneum of nude mice. Minute amounts of hAAT could be detected in the serum of recipient animals, and biochemical studies confirmed that this protein was functional. Our laboratory subsequently demonstrated that this ex vivo approach could also be used successfully with hepatocytes in normal dogs. A recombinant retrovirus expressing hAAT was used to transduce canine primary hepatocytes in culture. Up to 10^9 cells were autologously reimplanted into the respective donor animals, with the resulting production of 4 μg/ml of hAAT in the serum (43). The hAAT levels in the dogs receiving autologous transplants transduced by the recombinant retrovirus LN/CMV-hAAT began to fall after 10 days and were undetectable 20 to 50 days after transplantation (Fig. 1). PCR experiments suggested that this decrease was not due to loss of transplanted hepatocytes, but rather was caused by the shutdown in vivo of the CMV promoter used to drive transgene expression in the recombinant retrovirus (43).

Although successful in achieving transgene expresion in a large animal model, the ex vivo approach is technically very complex and labor-intensive. An alternative approach is to deliver recombinant retroviral vectors directly to the liver cells in vivo et situ. However, because retroviral vectors can only transduce actively dividing cells with high efficiency, the administration of retroviral vectors must be preceded by a chemical or surgical partial hepatectomy (44,45) in order to achieve viral transduction of normally quiescent hepatocytes. To test the efficacy of this approach, recombinant retroviral vectors were infused into the portal vasculature of partially hepatectomized mice. In the initial experiments, a recombinant retrovirus expressing the *E. coli* β-galactosidase gene was injected into the portal vein of partially hepatectomized mice. This procedure transduced approximately 1% to 2% of all hepatocytes in vivo (46), which compares favorably to that achieved by the ex vivo approach. To determine whether direct retrovirus delivery in vivo would result in long-

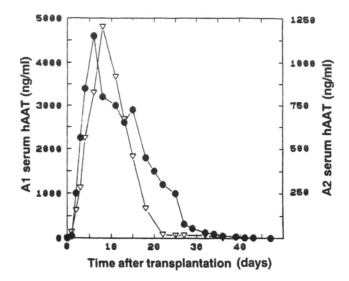

Figure 1 Serum hAAT levels in two dogs after transplantation of hepatocytes transduced with a recombinant retroviral vector expressing hAAT under the transcriptional control of the CMV promoter. Dog 1 (A1; solid circles) received 3.8×10^8 hepatocytes, while dog 2 (A2; open triangles) received 6.4×10^8 hepatocytes. Each point represents the mean of two determinations. (From Ref. 43.)

term expression, a recombinant retroviral vector expressing hAAT under the transcriptional control of the albumin promoter/enhancer element was constructed and infused into the portal vasculature of partially hepatectomized mice. Serum hAAT levels were then followed over time. hAAT levels in these animals ranged from 0.2 μg/ml to 1 μg/ml (Fig. 2). Despite this variation, the most important finding of these studies was the observation that transgene expression could be sustained in individual animals for more than 6 months (46).

Similar experiments were subsequently performed in the rat, where portal vein infusion of a recombinant retrovirus expressing β-galactosidase in partially hepatectomized animals resulted in the transduction of 10% to 15% of all hepatocytes (47). This level of hepatocyte transduction is about 10-fold higher than that observed in the mouse. Infusion of an LX/hAAT retroviral vector produced serum hAAT levels of approximately 2 μg/ml, a level of hAAT expression that was severalfold higher than that in the mouse using the same vector. Peak levels of hAAT persisted for the 4-month duration of the experiment, with no significant changes over time (47).

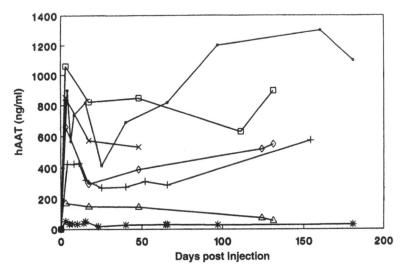

Figure 2 Serum hAAT levels in C57BL/6 mice after in vivo transduction of hepatocytes with a recombinant retroviral vector expressing hAAT under the transcriptional control of the albumin promoter. Each line represents an individual animal. Each point represents the mean of two determinations. (From Ref. 46.)

B. Adenovirus-Mediated Gene Therapy for hAAT Deficiency

While the persistence of hAAT expression following retrovirus-mediated gene transfer into hepatocytes was impressive, this approach has been able to achieve only 1/1000 of the serum hAAT level needed to be therapeutic. An alternative strategy to address this limitation is the use of recombinant adenoviral vectors. Several studies have now examined the potential of this vector system to deliver the hAAT gene, targeting a number of different tissues or cells including lung (48), liver (49,50), endothelial cells (51,52), ependymal cells (53), salivary glands (54), and keratinocytes (55). Studies of adenovirus-mediated gene transfer to the liver have shown that 50% to 100% of all hepatocytes of the mouse could be transduced following portal vein infusion of a recombinant adenoviral vector (56). This finding suggested that intraportal infusion of a recombinant adenoviral vector expressing hAAT might transduce a sufficiently large number of hepatocytes to produce therapeutic or physiological levels of hAAT. Adenoviral vectors containing the hAAT cDNA under the transcriptional control of either the Rous sarcoma virus long terminal repeat (RSV-LTR) or the phosphoglycerate kinase (PGK) promoter were constructed and introduced into C57BL/6 mice by portal vein infusion. Infusion of 10^{10} pfu of Ad.PGK-hAAT per mouse, either by portal or tail vein infusion, produced mean peak hAAT levels in serum of 10 to 100 μg/ml (50). Mice infused with 10^{10} pfu of Ad.RSV-hAAT produced mean peak hAAT levels in serum of about 500 μg/ml (50). These levels persisted for 2 to 3 months, followed by gradual decline over time (Fig. 3).

Despite this extraordinarily favorable result, there are major limitations associated with adenovirus-mediated gene transfer. The most critical problem is that of persistence. When E1-deleted adenoviral vectors were used for gene transfer into liver or other tissues, the level of transgene expression decreased by several orders of magnitude over a period of months in C57BL/6 mice (50), or weeks in other inbred and outbred strains of mice (57,58) and in outbred rabbits and dogs (59,60). Several studies have indicated that the transient expression of genes delivered by E1-deleted adenoviral vectors is primarily a consequence of a cytotoxic T cell-mediated immune response mounted in vivo against the virally transduced cells (61,62). These observations are supported by the fact that only slight decreases in the level of transgene expression were

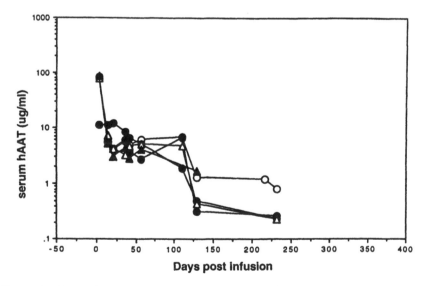

Figure 3 Serum hAAT levels in C57BL/6 mice after in vivo transduction of hepatocytes with a recombinant adenoviral vector expressing hAAT under the transcriptional control of the RSV-LTR promoter. Each line represents an individual animal. Each point represents the mean of two determinations. (From Ref. 50.)

observed for at least 6 months after adenovirus-mediated gene transfer in immune-deficient *nu/nu* (63) or *SCID* mice (58). Further support for this hypothesis is provided by the observation that administration of the immunosuppressive agent cyclosporine A can significantly prolong transgene expression following the infusion of E1-deleted adenoviral vectors (64–66), as can the depletion of CD4$^+$ cells by the anti-CD4 antibody GK1.5 (67,68) or the blockade of T-cell costimulatory molecules by administration of CTLA4Ig (69). In this latter study, hAAT levels of approximately 1 to 5 μg/ml were maintained in C3H mice for up to 1 year (69), but the mice were immunocompromised for at least 8 weeks after treatment.

While at least partially effective in prolonging transgene expression, each of the immunomodulatory treatments cited above has some practical limitations. For example, each of the treatments produces varying degrees of total rather than antigen-specific immunosuppression. Furthermore, none has been shown to be effective in maintaining therapeutic levels of transgene expression for an indefinite period of time. An alternative to immunomodulatory approaches to overcome the CTL response to viral proteins expressed in virally transduced cells is the further modification of the adenoviral vector to reduce or eliminate viral gene expression.

Several studies have examined whether modification (64,65,70) or deletion (71) of the E2 region can significantly prolong adenovirus-mediated transgene expression in lung or liver. In one study, tail vein infusion of a β-galactosidase-expressing adenoviral vector containing a temperature-sensitive ts125 mutation in the E2A gene resulted in transgene expression in 10% to 40% of the total liver area that persisted for up to 70 days (65). Animals similarly infused with β-galactosidase-expressing E1-deleted adenoviral vectors failed to maintain transgene expression for more than 14 days. However, other studies have failed to demonstrate statistically significant prolongation of transgene expression from adenoviral vectors containing this same ts125 mutation in immunocompetent mice or hemophilia B dogs, suggesting that the incorporation of this temperature-sensitive mutation into E2A alone may not be sufficient to significantly prolong persistence of transgene expression over that seen with E1-deleted recombinant viral vectors (70). An E1-deleted adenoviral vector containing a 1.3 kb deletion in the 1.6 kb ORF located in the E2 region has also been reported, along with the E1- and E2-complementing cell lines necessary to produce this vector (71). As yet, no data regarding the persistence of transgene expression following the in vivo infusion of this vector have been reported.

Other studies have reported the development of E1-deleted adenoviral vectors containing a variety of deletions in the E4 region (72–75), and the E1- and E4-complementing cell lines necessary for the production of these vectors (73–75). The E1/E4-deleted vectors produced by both the Yeh et al. (75) and

Wang et al. (74) were derived from H5dl1014 (76). This mutant contains two deletions, one from map units 92 to 93.8 and a second from map units 96.4 to 98.4. These deletions abolish the expression of all E4 genes except ORF4. Both studies report that these E1/E4-deleted vectors exhibit decreased cytopathic effects following the infection of cells in vitro, but neither of these two studies has yet reported the results of in vivo studies. In vivo administration of a β-galactosidase-expressing adenoviral vector similar to that described by Yeh et al. (75) did cause an inflammatory response in Balb/c mice, although this inflammatory response was less than that seen after infusion of the E1-deleted parent vector (77). As the β-galactosidase transgene expressed in these studies is itself capable of eliciting a CTL response (78), it is not yet clear whether the lack of observed persistence was the result of an immune response generated against ORF4, an immune response generated against or β-galactosidase, or an immune response against the residual amounts of late viral gene expression that may still occur in this vector. Thus, it remains to be seen whether recombinant adenoviral vectors deleted for E1 and E4 will yield more persistent transgene expression than that observed following the administration of E1-deleted vectors.

The ultimate form of adenoviral vector modification is an adenoviral amplicon deleted of all viral genes. The development of adenoviral amplicon vectors has been hampered by the inability to produce high titers of vectors and preparations that are free of wild-type helper virus. However, some progress has recently been made in this area. Mitani and co-workers (79) have reported the rescue, propagation, and partial purification of an E1-deleted helper virus-dependent adenoviral vector. This vector contains an SRα-βgeo expression cassette inserted in place of the viral genome from map unit 26.3 to 46.6. This replacement removes portions of the viral L1, L2, VAI, VAII, and pTP genes. Transfection of this plasmid into 293 cells fails to produce progeny virus. When a mixture of this plasmid and E1-deleted adenovirus DNA was transfected into 293 cells, plaque formation occurred and 1% to 5% of all plaques were positive for β-galactosidase activity. β-gal-positive plaques could be propagated in 293 cells with an E1-deleted helper virus and the recombinant vector could be partially purified from the helper virus by CsCl density-gradient centrifugation (79). Although the titer of this recombinant adenoviral amplicon vector was relatively low (about 1×10^7 β-gal-transducing particles/ml), the low sensitivity of the titering system leaves open the possibility that the actual titers may be much higher. This study suggests that this helper virus approach may permit the removal of most of the viral genome from adenoviral vectors. Furthermore, if the density of the recombinant vector can be made significantly different from that of the helper virus, repeated fractionation on CsCl gradients may yield preparations of recombinant viral vectors that are relatively free of helper virus.

Such a system has recently been developed. A plasmid containing the viral ITRs and the packaging signal flanking an expression cassette containing the dystrophin cDNA and a second expression cassette containing βgeo has been used in combination with a packaging-defective E1-deleted helper virus to produce a relatively high-titer ($>10^{10}$ infectious particles/ml) recombinant adenoviral vector. Although the size difference between the genomes of the recombinant vector and the helper virus is only about 5 kb, CsCl density gradient centrifugation can be used to obtain preparations of recombinant vector that contain less than 1% of the helper virus (80). Using a similar approach, Fisher and colleagues also reported the production of a recombinant adenoviral amplicon vector deleted of all adenoviral genes (81). Neither of these studies provides data regarding the long-term persistence of these vectors in vivo. Like the adenoviral vectors containing deletions of E2 and E4, the ultimate persistence and utility of adenoviral amplicons in vivo remains to be established.

A second problem associated with current recombinant adenoviral vectors is the inability to effectively readminister these vectors. If significant amounts of transgene expression could be periodically introduced, then the lack of persistence could at least be partially overcome. The cause of this problem is not a cellular immune response directed against viral proteins expressed from transduced cells, but rather a humoral immune response directed against the viral proteins present in the vector at the time of infusion (82). Our laboratory and others have shown that systemic readministration of adenoviral vectors in mice (57) and dogs (66) is ineffective for periods of at least 9 months, even in the presence of continuous immunosuppression with cyclosporine A (66). While the cellular immune response may be reduced or eliminated by vector modification, this approach cannot prevent the occurrence of a humoral immune response against the input of viral proteins that are necessary for entry of the vector into the target cell. Thus, various forms of host immunomodulation may still be useful to permit periodic readministration of vector to compensate for the loss of gene expression that may arise from gradual degradation of vector DNA or turnover of target cells. Two immunomodulatory agents that have been shown to permit some degree of reporter gene expression in lung and liver after readministration of adenoviral vectors are the CD4-depleting antibody GK1.5 (67,68,83) and the cytokine IL-12 (67,68). The administration of CD4-depleting antibodies immediately before, during, and after virus administration reduced the levels of neutralizing antiadenoviral antibodies in Balb/c (83) and C57BL/6 (67,68) mice and permitted some transgene expression following readministration of adenoviral vectors in both lung and liver. However, the efficiency of readministration in all of these studies was only a small fraction of that observed in naive animals. By blocking differentiation of Th2 cells, IL-12 administration was somewhat more successful than the

CD4-depleting antibody in permitting readministration of adenoviral vectors to the airway epithelium (67), but was largely ineffective in permitting readministration to the liver (68). Furthermore, the ability of IL-12 to permit reinfusion through its inhibition of Th2 cell differentiation appears to come largely at the expense of a heightened CTL response against virally-transduced cells due to the stimulatory effects of IL-12 on Th1 cell differentiation. While this result may be acceptable for readministration of viral vectors lacking expression of potentially immunogenic viral or transgenes, this result would ultimately limit the utility of this approach with existing E1-deleted adenoviral vectors. Finally, although promising, none of these studies has yet demonstrated the restoration of a phenotypic correction in an animal disease model following adenovirus-mediated transfer of a therapeutic gene, casting some doubt as to the ultimate significance and eventual utility of these procedures.

References

1. Crawford IP. Purification and properties of normal human alpha-1-antitrypsin. Arch Biochem Biophys 1973; 156:215–222.
2. Chan SK, Luby J, Wu YC. Purification and chemical compositions of human alpha1-antitrypsin of the MM type. FEBS Lett 1973; 35:79–82.
3. Carrell RW, Jeppsson JO, Laurell CB, et al. Structure and variation of human α1-antitrypsin. Nature 1982; 298:329–334.
4. Heimburger N, Heide K, Haupt H, Schultze HE. Bausteinanalysen von Humanserumproteinen. Clin Chem Acta 1964; 10:293–307.
5. Kueppers F. Alpha1-antitrypsin. Am J Hum Genet 1973; 25:677–686.
6. Hood JM, Koep LJ, Peters RL, et al. Liver transplantation for advanced liver disease with alpha-1-antitrypsin deficiency. N Engl J Med 1980; 302:272–275.
7. Cohen AB. Interrelationships between the human alveolar macrophage and alpha-1-antitrypsin. J Clin Invest 1973; 52:2793–2799.
8. Perlmutter DH, Coles FS, Kilbridge P, Rossing TH, Colten HR. Expression of the alpha₁-proteinase inhibitor gene in human monocytes and macrophages. Proc Natl Acad Sci USA 1985; 82:795–799.
9. Isaacson P, Jones DB, Milward-Sadler GH, Judd MA, Payne S. Alpha-1-antitrypsin in human macrophages. Lancet 1979; 2:964–965.
10. Kelsey GD, Povey S, Bygrave AE, Lovell-Badge RH. Species- and tissue-specific expression of human α_1-antitrypsin in transgenic mice. Genes Dev 1987; 1:161–171.
11. Sifers RN, Carlson JA, Clift SM, Demayo FJ, Bullock DW, Woo SLC. Tissue specific expression of the human alpha-1-antitrypsin gene in transgenic mice. Nucleic Acids Res 1987; 15:1459–1475.
12. Carlsson JA, Rogers BB, Sifers RN, Hawkins HK, Finegold MJ, Woo SLC. Multiple tissues express alpha1-antitrypsin in transgenic mice and man. J Clin Invest 1988; 82:26–36.

13. Travis J, Salvesen GS. Human plasma proteinase inhibitors. Annu Rev Biochem 1983; 52:655–709.
14. Cox DW. α₁-Antitrypsin deficiency. In: Scriver CR, Beaudet AL, Sly WS, Valle D, eds. The Metabolic and Molecular Bases of Inherited Disease, Vol. 3, 7th ed. New York: McGraw Hill, 1995:4125–4158.
15. Laurell CB, Eriksson S. The electrophoretic α1-globulin pattern of serum in α1-antitrypsin deficiency. Scand J Clin Lab Invest 1963; 15:132–140.
16. Sharp H, Bridges RA, Krivit W, Freier EF. Cirrhosis associated with alpha-1-antitrypsin deficiency: a previously unrecognized inherited disorder. J Lab Clin Med 1969; 75:934–939.
17. Jeppson JO, Laurell CB, Fagerhol MK. Properties of isolated α₁-antitrypsin Pi types M, S and Z. Eur J Biochem 1978; 83:143–153.
18. Olsen CN, Harris JO, Castle JR, Waldeman RH, Karmgard HJ. alpha-1-antitrypsin content in the serum, alveolar macrophages, and alveolar lavage fluid of smoking and nonsmoking normal subjects. J Clin Invest 1975; 55:427–430.
19. Tuttle WC, Jones MA. Fluorescent antibody studies of alpha-1-antitrypsin in adult human lung. Am J Clin Pathol 1975; 64:477–482.
20. Gadek JE, Fulmer JD, Gelfund JA, Frank MM, Petty TL, Crystal RG. Danzol-induced augmentation of serum alpha-1-antitrypsin levels in individuals with marked deficiency of this antiprotease. J Clin Invest 1980; 66:82–87.
21. Kuhn C III, Senior RM. The role of elastases in the development of emphysema. Lung 1978; 155:185–197.
22. Starkey PM, Barrett AJ. Human lysosomal elastase. Catalytic and immunological properties. Biochem J 1976; 155:265–271.
23. Hubbard RC, Fells G, Gadek J, Pacholok S, Humes J, Crystal RG. Neutrophil accumulation in the lung in alpha1-antitrypsin deficiency. Spontaneous release of leukotriene B4 by alveolar macrophages. J Clin Invest 1991; 88:891–897.
24. Fagerhol MK, Laurell CB. The Pi System—Inherited Variants of Serum α1-antitrypsin. In: Steinberg AG, Bearn AG, eds. Progress in Medical Genetics, Vol. 7. New York, Grune and Stratton, 1970:96–111.
25. Lai EC, Kao F-T, Law ML, Woo, SLC. Assignment of the α₁-antitrypsin gene and a sequence-related gene to human chromosome 14 by molecular hybridization. Somat Cell Genet 1983; 5:385–392.
26. Schroeder WT, Miller MF, Woo SLC, Saunders GF. Chomosomal localization of the human α₁-antitrypsin gene (Pi) to 14q31–32. Am J Human Genet 1985; 37: 868–872.
27. Long GL, Chandra T, Woo SL, Davie EW, Kurachi K. Complete sequence of the cDNA for human alpha1-antitrypsin and the gene for the S variant. Biochemistry 1984; 23:4828–4837.
28. Jeppson JO. Amino acid substitution Gly-Lys in α₁-antitrypsin Pi Z. FEBS Lett 1976; 65:195–197.
29. Yoshida A, Ewing C, Wessels M, Lieberman J, Gaidulus L. Molecular abnormality of Pi S variant of human alpha-1-antitrypsin. Am J Human Genet 1977; 29:233–239.
30. Kidd VJ, Wallace RB, Itakura K, Woo SLC. α₁-antitrypsin deficiency detection by direct analysis of the mutation in the gene. Nature 1983; 304:230–234.

31. Owen MC, Carrell RW. alpha-1-antitrypsin: molecular abnormality of S variant. Br Med J 1976; 1:130–131.
32. Fagerhol MK. The Pi system: genetic variants of serum α_1-antitrypsin. Ser Haematol 1968; 1:153–161.
33. Pierce JA, Eradio B, Dew TA. Antitrypsin phenotypes in St. Louis. JAMA 1975; 231:609–612.
34. Wewers MD, Casolaro MA, Sellers SE, et al. Replacement therapy for alpha$_1$-antitrypsin deficiency associated with emphysema. N Engl J Med 1987; 816:1055–1062.
35. Hubbard RC, Sellers S, Czerski D, Stephens L, Crystal RG. Biochemical efficacy and safety of monthly augmentation therapy for alpha$_1$-antitrypsin deficiency. JAMA 1988; 260:1259–1264.
36. Barker AF, Siemsen F, Pasley D, D'Silva R, Buist AS. Replacement therapy for hereditary alpha$_1$-antitrypsin deficiency. A program for long-term administration. Chest 1994; 105:1406–1410.
37. Hubbard RC, Crystal RG. Augmentation therapy of α_1-antitrypsin deficiency. Eur Respir J 1990; 9(suppl):44s–52s.
38. Hubbard RC, Crystal RG. Strategies for aerosol therapy of alpha1-antitrypsin deficiency by the aerosol route. Lung 1990; 168:565–578.
39. Casolaro MA, Fells G, Wewers M, et al. Augmentation of lung antineutrophil elastase capacity with recombinant human alpha-1-antitrypsin. J Appl Physiol 1987; 63:2015–2023.
40. Wright G, Carver A, Cotton D, et al. High level expression of active human α_1-antitypsin in the milk of transgenic sheep. Biotechnology 1991; 9:830–834.
41. Carlson JA, Rogers BB, Sifers RN, et al. Accumulation of PiZ α_1-antitrypsin causes liver damage in transgenic mice. J Clin Invest 1989; 83:1183–1190.
42. Garver RI, Chytil A, Courtney M, Crystal RG. Clonal gene therapy: transplanted mouse fibroblast clones express human α_1-antitrypsin gene in vivo. Science 1987; 237:762–764.
43. Kay MA, Baley P, Rothenberg S, et al. Expression of human alpha$_1$-antitrypsin in dogs. Proc Natl Acad Sci USA 1992; 89:89–93.
44. Ferry N, Duplessis O, Houssin D, Danos O, Heard JM. Retroviral-mediated gene transfer into hepatocytes in vivo. Proc Natl Acad Sci USA 1991; 88:8377–8381.
45. Kaleko M, Garcia JV, Miller AD. Persistent gene expression after retroviral gene transfer into liver cells in vivo. Human Gene Ther 1991; 2:27–32.
46. Kay MA, Li Q, Liu T-J, et al. Hepatic gene therapy: persistent expression of human α_1-antitrypsin in mice after direct gene delivery in vivo. Hum Gene Ther 1992; 3:641–647.
47. Kolodka T, Finegold M, Kay MA, Woo SLC. Hepatic gene therapy: efficient retroviral mediated gene transfer into rat hepatocytes in vivo. Somat Cell Mol Genet 1993; 19:5:491–497.
48. Rosenfeld MA, Siegried W, Yoshimura K, et al. Adenovirus-mediated transfer of a recombinant alpha-1-antitrypsin gene to the lung epithelium in vivo. Science 1991; 252:431–434.

49. Jaffe HA, Daniel C, Longenecker M, et al. Adenovirus-mediated in vivo gene transfer and expression in normal rat liver. Nature Genet 1992; 1:372–378.
50. Kay MA, Graham F, Leland F, Woo SLC. Therapeutic serum concentrations of human alpha-1-antitrypsin after adenoviral-mediated gene transfer into mouse hepatocytes. Hepatology 1995; 21:815–819.
51. Lemarchand P, Jaffe HA, Danel C, et al. Adenovirus-mediated transfer of a recombinant human α_1-antitrypsin cDNA to human endothelial cells. Proc Natl Acad Sci USA 1992; 89:6482–6486.
52. Lemarchand P, Jones M, Yamada I, Crystal RG. In vivo transfer and expression in normal uninjured blood vessels using replication-deficient recombinant adenovirus vectors. Circ Res 1993; 72:1132–1138.
53. Bajocchi G, Feldman SH, Crystal RG, Mastrangeli A. Direct in vivo gene transfer to ependymal cells in the central nervous system using recombinant adenovirus vectors. Nature Genet 1993; 3:229–234.
54. Mastrangeli A, O'Connel B, Aladib W, Fox PC, Baum BJ, Crystal RG. Direct in vivo adenovirus-mediated gene transfer to salivary glands. Am J Physiol 1994; 266:G1146–G1155.
55. Setoguchi Y, Jaffe HA, Danel C, Crystal RG. Ex vivo and in vivo gene transfer to the skin using replication-deficient recombinant adenovirus vectors. J Invest Dermatol 1994; 102:415–421.
56. Li Q, Kay MA, Finegold M, Stratford-Perricaudet LD, Woo SLC. Assessment of recombinant adenoviral vectors for hepatic gene therapy. Human Gene Ther 1993; 4:403–409.
57. Fang B, Eisensmith RC, Li XHC, et al. Gene therapy for phenylketonuria: phenotypic correction in a genetically deficient mouse model by adenovirus-mediated hepatic gene transfer. Gene Ther 1994; 1:247–254.
58. Barr D, Tubb T, Ferguson D, et al. Strain-related variations in adenovirally mediated transgene expression from mouse hepatocytes in vivo: comparisons between immunocompetent and immunodeficient inbred strains. Gene Ther 1995; 2:151–155.
59. Li J, Fang B, Eisensmith RC, et al. In vivo gene therapy for hyperlipidemia: phenotypic correction in Watanabe rabbits by hepatic delivery of the LDL receptor gene. J Clin Invest 1995; 95:768–773.
60. Kay MA, Landen CN, Rothenberg SR, et al. In vivo hepatic gene therapy: complete albeit transient correction of factor IX deficiency in hemophilia B dogs. Proc Natl Acad Sci USA 1994; 91:2353–2357.
61. Yang Y, Ertl HCJ, Wilson JM. MHC class I-restricted cytotoxic T lymphocytes to viral antigens destroy hepatocytes in mice infected with E1-deleted recombinant adenoviruses. Immunity 1994; 1:433–442.
62. Yang Y, Nunes FA, Berencsi K, Furth EE, Gönczöl E, Wilson JM. Cellular immunity to viral antigens limits E1-deleted adenoviruses for gene therapy. Proc Natl Acad Sci USA 1994; 91:4407–4411.
63. Yang Y, Nunes FA, Berencsi K, Gönczöl E, Englehardt JF, Wilson JM. Inactivation of E2a in recombinant adenoviruses improves the prospect for gene therapy in cystic fibrosis. Nature Genet 1994; 7:362–369.

64. Englehardt JF, Litzky L, Wilson JM. Prolonged transgene expression in cotton rat lung with recombinant adenoviruses defective in E2a. Human Gene Ther 1994; 5:1217–1229.

65. Englehardt JF, Ye X, Doranz B, Wilson JM. Ablation of E2A in recombinant adenoviruses improves transgene persistence and decreases inflammatory response in mouse liver. Proc Natl Acad Sci USA 1994; 91:6196–6200.

66. Fang B, Eisensmith RC, Wang H, et al. Gene therapy for hemophilia B: host immunosuppression prolongs the therapeutic effect of adenovirus-mediated factor IX expression. Human Gene Ther 1995; 6:1039–1044.

67. Yang Y, Trinchieri G, Wilson JM. Recombinant IL-12 prevents formation of blocking IgA antibodies to recombinant adenovirus and allows repeated gene therapy to mouse lung. Nature Med 1995; 1:890–893.

68. Yang Y, Greenough K, Wilson JM. Transient immune blockade prevents formation of neutralizing antibody to recombinant adenovirus and allows repeated gene transfer to mouse liver. Gene Ther 1996; 3:412–420.

69. Kay MA, Holterman AX, Meuse L, et al. Long-term hepatic adenovirus-mediated gene expression in mice following CTLA4Ig administration. Nature Genet 1995; 11:191–197.

70. Fang B, Wang H, Gordon G, et al. Lack of persistence of E1⁻ recombinant adenoviral vectors containing a temperature-sensitive E2A mutation in immunocompetent mice and dogs. Gene Ther 1995; 3:217–222.

71. Zhou H, Beaudet AL. Progress towards development of an adenoviral vector with deletions of E1 and E2a. J Cell Biochem 1995; 21A(suppl):434.1

72. Armentano D, Sookdeo CC, Hehir KM, et al. Characterization of an adenovirus gene transfer vector containing an E4 deletion. Human Gene Ther 1995; 6:1343–1353.

73. Krougliak V, Graham FL. Development of cell lines capable of complementing E1, E4 and protein IX defective adenovirus type 5 mutants. Human Gene Ther 1995; 6:1575–1586.

74. Wang Q, Jia X-C, Finer MH. A packaging cell line for propagation of recombinant adenovirus vecotrs conatining two lethal gene-region deletions. Gene Ther 1995; 2:775–783.

75. Yeh P, Dedieu JF, Orsini C, Vigne E, Denefle P, Perricaudet M. Efficient dual transcomplementation of adenovirus E1 and E4 regions from a 293-derived cell line expressing a minimal E4 functional unit. J Virol 1996; 70:559–565.

76. Bridge E, Ketner G. Redundant control of adenovirus late gene expression by early region 4. J Virol 1989; 63:631–638.

77. Lee MG, Haddada H, Perricaudet M. Reducing the immune response against adenoviral vectors. J Cell Biochem 1995; 21A(suppl):359.

78. Juillard V, Villefroy P, Godfrin D, Pavirani A, Venet A, Guillet JG. Long-term humoral and cellular immunity induced by a single immunization with replication-defective adenovirus recombinant vector. Eur J Immunol 1995; 25:3467–3473.

79. Mitani K, Graham FL, Caskey CT, Kochanek S. Rescue, propagation, and partial purification of a helper virus-dependent adenovirus vector. Proc Natl Acad Sci USA 1995; 92:3854–3858.

80. Kochanek S, Clemens PR, Mitani K, Chan S, Caskey CT. A new adenoviral vector for gene therapy: replacement of all viral coding sequences with 28 kb of foreign DNA. Am J Human Genet 1995; 57(suppl):A244.
81. Fisher KJ, Choi H, Burda J, Chen S-J, Wilson JM. Recombinant adenovirus deleted of all viral genes for gene therapy of cystic fibrosis. Virology 1996; 217:11–22.
82. Mittal SK, McDermott MR, Johnson DC, Prevec L, Graham FL. Monitoring foreign gene expression by a human adenovirus-based vector using the firefly luciferase gene as a reporter. Virus Res 1993; 28:67–90.
83. Kolls JK, Lei D, Odom G, et al. Use of CD4 lymphocyte depletion to prolong transgene expression of E1-deleted adenoviral vectors. Human Gene Ther 1996; 7:489–497.

14

Gene Therapy for Chronic Inflammatory Diseases of the Lungs

ANGELO E. CANONICO

Vanderbilt University School of Medicine
Nashville, Tennessee

I. Introduction

Gene therapy has potential as a future medical treatment for chronic inflammatory lung diseases. Most of the research in gene therapy for chronic inflammatory lung disease has focused on the inherited disorders alpha$_1$ antitrypsin deficiency (α1AT) and cystic fibrosis. Although the genetic bases for these diseases are different, both disorders are characterized by a chronic, neutrophil-dominated, inflammatory lung process. The presence of this persistent inflammatory environment in the lung is felt to contribute significantly to the progressive destruction of lung parenchyma and ultimately to respiratory failure. In α1AT deficiency, insufficient amount of the antiprotease alpha$_1$ antitrypsin results in a protease-antiprotease imbalance in the lung. For CF, the contribution of the genetic defect to the inflammatory state is more complicated and incompletely understood. Whether correcting the basic genetic defect will correct the inflammatory component if the disease, especially once the inflammatory cycle has been established, is unknown.

α1AT deficiency and CF are inherited, monogenetic disorders where no effective alternative therapies exist. However, other chronic inflammatory lung diseases may not have unique or identifiable genetic lesions responsible for the disease process. Nonetheless, gene therapy, viewed as a therapeutic intervention, has the potential to modify or attenuate the inflammatory component of these diseases. By targeting components of the inflammatory cascade, gene therapy can be utilized as a type of drug, inhibiting specific proinflammatory components of the disease process or augmenting the antiinflammatory defenses. Two examples of chronic inflammatory lung disease in which gene therapy may have potential in modifying the inflammatory environment are idiopathic pulmonary fibrosis and asthma.

In this review, I will discuss the progress in gene therapy in α1AT deficiency and CF, concentrating on the concept role of gene therapy as an antiinflammatory intervention. With IPF and asthma as models for other chronic inflammatory lung diseases, I will discuss the potential role of gene therapy as an antiinflammatory therapeutic intervention.

II. Alpha₁ Antitrypsin

Alpha₁ antitrypsin (α1AT) deficiency was initially described by Laurell and Eriksson in 1963 (1). In their seminal observation, they noted the association of severe, early-onset emphysema with an absent α1AT band on protein electrophoresis. Since then, much research has been done to elucidate the biochemical and molecular biological nature of this genetic defect (2–4). α1AT deficiency is a codominantly expressed genetic disorder seen predominantly in individuals of European descent. With an incidence of 1/3000, α1AT deficiency is second only to cystic fibrosis (CF) as the most common lethal hereditary disorder in whites.

The α1AT gene is a 12.2-kb gene located on chromosome 14q9 and encodes for a 52-kDa secreted glycoprotein. The α1AT gene has a high level of pleomorphism, with over 70 alleles documented. The liver is the primary site of protein synthesis; the monocyte-macrophage system contributes a small percentage of the total amount of α1AT. Although the liver is the primary site of synthesis, liver disease is less frequently seen than lung disease in α1AT deficiency and is not associated with all mutations. Disease manifestation occurs predominantly in the lung, where α1AT is responsible for over 95% of the antiprotease protection in the lower respiratory system. The normal variant, designated the M allele, is present in the majority of individuals. Because the gene is codominantly expressed, the presence of a normal M variant maintains adequate physiologic α1AT levels and obviates any risk for emphysema.

The most common genetic abnormality associated with premature emphysema is the Z allele. In this mutant allele, a lysine is substituted for glutamic acid at position 342. This substitution alters the three-dimensional configuration of the protein, which causes aggregation and accumulation in the rough endoplasmic reticulum of the hepatocyte. In addition to impaired protein secretions, the Z phenotype does not function normally as an inhibitor of neutrophil elastase. Other, rarer mutant alleles are responsible for reduced amounts or unstable α1AT protein, defective mRNA transcription, or intracellular destruction of the protein shortly after synthesis (4–7). Regardless of the genetic abnormality, a critical threshold of α1AT serum level less than 10 μM appears necessary for an individual to develop pulmonary emphysema (4).

In α1AT-deficient individuals, the hepatocytes are unable to produce or secrete adequate amounts of the antiprotease. Serum levels, and consequently epithelial lining fluid (ELF), are insufficient to provide adequate antiprotease protection in the lung. Unopposed elastolytic destruction of the lung parenchyma and subsequent panacinar emphysema ensues. The observations that α1AT deficiency causes premature emphysema coupled with animal models of protease-induced emphysema (8) spawned the protease theory of emphysema; that is, chronic, progressive destruction of the alveolar structures occurs because there is an imbalance between proteases and antiproteases in the lower respiratory tract. Neutrophils are recruited into the lungs by various stimuli and release elastase and other potentially toxic substances. In α1AT deficiency, the absence of this antiprotease permits unopposed elastolytic destruction of the lung parenchyma. Interestingly, the cause of liver disease in α1AT deficiency is not felt to be due to a protease imbalance but rather the intracellular accumulation of a defective protein (9). Unlike the risk for emphysema in individuals with critically low levels of α1AT, the risk for liver disease in confined almost entirely to Z homozygotes and other rare reported cases.

III. Gene Therapy for α1AT Deficiency

α1AT deficiency fulfills many of the criteria applicable to a gene therapy-based intervention (10). The monogenetic basis of the disease is understood, and the gene involved in the disease process is identified and characterized. A clear threshold for disease manifestation exists, and strict gene regulation does not appear necessary. Further, no ideal alternative therapy exists. Although α1AT protein augmentation is an approved therapy for deficient patients, weekly to monthly infusions are necessary to maintain adequate serum levels (11). In addition, α1AT protein used for intravenous augmentation is derived from human plasma, which exposes an individual to risks inherent in the transmission of bloodborne infectious agents.

Investigators have employed several different approaches to gene therapy for α1AT deficiency. Several factors relevant to α1AT protein and α1AT deficiency support investigating numerous strategies. First, the cellular biosynthesis does not require any specialized processing, and α1AT production can be achieved by numerous types of cells. Second, because it is a secreted protein, it can access the vascular compartment from different sites and localize in the lower respiratory tract. Third, since the protein is made predominantly in the liver but its site of protection is in the lung, it is logical to target specifically these two organs for in vivo gene expression.

A. Liver-Directed α1AT Gene Therapy

Because α1AT is normally synthesized in the liver, hepatocytes are an obvious cell target for α1AT gene therapy. Various strategies and delivery systems have successfully transferred the human α1AT gene to hepatocytes in vivo. Kay and colleagues harvested hepatocytes from dogs, transduced them ex vivo with an retrovirus vector containing the α1AT cDNA, and reinfused the transduced, autologous hepatocytes via the portal circulation back into the animal (12). Serum α1AT levels were detected for one month following gene transfer (Fig. 1). Because of the technical complexity and labor intensiveness of this procedure, these investigators researched alternative methods of hepatic gene delivery. Using a mouse model, Kay et al. performed partial hepatectomies followed by portal vein infusion of an recombinant retrovirus-α1AT vector (13). After portal vein infusion of recombinant retrovirus, 1% to 2% of hepatocytes were transduced and α1AT expression persisted for at least 6 months. This technique has also been successfully performed in rats using various liver-specific promoter elements (14).

Other gene delivery systems have also been used for liver-directed α1AT gene therapy. Direct injection of an α1AT cDNA containing recombinant adenovirus vector into the portal circulation of rats resulted in detectable serum levels of human protein for 4 weeks (15). Albino et al., employing plasmid DNA encapsulated into small liposomes, delivered the α1AT cDNA intravenously to the liver of mice and detected immunoreactive human α1AT in hepatocytes for 7 days after gene administration (16). Direct injection of naked plasmid DNA into the liver of animals also resulted in detectable levels of human α1AT in the serum (17).

Although successful liver-directed a1AT gene therapy is achieved by various strategies, serum α1AT levels in all these systems were far below what would be necessary for physiologic corrections. This obstacle has spurred many investigators to develop strategies to target the lower respiratory tract directly. Since ELF levels of α1AT required to maintain effective antielastase activity are much lower than serum levels, lung-directed gene therapy may circumvent the problems of low transgene expression.

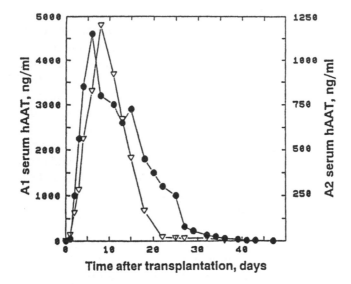

Figure 1 In vitro hAAT production in dogs A1 and A2 after transplantation of transduced hepatocytes. The serum concentrations of hAAT were determined before and after transplantation of 3.8×10^8 and 6.4×10^8 in animals A1 (solid circles) and A2 (open triangles), respectively. Each sample was analyzed in duplicate. (From: Kay MA, Baley P, Rothenberg S, et al. Proc Natl Acad Sci USA 1992; 89:89–93. Reprinted with permission.)

B. Lung-Directed α1AT Gene Therapy

One limitation of the current generation of gene delivery systems is inefficient transgene expression. The liver-directed gene therapy models previously described demonstrated successful gene transfer and persistent gene expression. Unfortunately, α1AT serum levels were far below the necessary amount for physiologic correction. In contrast, ELF levels of the protein are much lower than serum levels. By targeting the lung directly, local transgene production may be adequate to maintain antielastase activity in the ELF and to neutralize the protease burden present in the lower airways of α1AT deficient patients. Respiratory epithelial cell cultures, transduced with an adenovirus vector containing α1AT cDNA, secrete the protein to both the apical and basolateral surfaces of the epithelial cell (18), which suggests that local production of α1AT in respiratory epithelial cells can provide antielastase activity to both the ELF and the interstitium. Further, local expression of α1AT may prevent elastolysis within the neutrophil-epithelial cell interface, a microenvironment that excludes exogenously administered protein (19).

Rosenfeld and colleagues intratracheally delivered a recombinant adenovirus vector containing the α1AT cDNA to the lungs of cotton rats and demonstrated successful transgene expression (20). In their study, human α1AT mRNA transcripts were detected for 2 days following gene transfer in the respiratory epithelium, and human protein was detected in the ELF for 1 week. Using a plasmid-liposome complex (PLC) delivery system, we successfully delivered the α1AT cDNA directly into the lungs of New Zealand white rabbits (21). Transgene expression occurred by either intravenous or aerosol administration of the PLCs. RNA by Northern blot analysis revealed human α1AT transcripts present in the lung for at least 1 week, and organ cultures documented protein synthesis for at least 1 week as well. When the PLCs were delivered intravenously, protein was detected both in the pulmonary vascular endothelium and in the airway epithelium. In contrast, no protein was detected in the endothelium following aerosol delivery.

Because of the problems with adenoviral vector-induced inflammation (22–24), repetitive administration of this vector may be difficult. This may be less of a concern for PLC delivery systems. To address the potential toxicity of repeated administration of plasmid-liposome complexes to the lungs, we delivered intravenously or by aerosol plasmid-liposome complexes containing the human α1AT cDNA to New Zealand white rabbits at weekly intervals over a 4-week period (25). Protein expression persisted over this 4-week period (Fig. 2). No evidence of toxicity was documented by histology, pulmonary mechanics, or AaO_2 gradient (Fig. 3). From this and other preliminary data, we received approval and have begun a human gene therapy protocol studying PLC gene delivery of the α1AT cDNA into the airway epithelium of humans. This protocol consists of two arms. In the first arm, PLCs containing the human α1AT cDNA are administered directly to the nasal epithelium of α1AT deficiency individuals. In the second arm, the PLCs are bronchoscopically instilled into a region of normal lung which will subsequently be resected for other surgical indications. The first arm provides us an opportunity to assess directly transgene expression in α1AT-deficient individuals, and the second arm provides us an opportunity to assess the response to direct lower-airway administration of plasmid-liposome complexes.

C. Other Strategies for α1AT Gene Therapy

Because α1AT is a secreted protein and no unique intracellular processing is required to synthesize the protein, exploitation of other cells or body sites are potential approaches to α1AT gene therapy. Garver et al. transformed mouse fibroblasts ex vivo with a retroviral vector containing the α1AT cDNA and transplanted the fibroblasts into the peritoneal cavities of nude mice (26). Human protein was detected in both the sera and epithelial surface of the lungs

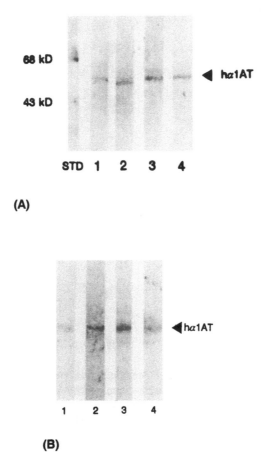

Figure 2 hα1-AT immunoreactive protein in lung organ cultures. (A) Intravenous administration of plasmid-liposome complexes. (B) Aerosol administration of plasmid-liposome complexes. STD, protein standard; lanes 1–4, weeks 1–4, respectively. (From: Canonico AE, Plitman JD, Conary JT, et al. J Appl Physiol 1994; 77:415–419. Reprinted with permission.)

for 1 month. Transplanted fibroblasts recovered four weeks later were still producing human α1AT. In another study, the peritoneal cavity functioned again as a biosource for transgene expression. Setoguchi and colleagues intraperitoneally injected an adenovirus vector containing the human α1AT cDNA into cotton rats and detected human α1AT protein in the serum for up to 24

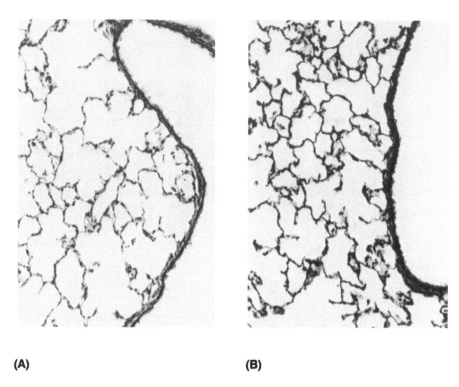

(A)   **(B)**

Figure 3 Light micrographs of alveolar region and bronchiolus after three or four weekly administrations of plasmid-liposome complexes. (A) Intravenous. (B) Aerosol. (C) Control. Hemotoxylin and eosin staining. Magnification ×320. (From: Canonico AE, Plitman JD, Conary JT, et al. J Appl Physiol 1994; 77:415–419. Reprinted with permission.)

days (Fig. 4) (23). Expression of the transgene remained localized to the peritoneal mesothelium, as PCR analysis did not detect expression of the exogenous gene in any other tissues evaluated.

In principle, α1AT deficiency is an ideal candidate for a gene therapy based intervention. However, the current generation of gene delivery systems is inadequate to produce sufficient serum levels of the antiprotease to achieve physiologic correction. Lung-directed α1AT gene therapy may overcome this limitation by directly augmenting antiprotease levels in the ELF and interstitium. Further, transgene expression of α1AT within specific cells may confer additional benefits and protection not seen with protein augmentation therapy (27). Nonetheless, further work is necessary to determine whether these promising observations will translate into successful medical therapy.

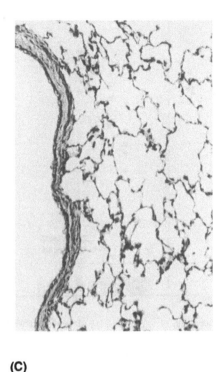

(C)

Figure 3 Continued

IV. Cystic Fibrosis

With an incidence of 1 in 2500 live births, CF is the most common lethal genetic disease in Caucasians (28). It rarely affects individuals of African or Asian ancestry. CF is characterized by a defect in epithelial chloride channel activity caused by mutations in the CF transmembrane conductance regulator (CFTR) gene (29-31). This gene, located on chromosome 7, consists of approximately 250,000 bp and encodes for an mRNA of 6.5 kb (32). Over 350 mutations have been identified; however, more than 70% of individuals afflicted with CF carry the delta F508 genetic defect.

The pathological manifestations of a defective CFTR are seen in cells of epithelial origin. Clinical manifestations of the disease occur primarily in the lung, liver, pancreas, and intestinal tract. Males are usually sterile. Although CF affects several organ systems, over 90% of deaths are the result of chronic obstructive pulmonary disease and respiratory failure. Consequently, most gene

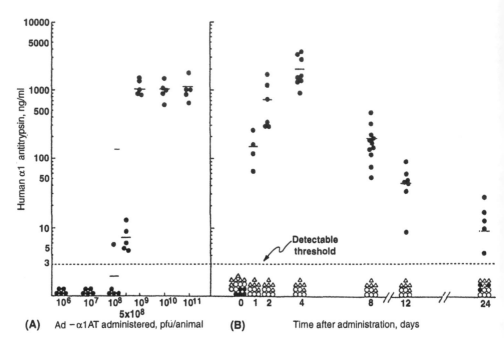

Figure 4 Serum human α1AT levels in cotton rats following intraperitoneal admini-stration of Adα1AT. Human α1AT levels were quantified by ELISA. Each data point represents the mean of duplicate determinations for one animal. The horizontal lines indicate the mean α1AT level of that group of determinations. The detectable thresh-old of the ELISA is 3 ng/ml (indicated by the dashed line). (A) Serum human α1AT levels following intraperitoneal administration of increasing amount of Adα1AT 4 days previously. pfu, plaque-forming units. (B) Serum human α1AT levels as a function of time following intraperitoneal administration of Adα1AT (closed circles), control vec-tor AdCFTR (open circles), or PBS (triangles). The zero from time point represents 1 hour after intraperitoneal administration of the vector. (From: Setoguchi Y, Jaffe HA, Chu C-S, Crystal RG. Am J Respir Cell Mol Biol 1994; 10:369–377. Reprinted with permission.)

replacement strategies have focused on correcting the genetic defect in the lungs.

A. Gene Therapy for CF

In the lung, CFTR has been localized to the airway epithelium and to the submucosal glands (33–36). CFTR functions as a regulated chloride channel, and, through mechanisms not completely understood, a defective CFTR results both in the lack of cyclic adenosine monophosphate (cAMP)-mediated chloride secretion and sodium hyperabsorption (37). Various pharmacological therapies attempt to correct the abnormalities caused by a defective CFTR (38). Rather than correcting the consequence of a defective CFTR, gene replacement therapy attempts to correct the underlying genetic abnormality. Utilizing different gene delivery systems, numerous in vitro studies have demonstrated successful transfer of a normal CF gene to airway epithelial cells (39–41). In vivo studies in several animal models have further confirmed that the human CFTR gene can be delivered to the airways and function as an ion transport channel (42–45). Based on extensive preclinical data, three gene delivery systems—the adenovirus, adeno-associated virus, and PLCs—have been approved for human gene therapy trials. Results from different adenovirus and PLC gene therapy trials have been reported (24,46–48). A consistent limitation in the reported human CF gene therapy trials is the low in vivo transduction efficiency for both delivery systems. Nonetheless, it is conceptually encouraging that a normal CF gene can be transferred in vivo to humans. As newer generations of gene delivery systems are developed with improved transduction efficiencies, CFTR gene therapy should become a viable therapeutic intervention in the future.

In addition to improving transduction efficiency, several issues regarding CFTR gene therapy remain unresolved. First, in addition to serving as a regulated chloride channel, additional data suggest that the CFTR is involved in other cellular functions (49–51); the loss of these functions to the pathogenesis of CF is unknown. Second, it is unclear which cells should ideally express the normal CFTR. If CF lung disease is a consequence of defective ion transportation, then delivery of a normal CF gene to the airway epithelium could correct these ion abnormalities and restore normal function and defense to the airways. However, if a normally functioning submucosal gland is necessary, then gene therapy strategies targeting the airway epithelium may be insufficient to prevent lung disease. Third, the percentage of cells expressing the normal gene required to correct the ion transport is unknown. In cell culture and bronchial xenografts, only 5% to 10% of cells must be transduced to correct the defective chloride transport (39,41), but correction of the sodium hyperabsorption requires a much higher rate of transduction (40,41). If sodium hyperabsorption is necessary, much higher transduction efficiency will be required. Finally, no study has correlated CFTR gene replacement with correction of the inflammatory process. It is unknown whether correction of the ion transport defect will be sufficient to ablate the inflammatory manifestations of the disease, especially once the inflammatory cycle is established.

Lung disease in CF is characterized by opportunistic bacterial infections—predominantly *Staphylococcus aureus* and *Pseudomonas aeruginosa*, neutrophil-dominated pulmonary inflammation, and a proinflammatory cytokine milieu. Although the mechanisms by which a defective CFTR results in lung disease are unknown, recent work has shed some light on how this defect causes lung disease. Smith et al. demonstrated that bactericidal activity of CF epithelia was impaired by the high NaCl concentration of the CF surface fluid (52). Pier et al. showed that cultured airway epithelial cells expressing the delta F508 allele were defective in their uptake of *P. aeruginosa* compared to cells expressing the wild-type allele (53). The abnormal CF epithelia may permit increased adherence of *P. aeruruginosa* to the cell (54) and the high level of CFTR expression in the submucosal glands may cause their secretions to be abnormal (55).

By mechanisms currently being elucidated, the defective CFTR results in a marked neutrophil dominated pulmonary inflammation. In patients with severe lung disease, a relationship has been established between lung inflammation and lung destruction. But even in patients with stable, clinically mild

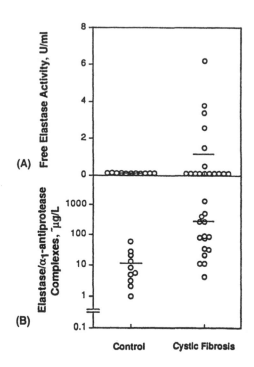

Figure 5   Free neutrophil elastase activity (A) and elastase/alpha$_1$-antiprotease levels (B) in the cell-free BALF of control infants and infants with CF. The bars represent the level of elastase and elastase/alpha$_1$-antiprotease complexes in each group. The levels of both free elastase and elastase/alpha$_1$-antiprotease complexes were significantly elevated in the CF group ($P = .03$ for elastase and $P < .001$ for elastase/alpha$_1$-antiprotease complexes.) (From: Khan TZ, Wagener JS, Bost T, Martinez J, Accurso FJ. Am J Respir Crit Care Med 1995; 151:1075–1082. Reprinted with permission.)

disease, bacteria reside in the lung and promote a persistent inflammatory environment (56). Further, although it has generally been hypothesized that the pulmonary inflammation is a result of bacterial infection, recent work suggests that inflammation may precede infection and contribute to lung destruction either directly or indirectly by establishing an environment in which bacterial pathogens flourish (Fig. 5) (57).

Chronic inflammation with large numbers of activated neutrophils in the lungs can cause pulmonary damage due to the release of neutrophil elastase (NE) and other potentially toxic substances (58). In addition to cleaving a variety of plasma proteins, NE is capable of degrading almost every connective tissue component of the lung. Normally, the lung contains sufficient quantities of α1AT to combat the effects of NE. However, in CF the antiprotease defense system may be overwhelmed, resulting in proteolytic damage to lung tissue (59–61).

In addition to direct proteolytic destruction, NE can affect lung destruction in CF indirectly. It can induce airway epithelium to produce several cytokines, including interleukin-8 (IL-8) (62). IL-8 is present in the bronchoalveolar lavage fluid of CF patients (63) and may contribute to the pathogenesis of neutrophil dependent airway inflammation. Excess NE impairs neutrophil killing of *P. aeruginosa* (64). NE also cleaves transferrin and lactoferrin, releasing iron which allows *Pseudomonas* to grow more readily (65). NE reduces the ciliary beat frequency and increases mucus production (66,67). Mucus hypersecretion, impaired ciliary clearance, and the accumulation of DNA from dead inflammatory cells may contribute to airflow obstruction in these patients (Fig. 6).

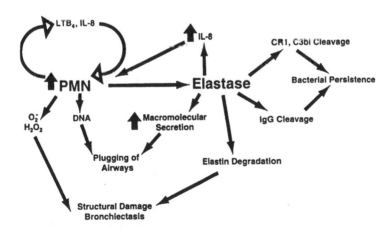

Figure 6 Neutrophil-mediated inflammation in CF lung disease. (From: Davis PB. Highlights [1994 CF Foundation Proceedings, Oct. 20–23] 1995; p. 6. Reprinted with permission.)

A growing body of evidence implicates NE as a primary mediator for destruction in CF. In CF, high levels of free NE have been measured in lung, and there is both biochemical and pathological evidence for active elastolysis (59–61,68). Although α1AT, the predominant antiprotease responsible for neutralizing NE, is present in the lungs from CF patients, most of it is in an inactive form (59,69). It is possible that α1AT could be oxidatively inactivated, but little data exist to support this idea. On the contrary, numerous studies have indicated that α1AT is proteolytically inactivated and this proteolysis is due to the overwhelming NE burden found in the lungs of CF patients (60,69).

Since the isolation and cloning of the CFTR gene, extensive work in gene therapy has focused on the possibility of CFTR gene replacement as treatment for CF. Although conceptually appealing, once the destructive cycle has begun, correction of the underlying genetic defect may not completely ablate the physiologic manifestations of the disease process. In addition to correcting the genetic defect, further therapy directly targeting the inflammatory component of the disease may be beneficial.

With the data accumulated implicating NE as a major agent of lung destruction in CF, antiprotease therapy seems rational. Available evidence suggests that a "protease-antiprotease" imbalance exists in the lungs of CF patients. In contrast to α1AT deficiency, in which the lung disease is the result of a systemic deficiency in antiprotease protection (2), lung disease in CF is the result of a local overwhelming protease burden. Therapies directed against the inflammatory component or against the products of inflammation in the CF airways have shown promising results. For example, ibuprofen, a cyclooxygenase inhibitor and anti-inflammatory agent (70), decreases the airway inflammation and the decline in pulmonary function in a subset of CF patients. Antielastase protein therapy with α1AT (71) or secretory leukoprotease inhibitor (SLPI) (63) decreases airway inflammation as well, and studies are under way to evaluate the role of antielastase protein therapy in preventing the ongoing lung destruction seen in CF.

Even though aerosol therapy with SLPI or α1AT protein corrects the protease-antiprotease imbalance in the ELF of CF patients (63,71), twice-daily therapy is necessary and is prohibitively expensive. α1AT gene therapy may permit administration of the transgene at weekly or possibly longer intervals with a more cost-effective profile. In addition, by targeting the epithelial cells, α1AT gene therapy may prevent elastolysis within the neutrophil-epithelial cell interface, a microenvironment that excludes the exogenously administered protein (19).

Current strategies for α1AT gene therapy to the lung have been previously described (see above). However, no in vivo studies with antiprotease gene therapy in CF have been performed. We examined the role of α1AT gene therapy in a CF bronchial epithelial cell line and its protective role against NE.

Using a PLC delivery system to transfer the human α1AT cDNA, we were able to demonstrate a protective benefit in CF cells transfected with α1AT (72). Release of neutrophil chemotactic factor(s) were blocked in CF cells transfected with α1AT and exposed to NE (Fig. 7). In addition, α1AT gene transfer prevented elastase-induced cell detachment (Fig. 8). This preliminary study suggests that α1AT gene therapy may be a therapeutic alternative to antiprotease protein augmentation of the lower airways in CF and may complement CFTR gene therapy in CF.

As newer generations of gene delivery systems are developed, CF gene therapy should become an acceptable therapeutic modality in the treatment of individuals afflicted with CF. Because inflammation is a significant contributor to lung destruction, anti-inflammatory therapy will be necessary to prevent the progressive lung destruction seen in this disease. To cure the lung disease

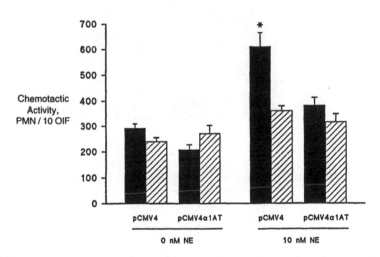

Figure 7 Rabbit and antihuman IL-8 antibody neutralizes chemotactic activcity in the supernatant of 2CFSMEo- cells after 24 hours of exposure to NE. Solid boxes: treatment with nonimmune rabbit serum. Patterned boxes: treatment with anti-IL-8 rabbit serum. The supernatant from cells transfected with pCMV4 and exposed to 10 nM NE and treated with nonimmune rabbit serum (*) had significantly greater chemotaxis than all other cells exposed to 10 nM NE ($P < .05$). Supernatant from cells transfected with pCMV4α1AT and exposed to 10 nM NE did not differ significantly in chemotaxis compared with supernatant from cells not exposed to NE. (From: Canonico AE, Brigham, KL, Carmichael LC, et al. Am J Respir Cell Mol Biol 1996; 14:348–355. Reprinted with permission.)

by transferring a normal CFTR gene to the airways of CF patients is conceptually appealing. Unfortunately, no data exist that correcting the defective CF gene will ablate the inflammatory component of the disease. Once the inflammatory cycle is established, additional therapy may be necessary to interrupt this cycle. Various strategies targeting the inflammatory component of CF are under investigation. Antiinflammatory gene therapy with α1AT or some other antiprotease is one strategy that may complement CFTR gene therapy and "cure" the deadly lung component of this disease.

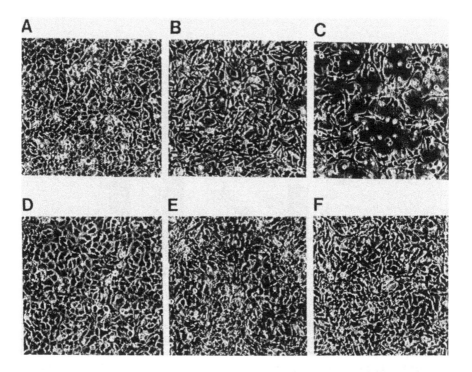

Figure 8 Transfer of the hα1AT gene to 2CFSMEo- cells prevents cell detachment following 24 hours of exposure to NE. Panels A to C: Cells transfected with 10 μg pCMV4. Panels D through F: Cells transfected with 10 μg pCMV4α1AT. Panels A and D: 0 nM NE; panels B and E: 10 nM NE; panels C and F: 25 nM NE. (From: Canonico AE, Brigham, KL, Carmichael LC, et al. Am J Respir Cell Mol Biol 1996; 14:348–355. Reprinted with permission.)

V. Other Chronic Inflammatory Lung Diseases Potentially Amenable to Gene Therapy

Most research in gene therapy for chronic inflammatory lung diseases has focused on α1AT deficiency and CF. However, idiopathic pulmonary fibrosis (IPF) and asthma are two other diseases in which gene therapy may play a potential role in the future.

IPF is a lethal disease with a medium time from diagnosis to death of 3 to 5 years (73,74). Current therapies for this disease have marginal effect on improved lung function or overall survival. Because of this, research into aggressive or novel therapeutic interventions seems justified. In IPF, an inflammatory response to an undefined insult or injury occurs followed by an exuberant fibrotic response. The initial inflammatory response is predominantly neutrophilic but evolves to a predominant lymphocytic and monocytic response. Although no specific genetic defect has been found, gene therapy could target specific sites in the disease pathway. Oxidants and proteases have been implicated in the progression of lung disease and could be a target for gene therapy.

Expression of tumor necrosis factor α and transforming growth factor β are increased in IPF (75,76). Treatment with the soluble TNF-α receptor has been shown to be beneficial in an animal model of pulmonary fibrosis (77). Instead of administering the receptor itself, expression of the soluble TNF-α receptor gene permits a localized, persistent expression of this inhibitor (78). Antisense gene therapy suppressing the production of these mediators may prevent progressive fibrosis (79). For example, antisense gene therapy could target specific growth factors or cytokines implicated in IPF. Another approach would be to express natural cytokine inhibitors such as interleukin-1 receptor antagonist or interleukin-4, which could block the effects of certain proinflammatory cytokines.

Another possibility involves hyperexpressing the cyclo-oxygenase gene. Wilborn et al. have shown that cultured fibroblasts from IPF patients are deficient in cyclo-oxygenase-2 (80). Transfer of the cyclo-oxygenase-2 gene could increase the amount of anti-inflammatory prostanoids PGE_2 and prostacyclin within the lung parenchyma thereby modulating the inflammatory and fibrotic component of the disease. In vivo gene therapy with the cyclooxygenase-1 gene has already been used successfully in an animal model of endotoxin lung injury (81) and has promise in the treatment of IPF.

Several of the same gene therapeutic strategies theoretically beneficial in IPF are applicable to asthma. Cyclo-oxygenase gene therapy and the consequent production of the PGE_2 could mediate the inflammation present in the airways of asthmatics (82). Antisense gene therapy directed against cellular adhesion molecules such as intercellular adhesion molecule-1 (ICAM-1) could modulate leukocyte trafficking into the lungs. Modulation of the cytokine cascade is another possibility for gene therapy. This could include antisense gene therapy against several cytokines such as TNF-α, IL-4, or IL-5 or possibly gene therapy with interferon γ or IL-10 (83).

VI. Conclusion

Most of the research in gene therapy for chronic inflammatory lung disease has focused on alpha$_1$ antitrypsin deficiency and cystic fibrosis. Significant advances have been made in the field of gene therapy for lung diseases. Current work has shown that exogenous DNA can be transferred to the lungs of animals and humans and function in a physiologically appropriate manner. Yet, much more research needs to be done. Most importantly, improvement in gene delivery vectors will be necessary for this new therapeutic intervention to realize its full potential.

In the future, routine therapy for chronic inflammatory lung diseases may include gene therapy. The exogenously administered gene may directly target the genetic cause of the inflammation such as alpha$_1$ antitrypsin deficiency or possibly cystic fibrosis. Alternatively, gene therapy may be utilized as a strategy against a component or components of the inflammatory process. This may involve antiprotease gene therapy in cystic fibrosis or idiopathic pulmonary fibrosis, antisense gene therapy, expression of antiinflammatory cytokines, or expression of cyclo-oxygenase, which exploits the antiinflammatory properties of its products.

Acknowledgements

This work was supported by grants no. HL 19153, HL 45151, and HL 07123 from the National Institutes of Health, National Heart, Lung, and Blood Institute; and the Cystic Fibrosis Foundation.

References

1. Laurell CB, Eriksson S. The electrophoretic $\alpha 1$-globulin pattern of serum in $\alpha 1$-antitrypsin deficiency. Scand J Clin Lab Invest 1963; 15:132–140.
2. Gadek JE, Crystal RG. $\alpha 1$-antitrypsin deficiency. In: Stanbury JB, Wyngaarden JB, Fredrickson DS, Goldstein JL, Brown MS, eds. The Metabolic Basis of Inherited Disease. New York: McGraw-Hill, 1982:1450–1467.
3. Crystal RG. The $\alpha 1$-antitrypsin gene and its deficiency states. TIG 1989; 5:411–417.
4. Crystal RG, Brantly ML, Hubbard RC, Curiel DT, States JT, Holmes MD. The alpha$_1$-antitrypsin gene and its mutations. Clinical consequences and strategies for therapy. Chest 1989; 95:196–208.

5. Curiel D, Brantly M, Curiel E, Stier L, Crystal RG. α1-antitrypsin deficiency caused by the α1-antitrypsin null$_{mattawa}$ gene. J Clin Invest 1989; 83:1144–1152.

6. Curiel D, Brantly M, Curiel E, Stier LE, Crystal RG. Molecular basis of α1- antitrypsin deficiency and emphysema associated with the α1-antitrypsin M$_{mineral springs}$ allele. Mol Cell Biol 1990; 10:47–56.

7. Curiel DT, Homes MD, Okayama H, et al. Molecular basis of the liver and lung disease associated with the α1-antitrypsin deficiency allele M$_{malton}$. J Biol Chem 1989; 264:13938–13945.

8. Gross P, Pfitzer EA, Tolker E, Babyak MA, Kaschak M. Experimental emphysema. Arch Environ Health 1965; 11:50–58.

9. Lomas DA, Evans DL, Finch JT, Carrell RW. The mechanism of Z alpha 1-antitrypsin accumulation in the liver. Nature 1992; 357:605–607.

10. Anderson WF. Human gene therapy. Science 1992; 256:808–813.

11. Wewers MD, Casolar MA, Sellers SE, et al. Replacement therapy for alpha$_1$-antitrypsin deficiency associated with emphysema. N Engl J Med 1987; 316:1055–1062.

12. Kay MA, Baley P, Rothenberg S, et al. Expression of human α1-antitrypsin in dogs after autologous transplanatation of retroviral transduced hepatocytes. Proc Natl Acad Sci USA 1992; 89:89–93.

13. Kay MA, Li Q, Liu TJ, et al. Hepatic gene therapy: persistent expression of human α1-antitrypsin in mice after direct gene delivery in vivo. Human Gene Ther 1992; 3:641–647.

14. Hafenrichter DG, Ponder KP, Rettinger SD, et al. Liver-directed gene therapy: evaluation of liver specific promoter elements. J Surg Res 1994; 56:510–517.

15. Jaffe HA, Danel C, Longenecker G, et al. Adenovirus-mediated in vivo gene transfer and expression in normal rat liver. Nature Genet 1992; 1:372–378.

16. Alino SF, Bobadilla M, Garcia-Sanz M, Lejarreta M, Unda F, Hilario E. In vivo delivery of human α1-antitrypsin gene to mouse hepatocytes by liposomes. Biochem Biophys Res Commun 1993; 192:174–181.

17. Hickman MA, Malone RW, Lehmann-Bruinsma K, et al. Gene expression following direct injection of DNA into liver. Human Gene Ther 1994; 5:1477–1483.

18. Siegfried W, Rosenfeld M, Stier L, et al. Polarity of secretion of α1-antitrypsin by human respiratory epithelial cells after adenoviral transfer of a human α1-antitrypsin cDNA. Am J Respir Cell Mol Biol 1995; 12:379–384.

19. Campbell EJ, Campbell MA. Pericellular proteolysis by neutrophils in the presence of proteinase inhibitors: effects of substrate opsonization. J Cell Biol 1988; 106:667–676.

20. Rosenfeld MA, Siegfried W, Yoshimura K, et al. Adenovirus-mediated transfer of a recombinant α1-antitrypsin gene to the lung epithelium in vivo. Science 1991; 252:431–434.

21. Canonico AE, Conary JT, Meyrick BO, Brigham KL. Aerosol and intravenous transfection of human α1-antitrypsin gene to lungs of rabbits. Am J Respir Cell Mol Biol 1994; 10:24–29.

22. Yang Y, Nunes FA, Berencsi K, Furth EE, Gonczol E, Wilson JM. Cellular immunity to viral antigens limits E1-deleted adenovirus for gene therapy. Proc Natl Acad Sci USA 1994; 91:4407–4411.

23. Setoguchi YH, Jaffe A, Chu CS, Crystal RG. Intraperitoneal in vivo gene therapy to deliver α1-antitrypsin to the systemic circulation. Am J Respir Cell Mol Biol 1994; 10:369–377, 1994.

24. Knowles MR, Hohneker KW, Zhou Z, et al. A controlled study of adenoviral-vector-mediated gene transfer in the nasal epithelium of patients with cystic fibrosis. N Engl J Med 1995; 333:823–831.

25. Canonico AE, Plitman JD, Conary JT, Meyrick BO, Brigham KL. No lung toxicity after repeated aerosol or intravenous delivery of plasmid-cationic liposome complexes. J Appl Physiol 1994; 77(1):415–419.

26. Garver RI Jr, Chytil A, Courtney M, Crystal RG. Clonal gene therapy: transplanted mouse fibroblast clones express human α1-antitrypsin gene in vivo. Science 1987; 237:762–764.

27. Persmark M, Canonico AE, Brigham KL, Stecenko AA. Inhibition of respiratory syncytial virus (RSV) infectivity by liposome-mediated antiprotease gene transfer. J Invest Med 1995; 43:220A.

28. Boat TF, Welsh MJ, Beaudet AL. Cystic fibrosis. In: Scriver CR, Beaudet AL, Sly WS, Valle D, eds. The Metabolic Basis of Inherited Disease. New York: McGraw-Hill, 1989:2649–2680.

29. Rommens JM, Iannuzzi MC, Kerem B-S, et al. Identification of the cystic fibrosis gene: chromosome walking and jumping. Science 1989; 245:1059–1065.

30. Riordan JR, Rommens JM, Kerem B-S, et al. Identification of the cystic fibrosis gene: cloning and characterization of complementary DNA. Science 1989; 245: 1066–1073.

31. Kerem BS, Rommens JM, Buchanan JA, et al. Identification of the cystic fibrosis gene: genetic analysis. Science 1989; 245:1073–1080.

32. Collins FS. Cystic fibrosis: molecular biology and therapeutic implications. Science 1992; 256:774–779.

33. Trapnell BC, Chu C-S, Paako P-K, et al. Expression of the cystic fibrosis transmembrane conductance regulator gene in the respiratory tract of normal individuals and individuals with cystic fibrosis. Proc Natl Acad Sci USA 1991; 88:6565–6569.

34. Sarkadi B, Bauzon D, Huckle WR, et al. Biochemical characterization of the cystic fibrosis transmembrane conductance regulator in normal and cystic fibrosis epithelial cells. J Biol Chem 1992; 267:2087–2095.

35. Engelhardt JF, Zepeda M, Cohn JA, Yankaskas JR, Wilson JM. Submucosal glands are the predominant site of CFTR expression in the human bronchus. Nature Genet 1992; 2:240–248.

36. Jacquot JE, Puchelle C, Hinnrasky J, et al. Localization of the cystic fibrosis transmembrane conductance regulator in airway secretory glands. Eur Respir J 1993; 6:169–176.

37. Boucher RC. Airway epithelial fluid transport. Am Rev Respir Dis 1994; 150:271–281, 581–593.

38. Wagne JA, Chao AC, Gardner P. Molecular strategies for therapy of cystic fibrosis. Annu Rev Pharmacol Toxicol 1995; 35:257–276.

39. Johnson LG, Olsen JC, Sarkadi B, Moore KL, Swanstrom R, Boucher RC. Efficiency of gene transfer for restoration of normal airway epithelial function in cystic fibrosis. Nature Genet 1992; 2:21–25.

40. Johnson LG, Boyles SE, Wilson J, Boucher RC. Normalization of raised sodium absorption and raised calcium-mediated chloride secretion by adenovirus-mediated expression of cystic fibrosis transmembrane conductance regulator in primary human cystic fibrosis airway epithelial cells. J Clin Invest 1995; 95:1377–1382.

41. Goldman M, Yang Y, Wilson JM. Gene therapy in a xenograft model of cystic fibrosis lung corrects chloride transport more effectively than the sodium defect. Nat Genet 1995; 9:126–131.

42. Flotte TR, Afione SA, Conrad C, et al. Stable in vivo expression of the cystic fibrosis transmembrane conductance regulator with an adeno-associated virus vector. Proc Natl Acad Sci USA 1993; 90:10613–10617.

43. Alton EWFW, Middleton PG, Caplen NJ, et al. Non-invasive liposome-mediated gene delivery can correct the ion transport defect in cystic fibrosis mutant mice. Nature Genet 1993; 5:135–142.

44. Hyde SD, Gill DR, Higgins CF, et al. Correction of the ion transport defect in cystic fibrosis transgenic mice by gene therapy. Nature 1993; 362:250–255.

45. Rosenfeld MA, Yoshimura K, Trapnell BC, et al. In vivo transfer of the human cystic fibrosis transmembrane conductance regulator gene to the airway epithelium. Cell 1992; 68:143–155.

46. Zabner J, Couture LA, Gregory RJ, Graham SM, Smith AE, Welsh MJ. Adenovirus-mediated gene transfer transiently corrects the chloride transport defect in nasal epithelia of patients with cystic fibrosis. Cell 1993; 5:207–216.

47. Caplan NJ, Alton EW, Middleton PG, et al. Liposome-mediated CFTR gene transfer to the nasal epithelium of patients with cystic fibrosis. Nature Med 1995; 1:39–46.

48. Crystal RG, McElvaney NG, Rosenfeld MA, et al. Administration of an adenovirus containing the human CFTR cDNA to the respiratory tract of individuals with cystic fibrosis. Nature Genet 1994; 8:42–51.

49. Bradbury NMA, Jilling T, Berta G, Sorscher EJ, Bridges RJ, Kirk KL. Regulation of plasma membrane recycling by CFTR. Science 1992; 256:530–532.

50. Barasch J, Kiss B, Prince A, Saiman L, Gruenert D, Al-Awqati Q. Acidification of intracellular organelles is defective in cystic fibrosis. Nature 1992; 352:70–73.

51. Mills CL, Pereira MMC, Dormer RL, McPherson, MA. An antibody against a CFTR derived synthetic peptide inhibits beta-adrenergic stimulation of mucin secretion. Biochem Biophys Res Commun 1992; 188:1146–1152.

52. Smith JJ, Travis SM, Greenberg EP. Cystic fibrosis airway epithelial fail to kill bacteria because of abnormal airway surface fluid. Cell 1996; 85:229–236.

53. Pier GB, Grout M, Zaide TS, et al. Role of mutant CFTR in hypersusceptibility of cystic fibrosis patients to lung infections. Science 1996; 271:64–66.

54. Saiman L, Prince A. *Pseudomonas aeruginosa* pili bind to asialoGM1 which is increased on the surface of cystic fibrosis epithelial cells. J Clin Invest 1993; 92:1875–1880.

55. Engelhardt JF, Yankaskas JR, Ernst SA, et al. Submucosal glands are the predominant site of CFTR expression in the human bronchus. Nature Genet 1992; 2:240–248.

56. Konstan MW, Hilliard KA, Norvell TM, Berger M. Bronchoalveolar lavage findings in cystic fibrosis patients with stable, clinically mild lung disease suggest ongoing infection and inflammation. Am J Respir Crit Care Med 1995; 150:448–454.

57. Khan TZ, Wagener JS, Bost T, Martinez J, Accurso FJ, Riches DW. Early pulmonary inflammation in infants with cystic fibrosis. Am J Respir Crit Care Med 1995; 151:1075–1082.

58. Weiss SJ. Tissue destruction by neutrophils. N Engl J Med 1989; 320:365–376.

59. Meyer KC, Lewandoski JR, Zimmerman JJ, Nunley D, Calhoun WJ, Dopico GA. Human neutrophil elastase and elastase/alpha₁-antiprotease complex in cystic fibrosis. Am Rev Respir Dis 1991; 144:580–585.

60. Cantin A., Lafrenaye S, Begin RO. Antineutrophil elastase activity in cystic fibrosis serum. Pediatr Pulmonol 1991; 11:249–253.

61. Bruce MC, Poncz L, Klinger JD, Stern RC, Tomashefski JF Jr, Dearborn DG. Biochemical and pathologic evidence for proteolytic destruction of lung connective tissue in cystic fibrosis. Am Rev Respir Dis 1985; 132:529–535.

62. Ruef C, Jefferson DM, Schlegel-Hauter SE, Suter S. Regulation of cytokine secretion by cystic fibrosis airway epithelial cells. Eur Respir J 1993; 6:1429–1436.

63. McElvaney NG, Nakamura H, Birrer P, et al. Modulation of airway inflammation in cystic fibrosis. J Clin Invest 1992; 90:1296–1301.

64. Tosi MF, Zakem H, Berger M. Neutrophil elastase cleaves C3bi on opsonized *Pseudomonas* as well as CR1 on neutrophils to create a functionally important opsonin receptor mismatch. J Clin Invest 1990; 86:300–308.

65. Britigan BE, Hayek MB, Doebbeling BN, Fick RB Jr. Transferrin and lactoferrin undergo proteolytic cleavage in the *Pseudomonas aeruginosa*-infected lungs of patients with cystic fibrosis. Infect Immun 1993; 61:5049–5055.

66. Amitani R, Wilson R, Rutman A, et al. Effects of human neutrophil elastase and *Pseudomonas aeruginosa* proteinases on human respiratory epithelium. Am J Respir Cell Mol Biol 1991; 4:26–32.

67. Sommerhoff CP, Nadel JA, Basbaum CB, Caughey GH. Neutrophil elastase and cathepsin G stimulate secretion from cultured bovine airway gland serous cells. J Clin Invest 1990; 85:682–689.

68. Goldstein W, Doring G. Lysosomal enzymes from polymorphonuclear leukocytes and proteinase inhibitors in patients with cystic fibrosis. Am Rev Respir Dis 1986; 134:49–56.

69. Suter S, Chevalier I. Proteolytic inactivation of α_1-proteinase inhibitor in infected bronchial secretions from patients with cystic fibrosis. Eur Respir J 1991; 4:40–49.

70. Konstan MW, Byard PJ, Hoppel CL, Davis PB. Effect of high-dose Ibuprofen in patients with cystic fibrosis. N Engl J Med 1995; 332:848–854.

71. McElvaney MG, Hubbard RC, Birrer P, et al. Aerosol α_1-antitrypsin treatment for cystic fibrosis. Lancet 1991; 337:392–394.

72. Canonico AE, Brigham KL, Carmichael LC, et al. Plasmid-liposome transfer of the $\alpha 1$ antitrypsin gene to cystic fibrosis bronchial epithelial cells prevents elastase induced cell detachment and cytokine release. Am J Respir Cell Mol Biol 1996; 14:348–355.

73. Turner-Warwick M, Burrows SB, Johnson A. Cryptogenic fibrosing alveolitis: clinical features and their effects on survival. Thorax 1980; 35:171–180.

74. Carrington CB, Gaensler EA, Coutu RE, Fitzgerald MX, Gupta RA. Natural history and treated course of usual and desquamative interstitial pneumonia. N Engl J Med 1978; 298:801–809.

75. Piguet PF, Collart MA, Grau GE, Kapanci Y, Vassalli P. Tumor necrosis factor/cachectin plays a key role in bleomycin-induced pneumopathy and fibrosis. J Exp Med 1989; 170:655–663.
76. Broekelmann TJ, Limper AH, Colby TV, McDonald JA. Transforming growth factor β_1 is present at sites of extracellular matrix gene expression in human pulmonary fibrosis. Proc Natl Acad Sci USA 1991; 88:6642–6646.
77. Piguet PF, Vesin C. Treatment by recombinant soluble TNF receptor of pulmonary fibrosis induced by bleomycin or silica in mice. Eur Respir J 1994; 7:515–518.
78. Rogy MA, Auffenberg R, Espat NJ, et al. Human tumor necrosis factor receptor (p55) and interleukin 10 gene transfer in the mouse reduces mortality to lethal endotoxemia and also attenuates local inflammatory responses. J Exp Med 1995; 181:2289–2293.
79. Itoh H, Mukoyama M, Pratt RE, Dzau VJ. Specific blockade of basic fibroblast growth factor gene expression in endothelial cells by antisense oligonucleotide. Biochem Biophys Res Commun 1992; 188:1205–1213.
80. Wilborn J, Crofford LJ, Burdick MD, Kunkel SL, Strieter RM, Peters-Golden M. Cultured lung fibroblasts isolated from patients with idiopathic pulmonary fibrosis have a diminished capacity to synthesize prostaglandin E_2 and to express cyclooxygenase-2. J Clin Invest 1995; 95:1861–1868.
81. Conary JT, Parker RE, Christman BW, et al. Protection of rabbit lungs from endotoxin injury by in vivo hyperexpression of the prostaglandin G/H synthase gene. J Clin Invest 1994; 93:1834–1840.
82. Christman BW, Christman JW, Dworski R, Blair IA, Prakash C. Prostaglandin E_2 limits arachidonic acid availability and inhibits leukotriene B_4 synthesis in rat alveolar macrophages by a nonphospholipase A_2 mechanism. J Immunol 1993; 151:2096–2104.
83. Wang P, Wu P, Siegel MI, Egan RW, Billah MM. IL-10 inhibits transcription of cytokine genes in human peripheral mononuclear cells. J Immun 1994; 153:811–816.

15

Gene Therapy for Acute Diseases of the Lungs

KENNETH L. BRIGHAM

Vanderbilt University School of Medicine
Nashville, Tennessee

I. Introduction

The possibility of transferring DNA, designed and constructed by man in vitro, to the lungs of man in a form in which the encoded protein is expressed, has captured the imagination of the scientific and lay public as well as the biotechnology industry. Most of the initial work in man, aimed at using this technology as a therapeutic tool, has focused rather narrowly on treatment of cystic fibrosis (CF) (1). Because the rationale for gene therapy in that disease is so simple and the disease is so tragic, investigation progressed rather rapidly to clinical studies (2). A great deal has been learned, both scientifically and socially, from those experiences. That knowledge base will be invaluable to further progress in the development of gene therapy for treatment of diseases of the lungs.

But, to limit the concept of gene therapy to diseases that are a consequence of the inheritance of a defect in a single gene is to restrict severely the potential value of this technology. The two most prevalent such lung diseases are CF and familial emphysema (alpha-1 antitrypsin deficiency), and together those two diseases account for but a small fraction of morbidity and mortality due to diseases of the lungs in man. It is likely that as understanding of mechanisms of disease increases, we will find that a number of diseases will be either a direct or an indirect consequence of multiple genetic defects or of unique interactions between genetic and environmental factors. Although the reasoning is less linear and the strategies are more complex, those diseases, too, might lend themselves to treatment or prevention by introduction of appropriate foreign DNA to the right place.

However, the overwhelming majority of diseases of the lungs is not a consequence of a genetic defect. Disease is a host response to a toxic stimulus, but the host response need not be wrong to result in disease. The usual case is that the stimulus outsmarts an appropriate host response. A special opportunity afforded by gene therapy is the potential for combining an understanding of disease pathogenesis at a molecular level with the technology for manipulating DNA to restore host homeostasis in the face of even the most clever toxic stimulus.

II. Rationale for Gene Therapy for Nongenetic Diseases

The rationale for using genetic techniques to treat diseases that are a consequence of an inherited defect in a single gene is straightforward. If an appropriate normal gene can be delivered in a functioning form in an adequate dose to the right cells, the disease (i.e., the structural and functional consequences of the genetic defect) should be cured. To that rationale should be added that the normal gene must be delivered early enough in the natural history of the disease that structural and functional consequences of the genetic defect and the abnormal host responses that result have not become irreversible. For example, once a patient with either CF or alpha-1 antitrypsin deficiency has destroyed, fibrotic lung, it is unlikely that restoring the cystic fibrosis transmembrane conductance regulator or the alpha-1 antitrypsin gene even to normal functioning levels in the appropriate cells will cure the lung disease.

To use gene therapy to treat acute lung disease, acquired rather than inherited, the rationale is more complex, and the strategies, in some cases, are more convoluted. First the pathogenesis of the disease must be understood in sufficient detail to identify a protein that, if increased (or decreased) at an appropriate site, will alter the course of the disease in favor of the host. Having done that, a DNA construct and a delivery system must be designed that delivers the therapeutic gene to the right site and assures that it functions for a limited period of time to an appropriate degree (3). The therapeutic gene need not qualitatively alter the host response but, like more conventional drugs, simply tilt the host-toxin interaction in the host's favor.

III. Design of DNA Delivery Systems for Treatment of Acute Diseases of the Lungs

The ultimate goal for gene therapy of inherited diseases would be to deliver the normal gene in a manner that assured its permanent expression. In contrast, DNA therapy for acute diseases requires that the transgene be expressed only transiently. Since one of the difficulties encountered in in vivo DNA delivery has been a short duration of expression of the transgene, this requirement of treating acute diseases may be easily met. The two general delivery systems with which there has been experience in man are viral vectors and plasmid/cationic liposome complexes. The predominant clinical experience has been with replication-deficient adenoviral vectors used to deliver the CFTR in patients with CF (4,5). These vectors do not appear to integrate into the host genome but rather function episomally, and thus might be useful for treatment of diseases in which permanent transgene expression was undesirable. However, the current generation of adenoviral vectors appears to be limited in clinical usefulness by the fact that they trigger both an acute inflammatory response and a host immune response (5). Other viral vectors, especially those based on adeno-associated viruses, are under investigation and appear to cause less of an undesirable host response (6). But, transgenes delivered in adeno-associated viral vectors appear to integrate into the host genome, a characteristic that would render these vectors useless for treating acute disease.

Cationic liposome/plasmid delivery systems have some significant advantages for therapy of acute diseases. Delivered to the lungs of experimental animals, either by intravenous injection or by aerosol, cationic liposome/plasmid complexes appear to be safe. Figures 1 and 2 illustrate physiologic data

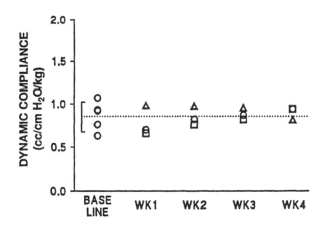

Figure 1 Dynamic lung compliance in rabbits following repeated administration of plasmid-liposome complexes (circles, untreated animals; triangles, intravenous plasmid-liposome; squares, aerosol plasmid-liposome). Plasmid-liposome was delivered weekly for 4 weeks. There was no detectable effect of the intervention on lung compliance. (Reprinted with permission from Canonico AE. Am J Physiol 1994; 77:415–419.)

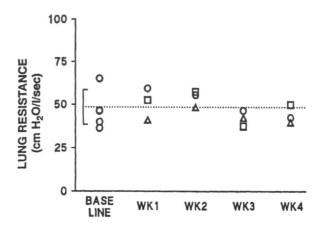

Figure 2 Lung resistance in rabbits following repeated administration of plasmid-liposome complexes (circles, untreated animals; triangles, intravenous plasmid-liposome; squares, aerosol plasmid-liposome). Plasmid-liposome was delivered weekly for 4 weeks. There was no detectable effect of the intervention on lung resistance. (Reprinted with permission from Canonico AE. J Appl Physiol 1994; 77:415–419.)

from rabbits given a plasmid containing the human alpha-1 antitrypsin gene driven by a cytomegalovirus promoter and complexed to cationic liposomes. The material was given weekly, either intravenously or by aerosol, for 4 weeks (7). There was no apparent effect on lung mechanics or gas exchange. In addition, histological examination of the lungs showed no difference between untreated controls and the treated animals.

Cationic liposome/plasmid complexes can deliver functioning transgenes to the lungs following either intravenous or airway administration (8–10). Such a system is particularly attractive for situations in which only transient expression of the transgene is desired since the transgene is contained in a plasmid that does not ordinarily integrate into the host genome or replicate in mammalian cells.

In mice, we initially demonstrated delivery of a chloramphenicol acetyltransferase reporter gene (8) to the lungs using a plasmid/cationic liposome delivery system; those data are summarized in Table 1. Subsequently, this technology has been shown capable of delivering other reporter genes (11), the human growth hormone gene (9), the CFTR gene (12), the alpha-1 antitrypsin gene (10), and the prostaglandin synthase gene (13) to the lungs of experimental animals. The CFTR gene has also been effectively delivered to the nasal mucosa of patients with cystic fibrosis using this delivery system (14). We have begun initial clinical investigations with this system in patients with alpha-1 antitrypsin deficiency.

Table 1 CAT Activity in Lungs, Liver, and Kidneys 72 Hours Following DNA-Liposome Injection into Mice

Route	% Chloramphenicol acetylation per hour per μg protein $\times$ 10^{-2}		
	Lungs	Liver	Kidneys
Intravenous			
30 μg DNA per mouse	24.5	0	0
15 μg DNA per mouse	10.1	0	0
Intratracheal			
30 μg DNA per mouse	35.7	0	0
Intraperitoneal			
30 μg DNA per mouse	0	0	0

Source: Reprinted with permission from Brigham KL. Am J Med Sci 1989; 298:278–281.

An appropriate delivery system may depend not only on getting enough functioning gene into the lungs, but also on getting the gene in the appropriate subpopulation of lung cells; different delivery technologies might be required to target different cell populations. Cell targeting will be especially important if the therapeutic transgene product acts only in the cells in which it is produced. For example, prevention of production and release of pro-inflammatory cytokines from airway epithelial cells might be achieved by genetically engineering those cells so that the transcription factor, NfκB, could not be translocated from the cytoplasm to the nucleus—a common mechanism for inducing cytokine production (15). In this case, only the cells expressing the transgene would be affected and the delivery system would need to target appropriate cells efficiently. The analogy to gene therapy for a genetic diseases would be cystic fibrosis. In that case, the gene product (CFTR) is an integral membrane protein, so only the cells expressing a normal CFTR gene will have their chloride transport defect corrected, and the delivery system needs to target respiratory epithelial cells.

If the therapeutic transgene product is a secreted protein or is an enzyme that catalyzes production of a therapeutic substance that exits the cell and acts on other cells, then the delivery system should maximize production of the transgene product, but the population of cells expressing the transgene is less critical. In fact, the first human trials of gene therapy were done by removing cells from the patient, engineering them in vitro to express the normal adenosine deaminase gene (a secreted protein) and then returning the cells to the patient (16). In the lung, the analogous genetic disease is alpha-1 antitrypsin deficiency. In that case, if sufficient extracellular concentrations of the deficient protein could be achieved, it might be relatively unimportant where the protein was produced (17).

As will be detailed below, alpha-1 antitrypsin gene therapy may also have use in acute lung diseases. In addition, increased expression of the enzyme, prostaglandin synthase, in the lungs can result in increased production of PGE2, an anti-inflammatory prostanoid, and can protect the lungs from acute injury in some experimental situations (18). In each of these cases, efficacy of gene therapy will depend mainly on maximizing expression of the transgene in the lungs, regardless of the cell population(s) targeted.

IV. Gene Therapy for Respiratory Viral Infections

Because viral infections of respiratory cells are a consequence of several genetically controlled events, it is tempting to think that cells could be rendered resistant to viral infection by appropriate genetic manipulation. One possible genetic manipulation is to enhance antiprotease activity in cells so that cell-associated proteolytic events essential for virus to invade cells and replicate are inhibited. Respiratory syncytial virus (RSV) is an example of a pathogen that might be amenable to this kind of therapy.

RSV infection is common in infants and children in North America. While overall mortality from RSV infection is low, mortality in children requiring hospitalization is substantial. RSV infection also occurs in adults, may be responsible for exacerbations of chronic lung disease (19), and is a special problem in immunocompromised patients following organ transplant (20). There is no effective prevention or treatment of RSV infection.

Proteolytic events may be essential to entry and replication of RSV in respiratory epithelium (21). Infectivity of RSV is mediated by a fusion glycoprotein (F) on the viral envelope. The F protein is synthesized during viral replication in a form that requires cleavage by a cell associated trypsinlike protease before it confers infectious potential. We reasoned that if we could express the gene encoding alpha-1 antitrypsin in respiratory epithelial cells, the cells would be rendered resistant to RSV infection.

To test this hypothesis, we used a continuous respiratory epithelial cell line known to be susceptible to RSV. We inserted the human alpha-1 antitrypsin transgene into a plasmid with a cytomegalovirus promoter and delivered this construct to the cells using a cationic liposome vehicle (10). The transfected cells generated increased levels of human alpha-1 antitrypsin both in the culture medium and inside the cells. We compared RSV infectivity in these transfected cells with infectivity in identical cells transfected in an identical manner with the same amount of a plasmid containing the reporter gene, chloramphenicol acetyltransferase (CAT) (22). As shown in Figure 3, cells expressing the AAT gene were relatively resistant to RSV infection. Addition of the human AAT protein to the culture medium, even in concentrations far in

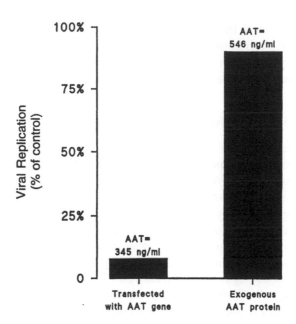

Figure 3 Inhibition of RSV infection of susceptible cells by transfection with a plasmid expressing the alpha-1 antitrypsin gene. The inhibitory effect of alpha-1 antitrypsin gene expression was not mimicked by addition of exogenous alpha-1 antitrypsin protein even at high concentrations. (Canonico AE, Stecenko AA, et al., unpublished observations.)

excess of those produced by the AAT-transfected cells, had no effect on RSV infectivity. These data suggest that antiprotease gene therapy may be effective in preventing RSV infection of respiratory epithelium and that this efficacy results from a cell-associated location of the gene product which is not attainable by administration of the protein extracellularly. A protein synthesized by the cell may have different therapeutic potential than even the same protein delivered exogenously.

In earlier studies, we showed that delivery of the human alpha-1 antitrypsin gene in an expression vector complexed to cationic liposomes either intravenously or by aerosol resulted in demonstrable expression of the transgene in lung endothelial and epithelial cells (10). We also showed that repeated intravenous or aerosol delivery of this formulation to animals weekly for 4 weeks caused no detectable alterations in lung function or structure (7). We have begun the first study using this system to deliver the alpha-1 antitrypsin gene to the nasal mucosa of humans with alpha-1 antitrypsin deficiency. Since

RSV infection starts in the upper respiratory tract and progresses downward, delivery of the alpha-1 antitrypsin gene effectively to upper-airway epithelium might prevent progression of the disease. With an effective delivery system, such a hypothesis would be easily tested in humans.

V. Gene Therapy for Acute Lung Injury

The entity called the adult respiratory distress syndrome (ARDS) is a consequence of diffuse inflammation in the lungs, often in the setting of sepsis or the sepsis syndrome (23). Although major strides have been made in understanding the pathogenesis of this disease and in describing quantitatively its pathophysiology, no clearly effective pharmacologic therapy has yet emerged.

In general, acute diffuse lung injury is thought to result from generation of chemotaxins in the lungs (probably at least from macrophages and airway epithelial cells), resulting in an influx of neutrophils that generate and release proteolytic enzymes (e.g., neutrophil elastase), and aggressive oxygen species that injure lung cells. There is abundant evidence in both animal models of lung injury and in the human syndrome to support this general pathogenetic hypothesis (24).

Several sites in the pathogenetic sequence by which diffuse inflammation in the lungs results in lung injury and dysfunction might lend themselves to manipulation by gene therapy. Such opportunities include manipulation of generation of either pro- or anti-inflammatory cytokines, enhancement of endogenous antioxidant defenses, alterations in expression of cell surface adhesion molecules, increasing endogenous antiproteases, increasing endogenous generation of protective prostanoids, and probably a very long list of other possibilities many of which are as yet unknown (24).

As an example of how such a therapy could develop, I will use our own work in which increased expression of the proximal enzyme in the synthesis of prostanoids from arachidonic acid (prostaglandin synthase, or cyclooxygenase) in lung endothelium is being tested as a possible gene therapy for ARDS (13). Development of other gene therapies might take a similar course.

The prostanoids prostacyclin and prostaglandin E2 (PGE2) have physiologic effects that could prevent or reverse acute lung injury and its consequent pathophysiology. Prostacyclin is a pulmonary vasodilator and prevents platelet activation. PGE2 can inhibit neutrophil aggregation, prevent induction of proinflammatory cytokines, and prevent airway constriction. Studies in animal models of acute lung injury show that exogenous delivery of pharmacologic amounts of these anti-inflammatory prostanoids can ameliorate lung injury and dysfunction (18).

Since prostacyclin and PGE2 are the principal prostanoids generated by vascular endothelium, we reasoned that if we could hyperexpress the prostaglandin synthase gene in lung endothelium, this would result in increased amounts of prostacyclin and PGE2 in the lungs and render the organ resistant to inflammatory injury. Further, the fact that the protective chemical species would not be the transgene product itself, but substances the production of which is catalyzed by the gene product, might amplify the physiologic consequences of in vivo gene transfer and therefore minimize efficiency requirements for the gene delivery system.

Figure 4 shows the synthesis of prostacyclin and PGE2 in lungs of rabbits 24 hours after intravenous delivery of the PG synthase gene in an expression plasmid complexed to cationic liposomes (13). In vivo expression of the prostaglandin synthase transgene is demonstrated by increased production of the endothelial-derived prostanoids, prostacyclin, and PGE2, in whole-lung organ cultures. Increased urinary concentrations of the stable prostacyclin metabolite also indicate increased in vivo generation of prostanoids in animals receiving the prostaglandin synthase gene.

Twenty-four hours after transfecting animals with either the prostaglandin synthase-containing plasmid or the same plasmid without the transgene,

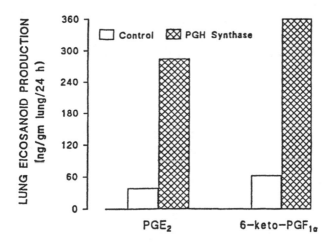

Figure 4 Production of prostaglandin E_2 and prostacyclin (measured as the stable metabolite 6-keto $PGF1\alpha$) by lungs removed from rabbits transfected 24 hours earlier by intravenous delivery of a plasmid containing the prostaglandin synthase gene complexed to cationic liposomes. Lung minces were incubated for 24 hours and eicosanoids measured in the supernatant by gas chromatography/mass spectrometry. Control animals were given a plasmid without the transgene. Both of the endothelial-derived prostanoids were produced in large quantities by lungs from animals that received the prostaglandin synthase transgene. (Reprinted with permission from Conary JT. J Clin Invest 1994; 93:1834–1840.)

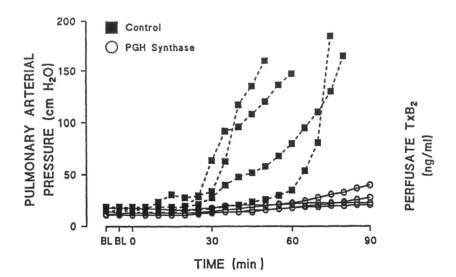

Figure 5 Effects of intravenous transfection with a plasmid containing the prostaglandin synthase gene complexed to cationic liposomes on the pulmonary vascular response to endotoxin in rabbits. Control animals received the same plasmid without the transgene. Animals receiving the prostaglandin synthase gene did not show the marked increase in pulmonary vascular resistance typical of the endotoxin response. (■) Control. (○) PGH synthase. (Reprinted with permission from Conary JT. Clin Invest 1994; 93:1834–1840.)

we measured responses of the lungs to endotoxemia. Figure 5 shows responses of pulmonary artery pressure in a preparation where pulmonary blood flow is constant in both groups of animals. Endotoxin normally causes pulmonary vasoconstriction in a variety of animal preparations. The same is true in this preparation in animals that did not receive the prostaglandin synthase gene. However, in animals receiving the gene and producing increased amounts of endothelial-derived prostanoids, the pulmonary vasoconstrictor response is prevented. In those same studies, we also showed significantly less edema in the lungs from animals receiving the prostaglandin synthase gene (13). These studies demonstrate that it is possible using available technology to transfer sufficient functioning transgene to the lungs of a living animal to alter a pathophysiologic response.

There is a sufficient knowledge base available to provide a rationale for antiprotease gene therapy for treatment of acute lung injury as well. Not only do activated neutrophils sequestered in the lungs release proteolytic enzymes that might directly injure lung cells, but antiproteases can have more general anti-inflammatory effects (24). We have constructed expression plasmids containing the human alpha-1 antitrypsin gene and demonstrated that the gene can be made to express in the lungs of rabbits following delivery of plasmid-cationic liposome complexes either intravenously or by aerosol (10). Studies under way confirm these observations in piglets. Intravenous delivery of the gene results in accumulation of the human alpha-1 antitrypsin protein in both lung endothelium and airway epithelium, whereas aerosol delivery results in the protein appearing primarily in respiratory epithelium (10). We could detect no abnormal physiologic or structural consequences of delivery of the plasmid-liposome complexes by either route (7). Studies are under way in piglets to determine whether animals expressing the human alpha-1 antitrypsin gene in their lungs are resistant to the effects of endotoxin.

Recent work reported by Canonico and associates demonstrates the potential for alpha-1 antitrypsin gene therapy as more general anti-inflammatory therapy. They showed that the release of neutrophil chemoattractants from human airway epithelial cells exposed to neutrophil elastase was prevented if the cells were transfected with the human alpha-1 antitrypsin gene prior to

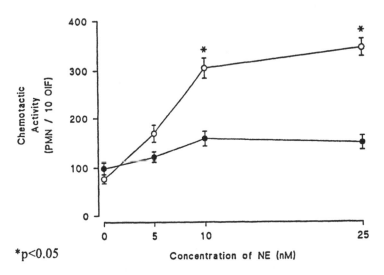

$*p < 0.05$

Figure 6   Effects of transfection of human cystic fibrosis airway epithelial cells with a plasmid containing the alpha-1 antitrypsin gene complexed to cationic liposomes on neutrophil elastase-induced release of neutrophil chemoattractants. Control cells were transfected with a plasmid without the alpha-1 antitrypsin transgene. Cells expressing the alpha-1 antitrypsin gene did not release chemotaxins with addition of even relatively high concentrations of elastase. (Reprinted with permission from Canonico AE. Am J Respir Cell Mol Biol 1996; 14:348–355.)

elastase exposure (25). Figure 6 (from their paper) summarizes those data and demonstrates the specificity of the effect (i.e., cells transfected with empty vector were not protected). There are a number of processes involved in the inflammatory response that involve proteolytic events inside cells. Addition of high-molecular-weight proteases to the extracellular space may not permit access to the appropriate microenvironment for exerting maximum anti-inflammatory activity. Engineering cells to synthesize the potentially therapeutic protein could be a useful strategy.

VI. Is DNA Really Just Another Drug?

Giving foreign genes to people has triggered both social and scientific ambivalence. The very potential of this technology is inextricable from the fact that DNA is the stuff of life—that is why manipulating DNA can have such profound curative powers. Society wants to avail itself of those powers without risking their going awry. Science wants, ultimately, to expose DNA's most private secrets because science believes that completely understanding the human organism is a noble goal, whether or not there are immediate therapeutic benefits. In fact, there is a respectable sector of science that does not believe that great effort should be spent on developing the technology for delivering genes to people until the knowledge base is more complete.

However, a case could be made that the challenges in developing gene therapy are not qualitatively different from those in developing any drug. This is especially true for somatic gene therapy using DNA constructs with minimal risk of reaching germ cell lines or integrating into the host genome. The development of DNA as a drug should synergize with basic molecular biology as development of chemical drugs synergizes with basic pharmacology.

Genes are molecules of DNA which, when delivered appropriately, instruct the cells which they transfect to produce a specific protein. The large and immediate issues are whether those processes are inherently toxic and whether there are unacceptable side effects. With other kinds of drugs, those issues are addressed in thorough basic and preclinical studies. Initial human studies are commonly aimed at defining pharmacodynamics and safety in dose ranging studies in normal volunteers. Following that, initial efficacy and safety are evaluated in a small group of patients in a clinical target population, and then investigation proceeds to a Phase III study of efficacy in large populations of patients where study design permits rigorous statistical analysis of the data.

Development of gene therapy has not followed this pattern. No exogenous genes have been delivered to normal humans. Very small groups of patients have been given foreign DNA with, at best, anecdotal conclusions about either toxicity or efficacy. While it is true that there are many unknowns about the long-term effects of delivering foreign DNA in vivo, the same is true for many new drugs. Even the possibility of affecting future generations is not unique to gene therapy, as recent drug development history attests.

The concept of gene therapy for acute acquired diseases of the lungs provides a particularly fertile opportunity for progressing to the next stage of development of DNA as a drug. The concept not only permits but requires use of delivery systems and DNA constructs that express only transiently and thus do not integrate into the host genome. The lung airways and the upper airway epithelium are easily accessible for drug delivery topically, which should minimize risk of delivery to other organs and permit direct assessment of effects of the therapy on lung cell structure and function. In ARDS, there is a highly lethal clinical syndrome for which there is no known specific pharmacologic therapy and with an incidence high enough to provide a target population of patients that could be studied.

Acknowledgments

This work was supported by NIH grants HL 45151 and HL 15153. Dr. Brigham serves as a paid consultant to GeneMedicine, Inc. Dr. Brigham is Ralph and Lulu Own Professor of Pulmonary Medicine at Vanderbilt University School of Medicine.

References

1. Korst R, McElvaney N, Chu C-S, et al. Gene therapy for respiratory manifestations of cystic fibrosis. Am Rev Respir Crit Care Med 1995; 151:575–587.
2. Crystal R, McElvaney N, Rosenfeld M, et al. Administration of an adenovirus containing the human CFTR cDNA to the respiratory tract of individuals with cystic fibrosis. Nature Genet 1944; 8:42–51.
3. Brigham KL, Stecenko AA. Gene therapy in acute critical illness. New Horizons 1995; 79:1163–1172.
4. Rosenfeld M, Yoshimora K, Trapnell B, et al. In vivo transfer of the human cystic fibrosis transmembrane conductance regulator gene to the airway epithelium. Cell 1992; 68:143–155.
5. Simon R, Engelhardt J, Yang Y, et al. Adenovirus mediated transfer of the CFTR gene to lung of nonhuman primate: toxicity study. Human Gene Ther 1993; 4:771–780.
6. Flotte T, Afione S, Conrad S, et al. Stable in vivo expression of the cystic fibrosis transmembrane conductance regulator with an adeno-associated virus vector. Proc Natl Acad Sci USA 1993; 90:10613–10617.
7. Canonico AE, Plitman JD, Conary JT, Meyrick BO, Brigham KL. No lung toxicity after repeated aerosol or intravenous delivery of plasmid-cationic liposome complexes. Am J Physiol 1994; 77:415–419.

8. Brigham K, Meyric B, Christman B, Magnuson M, King G, Berry LC Jr. Rapid communication: in vivo transfection of murine lungs with a functioning prokaryotic gene using a liposome vehicle. Am J Med Sci 1989; 298:278–281.

9. Brigham KL, Meyrick B, Christman B, et al. Expression of human growth hormone fusion genes in cultured lung endothelial cells and in the lungs of mice. Am J Respir Cell Mol Biol 1993; 8:209–213.

10. Canonico AE, Conary JT, Meyrick BO, Brigham KL. Aerosol and intravenous transfection of human α1-antitrypsin gene to lungs of rabbits. Am J Respir Cell Mol Biol 1994; 10:24–29.

11. Stribling R, Brunette E, Liggitt D, Gaensler K, Debs R. Aerosol gene delivery in vivo. Proc Natl Acad Sci USA 1992; 89:11277–11281.

12. Lecocq JP, Crystal RG. Expression of the human cystic fibrosis transmembrane conductance regulator gene in the mouse lung after in vivo intra-tracheal plasmid-mediated gene transfer. Nucleic Acids Res 1992; 20:3233–3240.

13. Conary JT, Parker RE, Christman BW, et al. Protection of rabbit lungs from endotoxin injury by in vivo hyperexpression of the prostaglandin G/H synthase gene. J Clin Invest 1994; 93:1834–1840.

14. Caplen N, Alton E, Middleton P, et al. Liposome-mediated CFTR gene transfer to the nasal epithelium of patients with cystic fibrosis. Nature Med 1995; 1:39–46.

15. Duckett CS. Structure and function of the NF-*k*B family of transcription factors. Promega Notes 1995; 44:40–43.

16. Anderson WF. Human gene therapy. Science 1992; 357:455–460.

17. Garver RI, Chytil A, Courtney M, Crystal RG. Clonal gene therapy: transplanted mouse fibroblast clones express human α-1 antitrypsin gene in vivo. Science 1987; 237:762–764.

18. Brigham KL, Serafin W, Zadoff A, Blair I, Meyrick B, Oates JA. Prostaglandin E_2 attentuation of sheep lung responses to endotoxin. J Appl Physiol 1988; 64:2568–2574.

19. Falsey AR, Cunningham CK, Barker WH, et al. Respiratory syncitial virus and influenza A infections in the hospitalized elderly. J Infect Dis 1995; 172:389–394.

20. Englund JA, Anderson LJ, Rhame FS. Nosocomial transmission of respiratory syncytial virus in immunocompromised adults. J Clin Microbiol 1991; 29:115–119.

21. Scheid A, Choppin PW. Identification of biological activities of paramyxovirus glycoproteins. Virology 1974; 57:475–490.

22. Persmark M, Canonico A, Brigham KL, Stecenko AA. Inhibition of respiratory syncytial virus (RSV) infectivity by liposome-mediated antiprotease gene transfer. J Invest Med 1995; 43:220A.

23. Bernard GR, Brigham KL. Increased lung vascular permeability. In: Ayers SM, ed. Textbook of Critical Care. Philadelphia: WB Saunders, 1995:674–680.

24. Canonico AE, Brigham KL. Biology of acute lung injury. In: Crystal R, West J, Weibel E, Barnes P, eds. The Lung: Scientific Foundations, 2nd ed. Philadelphia: Lippincott-Raven, 1996:190.1–190.24.

25. Canonico AE, Brigham KL, Carmichael LC, et al. Plasmid-liposome transfer of the α1 antitrypsin gene to cystic fibrosis bronchial epithelial cells prevents elastase-induced cell detachment and cytokine release. Am J Respir Cell Mol Biol 1996; 14:348–355.

16

Gene Therapy for Lung Cancer

CHOON-TAEK LEE

Korea Cancer Center Hospital
Seoul, Korea

DAVID P. CARBONE

Vanderbilt Cancer Center
Nashville, Tennessee

I. Introduction

Lung cancer is the most common cause of cancer death in the United States and is becoming a major cause in many other countries. While advances in chemotherapy and radiation therapy have resulted in remarkable progress in some tumors, only incremental progress has been achieved in lung cancer. Furthermore, several recently introduced anticancer drugs and new techniques of radiation have failed to show significant improvements in the outcome of patients with lung cancer (1). The marginal improvements thus far achieved in lung cancer patient survival with standard treatment modalities suggests the need for totally new treatment modalities.

Gene therapy is one such novel therapeutic approach. Advances in molecular biology and immunology are revealing the basic mechanisms of carcinogenesis and the immunological differences between cancer cells and their normal counterparts. This knowledge should help to develop future therapeutic strategies including gene therapy.

Gene therapy is the treatment of cancer using genetic material as a therapeutic agent. Gene therapy can be broadly classified into two categories, according to the mechanism of action: immunity-inducing and direct-acting.

The latter can be divided into approaches that are simply toxic, and those that modulate normal cellular signaling and regulation for their anticancer effect (2).

II. Immunity-Inducing Gene Therapy

During carcinogenesis, it is clear that multiple genetic and protein structural changes occur. Also clear is the fact that tumor cells acquire immunologic adaptations which allow them to effectively avoid immune surveillance. The inactivation of tumor suppressor genes and activation of oncogenes through mutation and/or dysregulated expression are examples of these genetic alterations. A number of tumor-associated antigens have been identified and more have been postulated. These alterations represent the differences between cancer cells and normal cells. The normally functioning immune system has the ability to detect and eliminate cells identified as non-self- (immune-) surveillance. Most cancer cells are likely to have protein structural features or patterns of protein expression which should allow immune detection and elimination. In real situations, the development of clinically evident tumors implies a failure of the immune system to detect and reject cancer cells as foreign.

Several mechanisms by which tumors avoid immune surveillance have been identified or hypothesized: (1) Tumor specific antigens do not exist or are structurally unable induce an effective T-cell immune response (not within the available genetic repertoire or unable to bind to MHC); (2) defects of antigen processing, or transport to the cell surface (e.g., defects in TAP or $\beta2$ microglobulin) or loss of expression of the appropriate MHC molecule; (3) inadequate "help" (cytokine environment or costimulatory interactions via molecules such as B7); (4) tumor production of immune-suppressive factors (TGFβ, IGF-1, IL-10); (5) a relative increase in the Th2 cell subpopulation among CD4$^+$ T helper cells. IL-10 secreted from Th2 cell subpopulation is known to suppress the production of IL-2, IFN-γ, and TNF-α from the Th1 subpopulation, thought to be responsible for the development of antitumor cytotoxic T cells; and (6) defects in signal transduction by the T cell receptor (TCR) in cells from cancer patients, resulting in an inability to respond to appropriate signals.

Genetic therapies (and other novel approaches as well) can be designed to address the regulatory issues involved in the failure of effective anticancer immunity. Gene therapy can deliver biologically active genes intended to alter the local immunological microenvironment and increase the immunogenicity of cancer cells, to increase the responsiveness of T cells, to improve antigen presentation, or to provide missing paracrine factors. The resulting immune response has the potential to detect and kill non-gene-modified tumor cells.

This characteristic has the greatest potential for improving outcome in the many clinical situations where there is concern for distant, even unrecognized metastases.

A variety of genetic approaches have been used to induce immunity via gene therapy. Allogenic and syngenic major histocompatibility complex (MHC) class I and II genes, costimulatory molecule genes (e.g., B7), and cytokine genes have been inserted to alter the immunological environment and overcome immune defects of cancer patients. These approaches may be viewed as attempts to alter tumor cells so as to make them effective antigen-presenting cells (APC). This is probably oversimplified, but it is a useful model for this discussion.

A. MHC Molecules

The MHC is a region of highly polymorphic genes whose products are expressed on the surface of most cells. MHC class I molecules bind endogenously synthesized peptide fragments and present them on cell surface for recognition by the T-cell receptor (TCR) on CD8$^+$ T cells. MHC class II molecules are primarily expressed on "professional antigen-presenting cells" such as macrophages and dendritic cells that are thought to be primarily responsible for binding peptide fragments derived from extracellular proteins and presenting them to T helper cells. This interaction induces the production of cytokines necessary for the expansion of cytotoxic effectors. MHC molecules therefore play a key role in all phases of the immune response. Tumor cells usually have some level of expression of MHC class I molecules on their surface, but in many tumor cells expression of MHC may be low.

Allogenic MHC

Attempts to increase the immunogenicity of tumor cells by insertion of foreign genes was one of the first gene therapeutic strategies. Among foreign genes, allogenic MHC molecules are a classic example. Allogenic MHC class I genes introduced into tumor cells will induce allo-MHC-specific T-cells. These allo-MHC-specific T-cells presumably then assist in the generation of T-cells specific for previously silent tumor specific antigens. In a murine lung cancer system (3LL/3) from C57BL/6 mice (MHC type H-2^b), transfection of the allogeneic MHC molecule H-2L^d caused a reduction in tumorigenicity and protection against unmodified 3LL/3 (3). Plautz et al. (4) showed that expression of a murine class I H-2K^s gene in CT26 mouse colon adenocarcinoma (H-2K^d) or MCA 106 fibrosarcoma (H-2K^d) induced a cytotoxic T-cell response to H-2K^d and, more importantly, to other antigens present on unmodified tumor cells that had not been recognized previously. Recently, allogenic MHC transfection has been applied to humans with HLA-B7 gene transfer.

Nabel et al. (5) reported the reduction of tumor size in a melanoma patient after the direct gene transfer of an HLA-B7 gene in a liposome complex. Clinical protocols of HLA-B7 gene transfer by lipofection in advanced cancers, including melanoma, renal cell cancer, and colon cancer are under way.

Syngenic MHC

Therapeutic transfection of syngenic MHC class I and II molecules has been attempted in several different tumor model systems. Transfection of syngenic MHC class II (Ak) in sarcoma cells induced an immune response against both transfected tumor cells and nontransfected tumor cells (6). This finding suggests that transfected MHC class II molecules can induce T helper cells specific to nontransfected tumor cells as well.

Low MHC class I molecule expression in some tumors may lead to decreased presentation of tumor-specific antigens on the cell surface, and, as stated above, this could be one mechanism for their escape from immune surveillance. Transfection of syngenic MHC class I molecules could induce or increase the presentation of endogenous peptide fragment derived from tumor specific antigens, leading to the generation of $CD8^+$ cytotoxic T lymphocytes specific for parental tumor antigens. Once induced, this response could then potentially cause recognition and lysis of non-gene-modified cells with lower levels of MHC expression. Increased MHC class I expression on tumor cells may thus result in enhanced immunogenicity and decreased tumorigenicity of tumor cells. Plaskin et al. (7) showed that high expression of a transfected $H-2K^b$ MHC gene in highly metastatic Lewis lung carcinoma cell (3LL) resulted in the conversion of a high-metastatic to a non- or low-metastatic phenotype and protection against the metastatic spread of 3LL.

B. Accessory Molecules (B7)

Effective antitumor immunity is usually dependent on T-cell-mediated responses. Two kinds of signals are required for the activation of T cells. As described above, the first signal is the antigen-specific binding of a peptide antigen-MHC complex on the surface of antigen presenting cells with antigen specific T-cell receptors. This interaction is necessary but not sufficient to induce primary T-cell activation and production of essential cytokines such as IL-2. It is hypothesized that the presence of this first signal alone may induce a state of readiness to respond to the second signal, known as costimulation. The second signal is transmitted by the antigen-independent binding of co-stimulatory molecules on APC with their corresponding receptors on T-cells. This signal is required to induce primary T-cell proliferation and other effector functions of the immune response. Without this second signal, the binding of antigen with TCR alone may cause prolonged unresponsiveness or specific

T-cell anergy. A number of molecules have been found to mediate this second signal, including the B7 family (unrelated to the HLA-B7 class I molecule described above), which interact with the CD28 receptor on T cells (8,9). B7 is expressed on activated B cells, macrophages, and dendritic cells and is the ligand for CD28 and CTLA-4 on T-cells (10,11).

Tumor cells may have tumor-specific antigens and MHC class I expression, but they only rarely express B7. The rationale for transfection of B7 into tumor cells is to provide or supplement the necessary costimulation for effective immune induction. B7-transfected tumor cells can directly stimulate CD8$^+$ cytotoxic T lymphocytes (CTL) without the help of APC and CD4$^+$ T helper cells. Townsend and Alison (12) demonstrated that B7 transfection into MHC class I and II positive melanoma induced the rejection of tumors in vivo. This rejection was mediated by CD8$^+$ T cells and did not require CD4$^+$ T cells. Chen et al. (13) showed that B7-transfection in HPV-16 E7 transfected tumors resulted in the loss of tumorigenicity and regression of E7$^+$B7$^-$ tumors by the immune response mediated by CD8$^+$ CTL. The costimulatory effect of B7 in antitumor immunity is dependent on tumor immunogenicity. Chen et al. (14) showed that B7 transduction by a retroviral vector decreased the tumorigenicity and induced protective immunity against immunogenic tumors. In nonimmunogenic tumors, B7 did not show any antitumor immunity. In contrast to defined antigen vaccine approaches, another important aspect of B7 gene therapy is the potential ability to increase the immunogenicity of all tumor-specific antigens, whether or not they have been identified. Many potential tumor-specific antigens have been postulated in tumor cells including viral antigens, fetal genes, and mutated oncogene or tumor suppressor gene products (15).

Carcinoembryonic antigen (CEA) is an oncofetal protein expressed by many adenocarcinomas, including those of lung origin. Recently, cotransduction of recombinant vaccinia virus-B7(rV-B7) and rV-CEA resulted in CEA specific T-cell responses and antitumor immunity against a murine carcinoma expressing CEA. These findings suggested that B7 increased the immunogenicity of CEA (16). Another recent study demonstrated the enhancement of the immunogenicity of a mutant p53 epitope after transduction of B7 via recombinant adenovirus (17). B7 transduction into murine fibrosarcomas expressing mutant p53 resulted in the induction of mutant p53-specific CTL and loss of tumorigenicity as well as protective immunity against challenge by untransduced tumor. Similar induction of immunity was found for the P1A epitope in the P815 (murine plasmacytoma) tumor model. Another recent study demonstrated that B7 and CD28 interaction provided costimulatory signals not only for T-cells but also for natural killer (NK) cells (18). Tumor cell resistance to NK cells is mediated by expression of MHC class I molecule, but B7:CD28 interaction induced by B7 transfection can overcome the MHC class I-mediated inactivation of NK cells, increasing their tumoricidal activity. All of these studies provide the experimental groundwork for attempts to utilize B7 in cancer gene therapeutic approaches.

C. Cytokine Gene Therapy

Rosenberg (19) was the first to report that systemic administration of IL-2 with or without in vitro expanded autologous lymphokine activated killer cells (LAK) to advanced cancer patients was associated with some significant responses (19). The toxicities of systemic cytokine administration interfered with the popularization of this approach. To avoid systemic side effects and more importantly to approximate normal physiological conditions, gene therapy has been used to insert cytokine genes into tumor cells and induce local production of cytokines in the vicinity of tumor. Local production of cytokines from tumors can modify the tumor's interactions with the host immune system. These altered immunological microenvironments may provide favorable conditions for host immune detection of tumor cells and previously unrecognized tumor-specific antigens. Effective recognition of these antigens and induction of a response can therefore allow detection and killing of non-gene-modified tumor cells with the same tumor-specific antigen even though they may be located far from the site of gene modification.

A number of cytokines have been tested for efficacy in animal models of cytokine gene therapy: IL-2, IL-4, IL-6, IL-7, IL-12, IFN-γ, TNF-α, G-CSF, and GM-CSF among others have been investigated. IL-2 in particular has been investigated extensively. Fearon et al. (20) demonstrated that IL-2 cDNA transfected into murine colon carcinoma (CT26) resulted in the loss of tumorigenicity and bypassed deficient T helper function in the generation of an antitumor response. Ley et al. (21) showed that IL-2-transfected murine mastocytoma (P815) cells induced P815-specific CTL, which led to regression of established tumors. Using a highly malignant and poorly immunogenic Lewis lung carcinoma, IL-2 production by retrovirally transduced tumor cells induced antitumor CTL and eliminated the generation of lung metastasis (22). In natural immune responses, IL-2 is produced by CD4$^+$ T cells activated by the binding of TCR and antigens presented by MHC class II molecules on APC. Secreted IL-2 will activate CD8$^+$ T cell precursors to cytotoxic T cells. IL-2 secretion by gene-modified tumor cells may activate CD8$^+$ T cells directly, bypassing T helper cells. IL-2 can overcome defects in signal transduction of T-cells of tumor-bearing patients caused by the decreased levels of p56^{lck}, and p59^{fyn} (23). Clinical trials of IL-2 cytokine gene therapy in lung cancer are planned at several sites, but preliminary data are not available.

IL-4 also shows antitumor effects in animal tumor models. IL-4 is produced by Th2 subset of helper T-cells and mast cells, and has many functions. It can induce LAK cells, stimulate B cell proliferation and maturation, and activate endothelial cells to express vascular cell adhesion molecules (24). Tepper et al. (25) demonstrated that IL-4 production from transfected tumor cells had antitumor effects in various tumor cell lines. This effect is blocked by anti-IL-4 antibody, related to the level of the production, and is evident in nude mice. Infiltration of the transduced tumor site with macrophages and eosinophils suggested that inflammatory mechanisms were involved.

IL-6 is the pleiotropic cytokine that can stimulate the differentiated functions of B-cells and T-cells and induce the production of mature myeloid cells and megakaryocytes (26). IL-6 has antitumor effects on murine models of lung cancer. The immunization of inactivated, IL-6-transfected Lewis lung carcinoma cells induced antitumor CTL and reduced the metastatic potential of these cells (27).

IL-7 is a bone marrow stromal cell-derived cytokine that stimulates pre-B-cell expansion. IL-7 also functions as a growth factor for thymocytes and CD4$^+$ and CD8$^+$ T cells. IL-7 can enhance the antigen presentation indirectly by increasing expression of costimulatory cell adhesion molecules such as ICAM-1 and inhibit the production of immunosuppressive cytokines such as TGF-β (28). The production of IL-7 by retroviral transduction in fibrosarcomas can decrease tumorigenicity and induce protective immunity against unmodified tumor challenge, and IL-7 transduction of immunogenic tumors can cause the regression of established lung metastasis (29). A clinical trial of cytokine gene therapy with IL-7 is planned in lung cancer, but no preliminary results are available.

Interferon-γ produced by activated T-cells can modulate the immune response in a number of ways. IFN-γ can induce MHC class I and II molecules, which will increase the presentation of antigen on the cell surface and can induce the activation of macrophages (30). Low expression of MHC molecules on tumor cells is one postulated mechanism by which tumors escape from host immune surveillance. Increased expression of MHC molecules by IFN-γ should increase the presentation of tumor-specific antigens and assist the induction of antigen-specific CTL. Transduction of a weakly immunogenic tumor (CMS-5) by retrovirus IFN-γ induced the abrogation of tumorigenicity and persistent and specific antitumor immunity against the unmodified CMS-5 challenge (31). The effect of IFN-γ has also been demonstrated in the 3LL mouse lung cancer model. Retroviral IFN-γ gene insertion into poorly immunogenic 3LL-D122 showed a significant decrease in tumorigenicity and metastatic potential, and induced tumor-specific CTL when modified tumor cells were injected after irradiation (32).

Tumor necrosis factor has direct cytolytic effects on tumor cells and attracts and augments the tumoricidal activity of macrophages (33). Systemic administration of TNF, however, was too toxic to achieve significant in vivo antitumor effects. Local production of TNF-α by retroviral transduction into the TNF-insensitive tumor cell line (J558L) drastically suppressed tumorigenicity in a syngenic animal, even though it did not appreciably affect growth in vitro (34). Administration of an anti-type 3 complement factor receptor to block the migration of inflammatory cells abolished the antitumor effects of TNF-α, which suggested the involvement of an inflammatory mechanism including the activation of macrophages (34). The antitumor effect of TNF-α was also proven in human lung cancer cell lines even though most human tumor cells are resistant to TNF-α. Insertion of human TNF-α cDNA into several human lung cancer cells resulted in decreased tumorigenicity in nude mice (35). Furthermore, injection of a mixture of 50% gene-modified and 50% parental cells also showed decreased tumor formation. These data suggest that local production of TNF-α can induce antitumor effects on human lung cancer cell growth, and every tumor cell need not be gene-modified to produce local antitumor effects.

GM-CSF appears to be one of the most active cytokines in the induction of antitumor immunity. In a comparison of the efficacy of a number of cytokines using retroviral vectors, GM-CSF demonstrated the most potent, specific, and long lasting antitumor immunity (36). The antitumor immunity induction after gene therapy with GM-CSF was dependent on both CD4$^+$ and CD8$^+$ T-cells. This activity may be related to its ability to promote the differentiation of hematopoietic precursors to dendritic cells and other professional antigen-presenting cells (37). Since culture and stable transduction of human tumors is problematic, "paracrine" GM-CSF release from gelatin-chondroitin microspheres mixed with irradiated tumor cells was tested and found to evoke antitumor immunity comparable to a GM-CSF-transduced tumor vaccine (38). We have designed and produced an adenovirus-GM-CSF vector which also overcomes many of the limitations of in vitro culture of primary human tumors. Transduction of 3LL with this adenovirus-GM CSF vector eliminated its tumorigenicity and induced tumor-specific CTL and the regression of established 3LL tumors (Lee et al., unpublished data). Furthermore, we showed that this was associated with an increased number of dendritic cells in the tumor vaccine injection site.

IL-12 is a new cytokine with dramatic antitumor effects and relatively modest systemic toxicity in animal models. IL-12 has the ability to promote the differentiation of uncommitted T-cells to Th1 cells, thought to be key to the production of effective antitumor immunity (39). IL-12 also activates the CD8$^+$ T cells and NK cells and generates LAK cells (40). Tahara and Lotze (41) demonstrated that IL-12-retroviral transduction of murine sarcomas showed suppression of tumorigenicity, induction of protective antitumor immunity, and suppression of preinjected nontransduced tumors. The efficacy of IL-12 was confirmed when delivered via recombinant vaccinia virus (42).

All of the data from cytokine gene-modified animal systems imply that locally released cytokines from tumor cells can alter local immunological microenvironment to help the host immune system detect previously unrecognized tumor-specific antigens and to induce the ability to kill the tumor cells. A number of human clinical trials utilizing cytokine gene modified cells are under way, and the results of these are expected in the near future (43).

To improve and facilitate the clinical application of cytokine gene therapy, several modifications of the published animal model tumor systems are being evaluated. Most animal studies, and some human trials have used in vitro cultured autologous tumor cells as targets for gene transfer. In practice this approach has serious limitations in a clinical setting in that generating an autologous tumor cell line from each patient's tumor is difficult, expensive, and time-consuming, and not possible in many circumstances. However, the key to cytokine gene therapy appears to be the production of appropriate cytokines in the vicinity of tumors, and not necessarily from tumor cells themselves. Several types of cells have been used for gene transfer including fibroblasts, tumor-infiltrating lymphocytes (TILs), and endothelial cells (44). Transduction of fibroblasts has many advantages over transduction of autologous cancer cells, as they can be obtained from skin biopsies, grow easily in culture for many passages, and can be efficiently transduced with viral vectors in vitro. Fakhrai et al. (45) showed that immunization with a mixture of irradiated tumor cells and IL-2-transduced fibroblasts induced protective antitumor immunity and remission of established tumors. A similar result was reported for IL-12 (46).

Genetic modification of immune effectors to increase cytotoxicity has also been explored. Tumor-infiltrating lymphocytes (TIL) are lymphocytes within a tumor mass that have homed to the tumor. Rosenberg et al. (47,48) have utilized TNF gene-transduced autologous TIL in advanced cancer patients and demonstrated enhanced antitumor effects with less toxicity than systemic administration. Transduction with chimeric T-cell receptor and antibody has the potential to improve targeting and killing by these effectors as well (48).

Another potential approach to improve the efficacy of immune-mediated gene therapy is to combine two different immunogenes to obtain a synergistic effects. The modification of tumor cells by MHC molecules and the B7 costimulatory molecule or by combinations of different cytokines with different mechanisms of action are possible approaches. Combination of B7-1-transduced tumors and concurrent administration of recombinant IL-12 can induce antitumor immunity synergistically (49). Combined gene therapy with B7 and IL-12 also showed an antitumor effect on lung metastases of the 3LL mouse lung carcinoma model (50). In this study, 3LL cells transfected with the B7 plasmid and transduced with retrovirus-IL12 induced 3LL specific CTL and significantly reduced the number of metastasis compared to parental 3LL (50). Combination of syngenic MHC class I transfection and IL-2 transduction in the mouse melanoma cells also showed synergistic effects on the eradication of established lung metastasis by the combined effects of efficient CTL induction and NK/LAK activity compared with single gene modified tumor cells (51).

D. Genetic Immunization

For defined target antigens, induction of epitope-specific immunity has been typically accomplished using synthetic peptides (52,53). The use of peptides as immunogens is complicated, however, by their weak inherent immunogenicity and variable chemical and physical properties, so a variety of strategies have been employed to enhance the efficacy of peptide-based vaccines (54–57). The chemical and physical problems of protein or peptide-based vaccines can be avoided by the use of "genetic vaccines," purified plasmid DNA expression vectors encoding the entire cloned open reading frame of proteins introduced into living animals. These DNA vaccine vectors may generate substantial humoral and cellular immunity with little or no toxicity (58–62). We have shown that the induction of T-cell epitope-specific (mutant p53) cellular immune responses and antitumor effects after introduction of a "genetic epitope" vaccine consisting of an expression cassette containing only an oligonucleotide coding for the desired epitope (63). In this study, we utilized a particle gun, which nontraumatically delivers microscopic gold particles coated with the plasmid DNA into the shaved skin of living animals. A plasmid vector containing the adenovirus E3 leader sequence was constructed, which facilitates transport of the mutant p53 epitope into the endoplasmic reticulum and shows that it can be important for optimal CTL induction and tumor-protective immunity. The use of epitope-minigene genetic vaccines may thus have significant potential for the induction of responses against identified T-cell epitopes in tumors.

III. Directly Cytotoxic Gene Therapy

Introduced genetic material can directly inhibit tumor growth by replacing a missing or damaged tumor suppressor gene, decreasing the expression of activated or overexpressed dominant oncogenes, or introducing an enzymatic activity that confers sensitivity to an otherwise nontoxic drug. Each of these will be discussed along with their actual or potential application to the gene therapy of lung cancer.

A. Tumor Suppressor Gene Replacement

Inactivation of tumor suppressor genes and the activation of oncogenes are important molecular events in carcinogenesis. Replacement of inactivated tumor suppressor genes by gene therapy has been extensively investigated, and several tumor suppressors are potential candidates for this approach. The inactivation of p53 may be the most common genetic alteration in human tumors, is found in human lung cancer with high frequency (64,65), and reintroduction inhibits in vitro growth (66). Replacement of p53 via retroviral vectors to human lung cancer cell lines with mutated p53 resulted in stable expression of p53 protein and apoptosis in vitro (67,68). In an orthotopic lung cancer model produced by intratracheal innoculation of human lung cancer cell lines into immunodeficient mice, direct intratracheal administration of retroviral-wt p53 expression vectors inhibited tumor growth (69). Wild-type p53 also has been introduced via recombinant adenovirus (70), with a pronounced antitumor effect after direct tracheal instillation in the same orthotopic lung tumor models (71). Direct peritumoral injection of adenovirus-p53 also inhibits solid tumor growth in nude mice (72). In these studies, the antitumor effects of p53 replacement by viral vectors appears to be more profound in p53-deleted or point-mutated tumors than in tumors with normal p53 (66). The mechanism of antitumor effects by p53 replacement appears to be primarily cell cycle arrest followed by the induction of apoptosis. Several human clinical trials using wild-type p53 are under way. A trial of a retrovirus expressing wild-type p53 directly injected into non-small-cell lung cancer endobronchial or intrapulmonary nodules has recently been completed at M.D. Anderson with some clinical responses observed (Roth, personal communication).

Replacement of p53 in p53-deleted lung cancer cell lines (NCI H358) appears to also increase chemosensitivity to cisplatin, a DNA-active agent. Direct injection of adenovirus-p53 into subcutaneous tumors in nude mice followed by IP cisplatin injection caused massive apoptotic death of tumors (73). Human clinical trials with adenovirus-p53 with and without cisplatin in human non-small-cell lung cancer have been started at the University of Texas M.D. Anderson Cancer Center.

Other tumor suppressor gene replacements have also been used therapeutically. These include the retinoblastoma protein (Rb), which is mutant in the majority of small-cell lung cancer and about 20% of non-small-cell lung cancer (74,75). Rb introduction has definite antitumor effects, which appear to depend on the extracellular environment of the tumor (76). A p53-inducible inhibitor of cyclin-dependent kinases, WAF/CIP1, is another candidate gene for the replacement therapy (77). A large part of the growth regulatory action of p53 appears to be mediated by this protein, also known as p21. Overexpression of these two genes can arrest the cell cycle in G1 phase (77). Adenovirus-p21 therapy of p53-deleted mouse prostate cancer cell lines induced growth arrest and resulted in a reduction of Cdk-2 kinase activity. And intratumoral injection with adenovirus-p21, but not with adenovirus-p53, prolonged the survival of tumor-bearing mice (78,79). Inhibitory activity was also found in human non-small-cell lung cancer cell lines (80).

The major conceptual problem with this sort of tumor suppressor gene replacement therapy is that all clonogenic tumor cells would theoretically need to receive the therapeutic tumor suppressor gene for there to be an observable clinical therapeutic impact. This goal is obviously very difficult or impossible to achieve. It is quite clear from chemotherapeutic studies that even single log reductions of tumor bulk (90% cell kill) are of only marginal clinical significance.

B. Drug-Sensitizing Gene Therapy

Since Moolten originally suggested the potential utility of herpes simplex virus thymidine kinase (HSV-TK) gene for cancer treatment (81), drug sensitivity gene therapies have been widely adopted in animal and human therapeutic approaches. All of these approaches attempt to confer tumor sensitivity to specific, otherwise nontoxic drugs by introduction of an activating enzyme followed by drug treatment. Tumor cells that express the activating enzyme are killed, and cells without the transduced gene and enzyme are not affected. Several drug resistance genes and drug combinations have been tested, including HSV-TK and ganciclovir (GCV), cytosine deaminase and 5-fluorouracil, and cytochrome P450 2B1 gene and cyclophosphamide.

HSV-TK and GCV combination has been the most widely investigated combination, and good responses have been observed in animal studies (reviewed in 82). Thymidine kinase is the enzyme responsible for phosphorylation of thymidine, and is involved in the salvage pathway for DNA synthesis. In contrast to cellular thymidine kinase, HSV-TK has an ability to phosphorylate not only thymidine but also several nucleoside analogs, including the guanosine analog ganciclovir. GCV can be phosphorylated by HSV-TK to the triphosphorylated form. This triphosphorylated GCV enters the DNA synthesis pathway instead of guanosine triphosphate, and blocks DNA synthesis, inducing cell death. In vitro tumor cell line experiments have shown that GCV sensitivity in HSV-TK gene-transduced tumors is at least 1000 times greater than control, nontransduced tumor cell lines. Furthermore, in animal tumor models, HSV-TK transduced tumors showed regression after GCV treatment (83).

An important advantage of this approach is the demonstrated existence of a "bystander effect." This means that not all tumor cells in a solid tumor need to be transduced to get 100% killing. In tumor models, HSV-TK-negative tumor cells in the vicinity of HSV-TK-positive tumor cells can be killed by exposure to GCV (84). The precise mechanisms of this bystander effect have not been fully elucidated, but there are several potential mechanisms. The transfer of apoptotic vesicles by endocytosis from killed HSV-TK-positive cells to adjacent HSV-TK-negative tumor cells has been identified in some cases. Recently, metabolic cooperation between cells via gap junctions has been demonstrated, making possible the cell-to-cell passage of small molecules, including toxic metabolites of GCV (85). Another mechanism that may play a role in some circumstances is the induction of an immune response by the recruitment of macrophages and CD4$^+$ and CD8$^+$ T cells to the tumor site.

Brain tumors may be ideal candidates for HSV-TK/GCV because of low metastatic potential and accessibility of direct tumoral injection. This is particularly true of retrovirally delivered TK, as retroviruses require cell cycling for integration, and the adult brain is comprised predominantly of noncycling cells. Thus there is also selective expression of TK in cycling tumor cells. Several clinical protocols have been completed or are under way in brain tumor patients, and tumor regression has been observed (Blaese, personal communication). This approach has been extended to various other tumors including hepatocellular carcinoma (86), ovarian carcinoma (87), mesothelioma (88), and lung cancer. Smythe et al. (88) demonstrated that transduction with an adenovirus expressing HSV-TK rendered human mesothelioma and lung cancer cell lines more sensitive to gancyclovir in vitro. This has been translated into a human clinical trial in mesothelioma, in which adenovirus TK is instilled into the pleural space through a chest tube, and 2 weeks of ganciclovir treatment is initiated 3 days later. Early patients in this study have tolerated this treatment well, and clinical results are awaited (Albelda, personal communication).

C. Antisense Treatment

In the process of carcinogenesis and maintenance of malignant phenotype, activation of oncogenes and continuous stimulation by certain growth factor signaling pathways are required. Some gene therapeutic strategies involve the inhibition of expression or function of these genes, which under certain circumstances can cause loss of the malignant phenotype and induction of apoptosis. Treatment with antisense nucleic acids attempts to block expression of these genes through the formation of translationally inactive RNA specific duplexes with the endogenous messenger RNA. This approach has been used to block the action of activated oncogenes, such as K-ras and c-myb, and growth factors or their receptors such as HER2/neu or insulin-like growth factors (IGF) and the IGF-I receptor.

Several antisense approaches have been investigated. One approach is through the use of antisense oligonucleotides, in which the phosphodiester backbone is modified to methylphosphonate or phosphorothioate to reduce degradation by nucleases. These modified antisense oligonucleotides can enter tumor cells by endocytosis and form DNA-RNA duplexes with endogenous sense mRNA, inhibiting translation. Another approach to inactivating mRNA is via ribozymes. Ribozymes are essentially antisense oligonucleotides that contain RNase active sites. A ribozyme possessing this activity allows catalytic gene ablation by sequence-specific cleavage of the target transcript. A third approach is to use plasmids or viral vectors for transferring an open reading frame fragment of the desired gene oriented backwards (3' to 5') behind a powerful promoter, resulting in the production of an antisense RNA. Antisense RNA transcribed from these constructs form a RNA duplex with sense mRNA inhibiting translation (89).

Blockade of activated oncogenes with antisense K-ras and c-myb has demonstrated antitumor activity. Intratracheal instillation of a retrovirus-containing antisense K-ras prevented tumor growth in an orthotopic lung cancer model with NCI H460a (90). A human clinical trial utilizing antisense Kras in lung cancer is under way at M.D. Anderson, but preliminary data are not available.

The growth of tumor depends on dysregulation in the regulation of cell cycle, and thus specific cell cycle regulatory proteins are reasonable targets for cancer gene therapy. The tumor suppressor genes p53 and Rb, discussed above, both directly down-modulate progression through the cell cycle, but a number of cyclins and growth factors act to promote proliferation, and are thus reasonable targets for gene-specific inhibition. Cyclin G1 (CYCG1) is one member of the G1 cyclins that is overexpressed in human osteosarcoma. Antisense cyclin G1 (CYCG1) delivered via a retroviral vector showed inhibition of growth of human osteogenic sarcoma cells (91).

A variety of growth factors and their receptors are also involved in cell cycle regulation (92). The insulin-like growth factors (IGF) and their receptors (IGF1r) can stimulate cellular proliferation, induce cellular differentiation, and have an important role in carcinogenesis and the maintenance of malignant phenotype (93). Mouse embryonic fibroblasts with a disruption of the IGF1r cannot be transformed by SV40 T antigen (94). Antisense IGF-1 suppresses the tumorigenicity of rat C6 glioblastomas and induces regression of established tumors (95). Stable transfection of antisense plasmids expressing the first 300 bp of IGF1r eliminates tumorigenicity of a variety of tumor cell lines and has been reported to induce systemic immune effects on established non-gene-modified tumors (96).

The IGFs and IGFr are also very important in lung development and the growth of cells in respiratory system (97). In most human lung cancer cells, IGF-1 and IGF-1r mediate autocrine proliferation, so lung cancer could be a good target for antisense IGF-1 and IGF-1r (98). We have constructed an adenovirus expressing antisense IGF-1r (Ad-IGF-1r/as) in an attempt to develop this observation into clinical therapeutic approach (99). A single transduction by Ad-IGF-1r/as decreased the receptor number by about 50% in human lung cancer cell lines, and decreased soft agar clonogenic ability almost 10- fold. In a model of an established intraperitoneal human lung cancer cell line in nude mice, intraperitoneal treatment by Ad-IGF-1r/as resulted in prolonged survival (99).

Ribozymes have been used in cancer therapeutic approaches to inhibit activated oncogenes (ras and bcr-abl), oncogenic viruses, and the MDR gene. A ribozyme designed to cleave only activated H-ras RNA abrogated the transformed phenotype of H-ras transfected murine NIH 3T3 cells (100). The effect of this anti-ras ribozyme was greater than that of a mutant ribozyme that could act only as antisense. To develop this strategy into a more practical therapeutic, a recombinant adenovirus containing the anti-H-ras ribozyme has been developed and demonstrated antitumor effects on EJ bladder carcinoma cells, expressing mutant H-ras (101). Since mutation of K-ras has been observed in about one third of lung adenocarcinomas (102,103), this ribozyme strategy may be applicable to the treatment of lung cancer.

Even though p53 is typically considered a tumor suppressor gene, the mutant p53 protein may not only lose tumor-suppressing ability but also can gain features of a dominant oncogene, either through a "dominant negative" effect where it binds to and inactivates wild-type p53 remaining within the cell, or gains an activity that independently promotes transformation. Reduction of mutant p53 protein expression in a tumor can thus have therapeutic effects, and can also be achieved by a retrovirally delivered anti-p53 ribozyme (104). Ribozyme-based techniques have also been investigated to block the expression of oncogenic viruses (human papilloma virus type 18) (105) and HIV (106).

D. Intracellular Antibody

Targeted abrogation of overexpressed oncoproteins in cancer cells can also be accomplished by the delivery of a gene encoding for what is known as an "intracellular antibody" (107). In this approach, the introduced gene encodes a single-chain Fv, which, when expressed and translated intracellularly, recognizes and entraps the oncoprotein, inhibiting its function. This has been successfully applied to erbB-2 in ovarian cancer (108,109), where dramatic growth inhibition and prolonged survival of human tumors in xenografted animals was observed after delivery of the gene with a recombinant adenovirus. As this oncoprotein is also expressed in breast and lung cancers (110–112), it has therapeutic potential for these tumor sites as well.

IV. Conclusion

Lung cancer is not an easy target for gene therapeutic approaches because it appears to be poorly immunogenic and often metastatic. Many genetic abnormalities have been found in lung cancer, however, and there are thus many potential targets for directed gene therapy approaches. Lung cancer also has the advantage of having a well-defined environmental etiology and discernible and accessible preneoplastic lesions. These preneoplastic lesions may ultimately be the most promising gene therapeutic targets.

Most of the gene therapeutic techniques described so far are potentially applicable to lung cancer. In mouse lung cancer models (Lewis lung carcinoma), several cytokines (GM-CSF, IL-2, IL-6) have shown antitumor effects, and these cytokines might be applied to human lung cancer. The relevance of this model, or any murine model of antitumor immunity, is at best uncertain, however.

p53 and K-ras mutations are frequently found in lung cancer, and replacement of p53 or blockade of K-ras are promising modalities currently being addressed in clinical trials in lung cancer. These approaches are not without problems either (113). Blocking key signal transduction pathways mediated by growth factors or their receptors (such as HER2/neu, IGF-1, and IGF-1r) by antisense, ribozyme, or intracellular antibodies is also promising in lung cancer, a disease commonly associated with autocrine growth stimulatory loops (98,114). Gene therapy using a drug-sensitizing gene (HSV-TK) could also be applied to inoperable, localized lung cancer or cancers localized to pleural space, a significant clinical problem. Novel approaches, such as the development of replication-competent, cytotoxic viruses that depend on the absence wild-type p53 or other tumor-specific features hold potential future promise.

In conclusion, molecular oncology and tumor immunology have revealed some of the secrets of cancer, and this knowledge has generated many interesting hypothetical therapeutic approaches, including gene therapeutic ones. The practical contribution of these to clinical oncology is still minimal, however, but we believe that gene therapy may ultimately represent a significant advance, or lead to as yet undiscovered novel approaches to the therapy of this frequently fatal disease.

References

1. Ginsberg RJ, Kris MG, Armstrong JG. Cancers of the Lung. In: DeVita VT Jr, Hellman S, Rosenberg RA, eds. Cancer: Principles and Practice of Oncology. Philadelphia: Lippincott, 1993:673–758.
2. Lee C-T, Chen HL, Carbone DP. Gene therapy for lung cancer. Ann Oncol 1995; 6:61–63.

3. Itaya T, Yamagiwa S, Okada F, et al. Xenogenization of a mouse lung carcinoma (3LL) by transfection with an allogenic class I major histocompatibility complex gene (H-2Ld). Cancer Res 1987; 47:3136–3140.
4. Plautz GE, Yang Z-Y, Wu B-Y, Gao X, Huang L, Nabel GJ. Immunotherapy of malignancy by in vivo gene transfer into tumors. Proc Natl Acad Sci USA 1993; 90:4645–4649.
5. Nabel GJ, Nabel EG, Yang ZY, et al. Direct gene transfer with DNA-liposome complexes in melanoma: expression, biological activity, and lack of toxicity in humans. Proc Natl Acad Sci USA 1993; 90:11307.
6. Ostrand-Rosenberg S, Thakur A, Clements V. Rejection of mouse sarcoma cells after transfection of MHC class II genes. J Immunol 1990; 144:4068–4071.
7. Plaskin D, Gelber C, Feldman M, Eisenbach L. Reversal of the metastatic phenotype in Lewis lung carcinomas after transfection with syngeneic H-2Kb gene. Proc Natl Acad Sci USA 1988; 85:4463–4467.
8. June CH, Bluestone JA, Nadler LM, Thompson CB. The B7 and CD28 receptor families. Immunol Today 1994; 15:321–331.
9. Guinan EC, Gribben JG, Boussiotis VA, et al. Pivotal role of the B7:CD28 pathway in transplantation tolerance and tumor immunity. The B7 and CD28 receptor families. Blood 1994; 84:3261–3282.
10. Linsley PS, Brady W, Grosmaire L, Aruffo A, Damle NK, Ledbetter JA. Binding of B-cell activation antigen B7 to CD28 costimulates T cell proliferation and interleukin-2 accumulation. J Exp Med 1991; 173:721–730.
11. Linsley PS, Greene JL, Tan P, et al. Coexpression and functional cooperation of CTLA–4 and CD28 on the activated T lymphocytes. J Exp Med 1992; 176:1595–1604.
12. Townsend SE, Alison JP. Tumor rejection after direct costimulation of CD8+ T cells by B7-transfected melanoma cells. Science 1993; 259:368–370.
13. Chen L, Ashe S, Brady WA, et al. Costimulation of antitumor immunity by the B7 counterreceptor for the T lymphocyte molecules CD28 and CTLA-4. Cell 1992; 71:1093–1102.
14. Chen L, McGowan P, Ashe S, et al. Tumor immunogenicity determines the effect of B7 costimulation on T cell-mediated tumor immunity. J Exp Med 1994; 179: 523–532.
15. Melief CJM, Kast WM. Potential immunogenicity of oncogene and tumor suppressor gene products. Curr Op Immunol 1993; 5:709–713.
16. Hodge JW, McLaughlin JP, Abrams SI, Shupert L, Schlom J, Kantor JA. Admixture of a recombinant vaccinia virus containing the gene for the costimulatory molecule B7 and a recombinant vaccinia virus containing a tumor-associated antigen gene results in enhanced specific T-cell responses and antitumor immunity. Cancer Res 1995; 55:3598–3603.
17. Lee CT, Ciernik IF, Wu S, et al. Increased immunogenicity of tumors bearing mutant p53 and P1A epitopes after transduction of B7-1 via recombinant adenovirus. Cancer Gene Ther 1996; 3:238–244.
18. Geldhof AB, Raes G, Bakkus M, Devos S, Thielemans K, De Baetselier P. Expression of B7-1 by a highly metastatic mouse T lymphoma induces optimal natural killer cell-mediated cytotoxicity. Cancer Res 1995; 55:2730–2733.

19. Rosenberg SA. Immunologic manipulation can mediate the regression of cancers in humans. J Clin Oncol 1988; 6:403–406.
20. Fearon ER, Pardoll DM, Itaya T, et al. interleukin-2 production by tumor cells bypasses T helper function in the generation of an antitumor response. Cell 1990; 60:397–403.
21. Ley V, Langlade DP, Kourilsky P, Larsson SE. Interleukin 2-dependent activation of tumor-specific cytotoxic T lymphocytes in vivo. Eur J Immunol 1991; 21:851–854.
22. Porgador A, Gansbacher B, Bannerji R, et al. Anti-metastatic vaccination of tumor-bearing mice with IL-2-gene-inserted tumor cells. Int J Cancer 1993; 53:471–477.
23. Salvadori S, Gansbacher B, Pizzimenti AM, Zier KS. Abnormal signal transduction by T cells of mice with parental tumors is not seen in mice bearing IL-2-secreting tumors. J Immunol 1994; 153:5176–5182.
24. Paul WE. Interleukin 4/B cell stimulatory factor 1: one lymphokine, many functions. FASEB 1987; 1:456–461.
25. Tepper RI, Pattengale PK, Leder P. Murine interleukin-4 displays potent anti-tumor activity in vivo. Cell 1989; 57:503–512.
26. Kishimoto T. The biology of interleukin 6. Blood 1989; 74:1–10.
27. Porgador A, Tzehoval E, Katz A, et al. Interleukin 6 gene transfection into Lewis lung carcinoma tumor cells suppresses the malignant phenotype and confers immunotherapeutic competence against parental metastatic cells. Cancer Res 1992; 52:3679–3686.
28. McBride WH, Dougherty GJ, Dubinett SM, Economou JS. Interleukin-7 mediated cancer gene therapy. In: Sobol RE, Scanlon KJ, eds. The Internet Book of Gene Therapy. Stamford, CT: Appleton and Lange, 1995: 181–187.
29. McBride WH, Thacker JD, Comora S, et al. Genetic modification of a murine fibrosarcoma to produce interleukin 7 stimulates host cell infiltration and tumor immunity. Cancer Res 1992; 52:3931–3937.
30. Collins T, Korman AJ, Wake CT, et al. Immune interferon activates multiple class-II major histocompatibility complex genes and the associated invariant chain gene in human endothelial cells and dermal fibroblasts. Proc Natl Acad Sci USA 1984; 81:4917–4921.
31. Gansbacher B, Bannerji R, Daniels B, Zier K, Cronin K, Gilboa E. Retroviral vector-mediated gamma-interferon gene transfer into tumor cells generates potent and long lasting antitumor immunity. Cancer Res 1990; 50:7820–7825.
32. Porgador A, Bannerji R, Watanabe Y, Feldman M, Gilboa E, Eisenbach L. Antimetastatic vaccination of tumor-bearing mice with two types of IFN-gamma gene-inserted tumor cells. J Immunol 1993; 150:1458–1470.
33. Beutler B, Cerami A. The biology of cachectin/TNF: a primary mediator of host response. Annu Rev Immunol 1989; 7:625.
34. Blankenstein T, Qin ZH, Uberla K, et al. Tumor suppression after tumor cell-targeted tumor necrosis factor alpha gene transfer. J Exp Med 1991; 173:1047–1052.
35. Han SK, Brody SL, Crystal RG. Suppression of in vivo tumorigenicity of human lung cancer cells by retrovirus-mediated transfer of the human tumor necrosis factor-alpha cDNA. Am J Respir Cell Mol Biol 1994; 11:270–278.

36. Dranoff G, Jaffee E, Lazenby A, et al. Vaccination with irradiated tumor cells engineered to secrete murine granulocyte-macrophage colony-stimulating factor stimulates potent, specific, and long-lasting anti-tumor immunity. Proc Natl Acad Sci USA 1993; 90:3539–3543.

37. Inaba K, Inaba M, Romani N, et al. Generation of large numbers of dendritic cells from mouse bone marrow cultures supplemented with granulocyte/macrophage colony-stimulating factor. J Exp Med 1992; 176:1693–1702.

38. Golumbek PT, Azhari R, Jaffee EM, et al. Controlled release, biodegradable cytokine depots: a new approach in cancer vaccine design. Cancer Res 1993; 53: 5841–5844.

39. Hsieh CS, Macatonia SE, Tripp CS, Wolf SF, O'Garra A, Murphy KM. Development of TH1 CD4+ T cells through IL-12 produced by Listeria-induced macrophages [see comments]. Science 1993; 260:547–549.

40. Scott P. IL-12: initiation cytokine for cell-mediated immunity [comment]. Science 1993; 260:496–497.

41. Tahara H, Lotze MT. Antitumor effects of interleukin-12 (IL-12): applications for the immunotherapy and gene therapy of cancer. Gene Ther 1995; 2:96–106.

42. Meko JB, Yim JH, Tsung K, Norton JA. High cytokine production and effective antitumor activity of a recombinant vaccinia virus encoding murine interleukin 12. Cancer Res 1995; 55:4765–4770.

43. Sobol RE, Fakhrai H, Shawler D, et al. interleukin-2 gene therapy in a patient with glioblastoma. Gene Ther 1995; 2:164–167.

44. Schmidt-Wolf GD, Schmidt-Wolf IG. Cytokines and gene therapy. Immunol Today 1995; 16:173–175.

45. Fakhrai H, Shawler DL, Gjerset R, et al. Cytokine gene therapy with interleukin-2-transduced fibroblasts: effects of IL-2 dose on anti-tumor immunity. Human Gene Ther 1995; 6:591–601.

46. Zitvogel L, Tahara H, Robbins PD, et al. Cancer immunotherapy of established tumors with IL-12. Effective delivery by genetically engineered fibroblasts. J Immunol 1995; 155:1393–1403.

47. Anonymous, Hwu P, Rosenberg SA. Immunization of cancer patients using autologous cancer cells modified by insertion of the gene for tumor necrosis factor. Human Gene Ther 1992; 3:57–73.

48. Hwu P, Rosenberg SA. The genetic modification of T cells for cancer therapy: an overview of laboratory and clinical trials. Cancer Detect Prev 1994; 18:43–50.

49. Coughlin CM, Wysocka M, Kurzawa HL, Lee WM, Trinchieri G, Eck SL. B7-1 and interleukin 12 synergistically induce effective antitumor immunity. Cancer Res 1995; 55:4980–4987.

50. Kato K, Yamada K, Wakimoto H, Okumura K, Yagita H. Combination gene therapy with B7 and IL-12 for lung metastasis of mouse lung carcinoma. Cancer Gene Ther 1995; 2:316s–317s.

51. Porgador A, Tzehoval E, Vadai E, Feldman M, Eisenbach L. Combined vaccination with major histocompatibility class I and interleukin 2 gene-transduced melanoma cells synergizes the cure of postsurgical established lung metastases. Cancer Res 1995; 55:4941–4949.

52. Ishioka GY, Colon S, Miles C, Grey HM, Chesnut RW. Induction of class I MHC-restricted, peptide-specific cytotoxic T cells by peptide priming in vivo. J Immunol 1989; 143:1094–1100.
53. Carbone F, Bevan M. Induction of ovalbumin-specific cytotoxic T cells by in vivo peptide immunization. J Exp Med 1989; 169:603–612.
54. Deres K, Schild H, Wiesmüller K-H, Jung G, Rammensee H-G. In vivo priming of virus-specific cytotoxic T lymphocytes with synthetic lipopeptide vaccine. Nature 1989; 342:561–564.
55. Gupta RK, Relyveld EH, Lindblad EB, Bizzani B, Ben-Efraim S, Kanta Gupta C. Adjuvants—a balance between toxicity and adjuvanticity. Vaccine 1993; 11:293–307.
56. Romero P, Cerottini J-C, Luescher I. Efficient in vivo induction of CTL by cell-associated covalent H-2Kd-peptide complexes. J Immunol Methods 1994; 171:73–84.
57. Berzofsky JA. Epitope selection and design of synthetic vaccines: molecular approaches to enhancing immunogenicity and crossreactivity of engineered vaccines. Ann NY Acad Sci 1993; 690:256–264.
58. Tang D, DeVit M, Johnston SA. Genetic immunization: A simple method for eliciting an immune response. Nature 1992; 356:152–154.
59. Ulmer JB, Donnelly JJ, Parker SE, et al. Heterologous protection against influenza by injection of DNA encoding a viral protein. Science 1993; 259:1745–1749.
60. Eisenbraun MD, Heydenburg Fuller D, Haynes JR. Examination of parameters affecting the eliction of humoral immune response by particle bombardment-mediated genetic immunization. DNA Cell Biol 1993; 12:791–797.
61. Wang B, Ugen KE, Srikantan V, et al. Gene inoculation generates immune response against human immunodeficiency virus type 1. Proc Natl Acad Sci USA 1993; 90:4156–4160.
62. Fynan EF, Webster RG, Fuller DH, Haynes JR, Santoro JC, Robinson HL. DNA vaccines: protective immunization by parental, mucosal and gene-gun inoculation. Proc Natl Acad Sci USA 1993; 90:11478–11482.
63. Ciernik IF, Berzofsky JA, Carbone DP. Induction of cytotoxic T lymphocytes and anti-tumor immunity with DNA vaccines expressing single T cell epitopes. J Immunol 1996; 156:2369–2375.
64. Chiba I, Takahashi T, Nau MM, et al. Mutations in the p53 gene are frequent in primary, resected non-small cell lung cancer. Oncogene 1990; 5:1603–1610.
65. D'Amico D, Carbone D, Mitsudomi T, et al. High frequency of somatically acquired p53 mutations in small cell lung cancer cell lines and tumors. Oncogene 1992; 7:339–346.
66. Takahashi T, Carbone D, Takahashi T, et al. Wild-type but not mutant p53 suppresses the growth of human lung cancer cells bearing multiple genetic lesions. Cancer Res 1992; 52:2340–2343.
67. Cai DW, Mukhopadhyay T, Liu Y, Fujiwara T, Roth JA. Stable expression of the wild-type p53 gene in human lung cancer cells after retrovirus-mediated gene transfer. Human Gene Ther 1993; 4:617–624.
68. Fujiwara T, Grimm EA, Mukhopadhyay T, Cai DW, Owen-Schaub LB, Roth JA. A retroviral wild-type p53 expression vector penetrates human lung cancer spheroids and inhibits growth by inducing apoptosis. Cancer Res 1993; 53:4129–4133.

69. Fujiwara T, Cai DW, Georges RN, Mukhopadhyay T, Grimm EA, Roth JA. Therapeutic effect of a retroviral wild-type p53 expression vector in an orthotopic lung cancer model. J Natl Cancer Inst 1994; 86:1458–1462.

70. Santoso JT, Tang DC, Lane SB, et al. Adenovirus-based p53 gene therapy in ovarian cancer [see comments]. Gynecol Oncol 1995; 59:171–178.

71. Zhang WW, Fang X, Mazur W, French BA, Georges RN, Roth JA. High-efficiency gene transfer and high-level expression of wild-type p53 in human lung cancer cells mediated by recombinant adenovirus. Cancer Gene Ther 1994; 1:5–13.

72. Wills KN, Maneval DC, Menzel P, et al. Development and characterization of recombinant adenoviruses encoding human p53 for gene therapy of cancer. Human Gene Ther 1994; 5:1079–1088.

73. Fujiwara T, Grimm EA, Mukhopadhyay T, Zhang WW, Owen-Schaub LB, Roth JA. Induction of chemosensitivity in human lung cancer cells in vivo by adenovirus-mediated transfer of the wild-type p53 gene. Cancer Res 1994; 54:2287–2291.

74. Harbour JW, Lai S-L, Whang-Peng J, Gazdar AF, Minna JD, Kaye FJ. Abnormalities in structure and expression of the human retinoblastoma gene in SCLC. Science 1988; 241:353–357.

75. Horowitz J, Park S-H, Bogenmann E, et al. Frequent inactivation of the retinoblastoma anti-oncogene is restricted to a subset of human tumor cells. Proc Natl Acad Sci USA 1990; 87:2775–2779.

76. Kratzke RA, Shimizu E, Geradts J, et al. RB-mediated tumor suppression of a lung cancer cell line is abrogated by an extract enriched in extracellular matrix. Cell Growth Differ 1993; 4:629–635.

77. el-Deiry WS, Tokino T, Velculescu VE, et al. WAF1, a potential mediator of p53 tumor suppression. Cell 1993; 75:817–825.

78. Eastham JA, Hall SJ, Sehgal I, et al. In vivo gene therapy with p53 or p21 adenovirus for prostate cancer. Cancer Res 1995; 55:5151–5155.

79. Yang C, Cirielli C, Capogrossi MC, Passaniti A. Adenovirus-mediated wild-type p53 expression induces apoptosis and suppresses tumorigenesis of prostatic tumor cells. Cancer Res 1995; 55:4210–4213.

80. Inoue F, Hamada K, Kataoka M, et al. Growth inhibition of non-small cell lung cancer cells by a recombinant adenovirus-mediated transfer of the p21 gene. Cancer Gene Ther 1995; 2:339s.

81. Moolten FL. Tumor chemosensitivity conferred by inserted herpes thymidine kinase genes: paradigm for a prospective cancer control strategy. Cancer Res 1986; 46:5276–5281.

82. Moolten FL. Drug sensitivity ("suicide") genes for selective cancer chemotherapy. In: Sobol RE, Scanlon KJ, eds. The Internet Book of Gene Therapy: Cancer Therapeutics. Stamford, CT: Appleton & Lange, 1995:119–127.

83. Culver KW, Ram Z, Wallbridge S, Ishii H, Oldfield EH, Blaese RM. In vivo gene transfer with retroviral vector-producer cells for treatment of experimental brain tumors. Science 1992; 256:1550–1552.

84. Freeman SM, Abboud CN, Whartenby KA, et al. The bystander effect: tumor regression when a fraction of the tumor mass is genetically modified. Cancer Res 1993; 53:5274–5283.

85. Bi WL, Parysek LM, Warnick R, Stambrook PJ. In vitro evidence that metabolic cooperation is responsible for the bystander effect observed with HSV tk retroviral gene therapy. Human Gene Ther 1993; 4:725–731.

86. Wills KN, Huang WM, Harris MP, Machemer T, Maneval DC, Gregory RJ. Gene therapy for hepatocellular carcinoma: chemosensitivity conferred by adenovirus-mediated transfer of the HSV-1 thymidine kinase gene. Cancer Gene Ther 1995; 2:191–197.

87. Rosenfeld ME, Feng M, Michael SI, Siegal GP, Alvarez RD, Curiel DT. Adenoviral-mediated delivery of the herpes simplex virus thymidine kinase gene selectively sensitizes human ovarian carcinoma cells to ganciclovir. Clin Cancer Res 1995; 1:1571–1580.

88. Smythe WR, Hwang HC, Amin KM, et al. Use of recombinant adenovirus to transfer the herpes simplex virus thymidine kinase (HSVtk) gene to thoracic neoplasms: an effective in vitro drug sensitization system. Cancer Res 1994; 54:2055–2059.

89. Mercola D, Cohen JS. Antisense approaches to cancer gene therapy. In: Sobol RE, Scanlon KJ, eds. The Internet Book of Gene Therapy: Cancer Therapeutics. Stamford, CT: Appleton & Lange, 1995:77–89.

90. Georges RN, Mukhopadhyay T, Zhang Y, Yen N, Roth JA. Prevention of orthotopic human lung cancer growth by intratracheal instillation of a retroviral antisense K-ras construct. Cancer Res 1993; 53:1743–1746.

91. Skotzko M, Wu L, Anderson WF, Gordon EM, Hall FL. Retroviral vector-mediated gene transfer of antisense cyclin G1 (CYCG1) inhibits proliferation of human osteogenic sarcoma cells. Cancer Res 1995; 55:5493–5498.

92. Baserga R, Rubin R. Cell cycle and growth control. Crit Rev Eukaryot Gene Expr 1993; 3:47–61.

93. Baserga R. The insulin-like growth factor I receptor: a key to tumor growth? Cancer Res 1995; 55:249–252.

94. Sell C, Rubini M, Rubin R, Liu JP, Efstratiadis A, Baserga R. Simian virus 40 large tumor antigen is unable to transform mouse embryonic fibroblasts lacking type 1 insulin-like growth factor receptor. Proc Natl Acad Sci USA 1993; 90:11217–11221.

95. Trojan J, Johnson TR, Rudin SD, Ilan J, Tykocinski ML, Ilan J. Treatment and prevention of rat glioblastoma by immunogenic C6 cells expressing antisense insulin-like growth factor I RNA. Science 1993; 259:94–97.

96. Resnicoff M, Sell C, Rubini M, et al. Rat glioblastoma cells expressing an antisense RNA to the insulin-like growth factor-1 (IGF–1) receptor are nontumorigenic and induce regression of wild-type tumors. Cancer Res 1994; 54:2218–2222.

97. Stiles AD, D'Ercole AJ. The insulin-like growth factors and the lung. Am J Respir Cell Mol Biol 1990; 3:93–100.

98. Nakanishi Y, Mulshine JL, Kasprzyk PG, et al. Insulin-like growth factor-I can mediate autocrine proliferation of human small cell lung cancer cell lines in vitro. J Clin Invest 1988; 82:354–359.

99. Lee C-T, Wu S, Gabrilovich D, et al. Antitumor effects of an adenovirus expressing antisense insulin-like growth factor I receptor on human lung cancer cell lines. Cancer Res 1996; 56:3038–3041.

100. Kashani-Sabet M, Funato T, Florenes VA, Fodstad O, Scanlon KJ. Suppression of the neoplastic phenotype in vivo by an anti-ras ribozyme. Cancer Res 1994; 54:900–902.
101. Feng M, Cabrera G, Deshane J, Scanlon KJ, Curiel DT. Neoplastic reversion accomplished by high efficiency adenoviral-mediated delivery of an anti-ras ribozyme. Cancer Res 1995; 55:2024–2028.
102. Mitsudomi T, Steinberg SM, Oie HK, et al. *ras* gene mutations in non-small cell lung cancers are associated with shortened survival irrespective of treatment intent. Cancer Res 1991; 51:4999–5002.
103. Mitsudomi T, Viallet J, Mulshine JL, Linnoila RI, Minna JD, Gazdar AF. Mutations of *ras* genes distinguish a subset of non-small-cell lung cancer cell lines from small-cell lung cancer cell lines. Oncogene 1991; 6:1353–1362.
104. Cai DW, Mukhopadhyay T, Roth JA. Suppression of lung cancer cell growth by ribozyme-mediated modification of p53 pre-mRNA. Cancer Gene Ther 1995; 2:199–205.
105. Chen Z, Kamath P, Zhang S, Weil MM, Shillitoe EJ. Effectiveness of three ribozymes for cleavage of an RNA transcript from human papillomavirus type 18. Cancer Gene Ther 1995; 2:263–271.
106. Sarver N, Cantin EM, Chang PS, et al. Ribozymes as potential anti-HIV–1 therapeutic agents. Science 1990; 247:1222–1225.
107. Deshane J, Loechel F, Conry RM, Siegal GP, King CR, Curiel DT. Intracellular single-chain antibody directed against erbB2 down-regulates cell surface erbB2 and exhibits a selective anti-proliferative effect in erbB2 overexpressing cancer cell lines. Gene Ther 1994; 1:332–337.
108. Deshane J, Cabrera G, Grim JE, et al. Targeted eradication of ovarian cancer mediated by intracellular expression of anti-erbB-2 single-chain antibody [see comments]. Gynecol Oncol 1995; 59:8–14.
109. Deshane J, Siegal GP, Alvarez RD, et al. Targeted tumor killing via an intracellular antibody against erbB-2. J Clin Invest 1995; 96:2980–2989.
110. Wodrich W, Volm M. Overexpression of oncoproteins in non-small cell lung carcinomas of smokers. Carcinogenesis 1993; 14:1121–1124.
111. Yu D, Wang SS, Dulski KM, Tsai CM, Nicolson GL, Hung MC. c-erbB–2/neu overexpression enhances metastatic potential of human lung cancer cells by induction of metastasis-associated properties. Cancer Res 1994; 54:3260–3266.
112. Tsai CM, Chang KT, Perng RP, et al. Correlation of intrinsic chemoresistance of non-small-cell lung cancer cell lines with HER-2/neu gene expression but not with ras gene mutations. J Natl Cancer Inst 1993; 85:897–901.
113. Carbone DP, Minna JD. In vivo gene therapy of human lung cancer using wild-type p53 delivered by retrovirus. J Natl Cancer Inst 1994; 86:1437–1438.
114. Cuttitta F, Carney DN, Mulshine J, et al. Bombesin-like peptides can function as autocrine growth factors in human small-cell lung cancer. Nature (Lond) 1985; 316:823–826.

Part Five

EPILOGUE

17

The Regulatory Process and Gene Therapy

SUZANNE EPSTEIN

Center for Biologics Evaluation and Research
Food and Drug Administration
Bethesda, Maryland

I. Introduction

Gene therapy is a rapidly evolving field in which a wide variety of therapeutic approaches are being considered and tested. Approaches to treatment of various lung diseases include study and use of adenovirus vectors (1–4), adeno-associated virus vectors (5,6), and DNA/liposome vectors (7,8) for delivery of therapeutic genes. Retroviral vectors have also been studied (9).

In regulating product development in such an area, it is important to be flexible and attentive to practical limitations. If the barrier to starting exploratory clinical trials is too high, then only a few therapies will be tried and the ones that might have succeeded may be missed. On the other hand, the public should be protected from unnecessary or unreasonable risk, and also from the raising of unrealistic hopes. In its oversight of biological products for gene therapy, the Center for Biologics Evaluation and Research (CBER) of FDA seeks a balance between these considerations. As time goes on and a longer track record accumulates, it will likely be possible to relax recommendations for some types of testing which pioneering investigators had to perform. Sponsors should be aware that change in regulatory policy occurs more quickly now than in the past, especially in an area like gene therapy with rapidly evolving technology. However, when products reach later stages of development, standards for approval call for demonstration of efficacy, as with any other product.

While FDA regulation applies to human clinical trials rather than to laboratory research, CBER staff can assist during the lab research phase of product development prior to submission of an investigational new drug (IND) application. Early interaction with CBER staff can be helpful, especially for those, such as academic scientists, not used to viewing their work as manufacturing. Pre-IND interaction can help sponsors anticipate questions, facilitate later stages of the process, benefit from accumulated experience of others, and avoid pitfalls. In addition to telephone consultation, pre-IND meetings can be arranged. Guidance documents are also available, as listed in Table 1.

The chapter title includes "regulatory process" to emphasize the idea of regulation as a dynamic process involving exchange and dialogue between the investigator/sponsor and CBER staff. It is inappropriate to think of regulation of biologics in terms of a check list of tests every product must undergo, especially in an area such as gene therapy in which systems differ greatly and available information changes so rapidly. The fundamentals have not changed: demonstration of the safety, identity, purity, potency, and ultimately efficacy of the product. However, the details depend on the experimental system, the patient population, and the knowledge base for related products. The goal should be a process during which the sponsor guides a product through development and clinical testing, consulting with CBER staff whenever appropriate.

II. Manufacturing and Quality Control of Gene Therapy Products

Production and testing of products for gene therapy overlap in many respects with those for other biologicals, as discussed previously (10). In particular, there are resemblances to the handling of other products produced by recombinant DNA technology and to handling of cell lines for use in making other biologicals. For this reason, guidance documents in these areas are helpful (Table 1). For example, manufacturing of gene therapy products may involve bacterial cultures producing plasmids, mammalian packaging cell lines, and producer cell lines making viral vectors. Cell banks of these types are to be characterized and documented as described in the "Points to Consider in the Characterization of Cell Lines Used to Produce Biologicals (1993)." As with other products, in process testing and product quality control testing should use quantitative assays of demonstrated sensitivity and specificity whenever possible.

Conventional concerns about toxic contaminants of course apply to these novel products, too. Unnecessary use of toxic materials during manufacturing

Table 1 Points to Consider Documents Relevant to Gene Therapy Issued by the Center for Biologics Evaluation and Research

Points to Consider in the Production and Testing of New Drugs and Biologicals Produced by Recombinant DNA Technology (1995)
Points to Consider in the Characterization of Cell Lines Used to Produce Biologicals (1993)
Points to Consider in the Manufacture and Testing of Monoclonal Antibody Products for Human Use (1995)
Points to Consider in the Collection, Processing, and Testing of Ex-Vivo-Activated Mononuclear Leukocytes for Administration to Humans (1989)
Points to Consider in Human Somatic Cell Therapy and Gene Therapy (1991)

should be avoided. If toxic solvents, metals, dyes, or other chemicals are used during production, the risk they pose should be considered, residual levels measured, and specifications set. One example would be chloroform used in manufacturing liposomes. Chloroform is readily removed, assays for residual levels are sensitive, and its removal should be verified by testing.

Gene therapy products also present novel issues. One example is the potential for recombination. Recombination can occur between sequences in a vector and other elements in a packaging cell or producer cell (11). Recombination may also occur between a vector and either endogenous sequences in a recipient (12) or viruses infecting the recipient. Recombination between two retroviruses infecting a single patient has been seen to lead to recombination (13). The significance of this type of concern would depend on the product and the patient group. For retroviral products, either transduced cells or retroviral vectors for direct administration to patients, one major concern is exclusion of replication-competent retroviruses (RCR). This concern was at first theoretical, as initial animal studies did not show adverse effects (14). While retroviral vectors have been used in humans for several years without apparent adverse effects (15,16), the concern was heightened by the development of fatal lymphomas in immunosuppressed monkeys exposed to replication-competent murine retrovirus (17). For this reason, testing for RCR in cell banks, testing of bulk and final products, and monitoring of patients are recommended. For detailed recommendations, consult current versions of "Points to Consider" documents or contact CBER for guidance.

For adenovirus vectors, the analogous issue of replication-competent adenovirus (RCA) is of concern. Adenovirus infections of the respiratory tract are often mild, but could be significant adverse events in patients already compromised by lung diseases. Quantitative testing for presence of RCA is needed, and levels of RCA in patient doses are to be minimized.

For plasmid gene therapies, one issue has been raised by the common use of ampicillin resistance as a selection marker. As stated in the "Points to Consider in the Characterization of Cell Lines Used to Produce Biologicals (1993)," penicillin and other beta-lactam antibiotics should be avoided in production cultures, due to the possibility of residual antibiotic or adducts in the final product and antibiotic hypersensitivity in some people. For licensed products, current regulations (21 CFR 610.61[m]) require product labeling to declare any antibiotics added during manufacturing, and use of penicillin may concern physicians. If a construct has already been made and contains the amp^r gene, it may be possible to use it in some situations if ampicillin is not used in making the final product. If the antibiotic is used in production, contact CBER concerning the nature of the protocol and the proposed patient population.

Investigators with products under development should be aware of this problem and consider alternative selection markers, especially since simply omitting ampicillin from production runs may lead to poor yields. Choosing a different selection marker avoids the difficulties of attempting later to prove absence of residual beta lactam and/or derivatized products, and of labeling issues if the product proceeds to licensure. Currently, other antibiotic selection systems can be used. A variety of nonantibiotic selection systems are being developed. Such approaches would have the added advantage of avoiding possible transfer of antibiotic resistance genes to other bacteria.

III. Preclinical Animal Studies

Animal studies are usually needed and informative in testing of new products prior to human use, even though they provide imperfect predictions about effects in humans (though see Sect. IV, below). An animal species should be chosen that can be infected by the vector class in use, and if possible an animal that provides a model of the biological activity of the product. It will sometimes be possible to achieve an animal model of the disease. The appropriate species will vary, and uniform use of particular animal species such as primates is not mandated by CBER. The route of administration intended for clinical trials should be used if possible, as well as intravenous administration in certain cases. The dose range tested should include at least one dose within and one dose above the proposed clinical range, and should attempt to determine a toxic dose.

One goal of animal studies is of course testing for toxicity. Formulation components can be responsible for toxicity, so the final product formulation intended for human use should be the one studied. For example, a variety of cationic lipids have been studied as possible delivery systems for plasmid gene therapy (18,19). One observed toxicity of DNA/liposome products is inflammation of the lungs, and this effect has been shown to be due to the lipid component (Dr. S. Cheng, Genzyme, personal communication with permission).

In addition to toxicity, the goals of animal studies should be to study where the vector localizes, where it is expressed, and effects of overexpression or ectopic expression. Tumorigenicity may also be a concern in some cases. Possible adverse events might include effects of expression at higher than normal levels, or effects of expression in unintended cells or tissues. Such effects could include immune responses to a gene product in hosts not normally expressing it, or in hosts expressing a mutated form of that product. Such responses could result in autoimmunity or in alteration of the safety and/or efficacy of repeat treatment.

Of special concern is whether vector localizes to particular sites, and whether it is found in the gonads. While the gonads and germ line are not impervious to viral attack and alteration (20), it may be that gene therapy vectors in current use will not alter the germ line. In one model system, plasmid DNA administered to pregnant mice was found in the progeny but had not been integrated into the germline (21). Nevertheless, due to the long term consequences of germ line alteration and the lack of societal consensus on germ line genetic intervention, preclinical animal testing is used to examine potential effects of gene therapy products. These questions call for well-designed animal trials analyzed with sensitive, specific assay methods such as validated PCR. Positive signals in gonadal tissue would call for follow-up studies excluding blood contamination and analyzing the cell type(s) involved. Note, however, that CBER's position is not an absolute one prohibiting clinical use of products with possible mutagenic effects. Even negative results from animal studies analzyed with the most sensitive techniques can only put an upper limit on the possible frequency of altered sperm, for example. Other accepted therapeutics such as some chemotherapy drugs are known to be mutagenic. The goal is to explore the issue carefully, to follow up on unexpected results, and to accumulate data that will allow full patient informed consent, including knowledgeable advice to patients in the future, and will permit an accurate risk-benefit assessment to be made.

Recombination of vectors with other viruses present in the host at the time of treatment or with viruses infecting the host later are possible sources of concern in gene therapy, and have been discussed earlier. Whether such possibilities can appropriately be addressed in vitro or in animal models depends on the system.

IV. Vector Modification

In the past, CBER has considered any change in a vector to result in a new product, necessitating a new IND. There is now an accumulating body of evidence that will permit flexibility in this policy, and in the future further changes in policy may be possible. The goal is to facilitate progress toward effective therapies by abbreviating testing and reducing documentation, when this can be done while preserving patient safety.

Ordinarily, a new biological product requires a new IND. For gene therapy vectors this is not always appropriate. Vector modification is a process involving many incremental changes potentially leading to safer and more effective vectors. Each modified vector must receive appropriate testing and must meet appropriate specifications. However, repeating all preclinical studies will not be necessary in all cases. If this were required, it could lead to inappropriate conservatism by investigators and to continued use of vectors that might not be as good as newer ones, but were considered by investigators to be "approved." Note that CBER does not approve vectors for investigational use in general, but rather permits particular clinical trial proposals to proceed based on adequate data. What is adequate may depend on the intended use.

How much of a change in a vector causes it to be classed as a new product subject to full retesting? This question must be judged case by case, and decisions will be based in part on the data accumulating concerning the effects or lack of effects of particular structural changes on localization, expression, etc. Still, comments can be made about some examples. If the only change in a series of closely related vectors is substitution of a different (especially if closely related) therapeutic gene as the insert, with no change in the viral or plasmid structures, the vectors may in some cases be considered members of a panel. Another case would be replacement of one antibiotic selection marker with another. Other minor changes in vector structure may also fall into this category. In such cases testing could be abbreviated. Preclinical animal testing such as tissue localization, germ line alteration, tumorigenicity, and pharmacology/ toxicology studies may not need to be repeated. Instead, safety testing for each vector will focus on safety issues presented by the insert and its expressed product, plus quality control testing. If the inserted genes in a series of vectors are the same or closely related and would be expected to have the same potential for adverse effects, even such testing could be abbreviated.

Other modifications may consist of changes in lipids or other formulation components used with vectors. Different lipid components, different ratios of two lipids, or different ratios of lipid to DNA may be used to formulate a vector for administration to patients. The properties of the particular lipids used will affect the activity of liposomes in enhancing transfection (22). Use of a new lipid requires safety information on that new lipid, but not necessarily in conjuction with the precise vector proposed for use if the new lipid has been used with other vectors. If tissue localization, germ line alteration, and classical animal pharmacology/toxicology data are compared for use of varied lipid ratios, compositions, or ratios to DNA, this data can be used to evaluate whether such studies are necessary for additional combinations.

Characterization of modified vectors should include confirmation of the intended changes in structural features. If the same facility is producing multiple related vectors, it would be advisable to perform a quality control test capable of distinguishing the various vectors, such as mapping with appropriately chosen restriction enzymes.

V. Clinical Trials

Gene therapy clinical trials involve special unresolved issues, as well as usual issues of design and ethics affecting all trials. The concerns come about owing to possible infectious spread or intrapatient recombination or replication of agents used in the trial, possible irreversible effects, potential for germ line alteration, and the uncertain time frame and nature of any adverse effects. Such issues have been analyzed by ethicists and by investigators in the field (23). Other issues arise mainly due to the excitement surrounding this technology and the potential for unrealistic hopes and exploitation of patients or their families. Some of these issues can be addressed by appropriate conduct of recruiting and the informed consent process, some by long-term patient monitoring, and some by additional attention to animal studies even while clinical trials are under way.

Regarding recruitment, an important issue is definition of diagnostic criteria to define patient subgroups in which genotype or diagnosis permits prediction of disease severity. Such criteria permit more realistic risk/benefit assessment. Another issue is the accuracy of laboratory tests used for diagnosis. Next, there is need for balanced presentation to patients or families of the options they have, whether or not these lie within the area of expertise of the investigator.

Access to clinical trials should be the same for patients of both genders unless there is a valid biological or medical reason to exclude one gender. Reproductive risks should also be handled comparably for both genders unless there are biological or medical reasons to do otherwise. Exclusion of pregnant women may be appropriate in many trials, while intentional in utero gene therapy brings up a host of issues that are currently under discussion nationally and internationally. However, requirements for use of birth control or for reproductive sterility will usually apply to both genders if they apply at all. Any hazard to germ cells may be greater in the male due to the rapid turnover of gametes. In addition to use of birth control, reproductive risks can be dealt with by informed consent procedures that honestly state the limits of our knowledge about risks to future offspring.

Patient monitoring in gene therapy trials will need to be long term, often lifelong. Adverse events cannot be assumed unrelated to therapy simply because they are not immediate. Many consequences, for example immunological ones, might be considerably delayed. Consent forms should mention the need for long-term follow-up, including the possible need for samples. Planning is under way for a national information network to track such data. Patients should be advised not to donate blood, organs, tissues, or gametes until more is known about genetic alterations. Consent can also include mention of the importance of autopsy, and autopsy protocols should include germ cell analysis.

If there is reasonable risk that a vector used in a clinical trial may be subject to environmental release and spread or may mutate or recombine to release an infectious agent, then clinical trial design should include provision for appropriate isolation of the subject and for monitoring for such agents.

Questions arise as to the risk of interactions if one patient is exposed to multiple gene therapy agents. As with interactions among other drugs, this possibility should be examined case by case. There is no general prohibition on patients who have already received one gene therapy later receiving another gene therapy intervention, particularly if cells have been treated with vector ex vivo.

VI. "Well-Characterized Products"

At a CBER-sponsored workshop, December 11–13, 1995, on well-characterized biotechnology products, a breakout session on plasmid DNA products was held. The recommendation of this session was that plasmid DNA products be considered eligible for designation as "well characterized." The agency is considering this policy decision. A "well-characterized product" would have well-defined physicochemical characteristics, could be manufactured reproducibly, and would not require a separate establishment license or lot release submission to CBER if licensed. Note that a product can be considered well characterized by virtue of its structure and mode of manufacture and testing, and yet the product is not eligible for licensure unless data collected under an IND demonstrate safety and efficacy in an appropriate patient population.

VII. Conclusion

Application of the new technology of gene therapy to treatment of lung diseases will yield the greatest benefits if approached on a sound scientific and medical basis. CBER welcomes suggestions from the research and manufacturing communities as to how this process can best be facilitated in the regulatory arena.

Guidance documents including Points to Consider documents and other policy documents are available from CBER as follows:

Division of Congressional and Public Affairs
CBER, HFM-12
1401 Rockville Pike, Suite 200N
Rockville, MD 20852-1448
Phone (301) 594-1800 or (800) 835-4709
CBER FAX Information System (301) 594-1939 (call from a FAX
machine with touchtone phone and then follow the instructions)
Or send E-mail to GSTA@A1.CBER.FDA.GOV
You will receive an automatic reply listing documents available by return
E-mail.

CBER staff are also available for questions about what products FDA
regulates, for arrangement of pre-IND meetings, or for any concerns relating
to gene therapy product development, at (301) 594-0830.

Note Added in Proof

See Federal Register, Vol. 61, No. 94, May 14, 1996, "Elimination of Establishment
Licence Application for Specified Biotechnology and Specified Synthetic Biological
Products." The term "well-characterized" has been superceded; plasmid DNA
therapeutic products are included as "specified products."

Note also that presence of vectors in the gonads without chromosomal integration could also have reproductive effects of concern.

References

1. Rosenfeld MA, Yoshimura K, Trapnell BC, et al. In vivo transfer of the human
   cystic fibrosis transmembrane conductance regulator gene to the airway epithelium.
   Cell 1992; 68:143–155.
2. Rosenfeld MA, Siegfried W, Yoshimura K, et al. Adenovirus-mediated transfer of
   a recombinant α1-antitrypsin gene to the lung epithelium in vivo. Science 1991;
   252:431–434.
3. Engelhardt JF, Simon RH, Yang Y, et al. Adenovirus-mediated transfer of the
   CFTR gene to lung of nonhuman primates: biological efficacy study. Human Gene
   Therapy 1993; 4:759–769.
4. Crystal RG, McElvaney NG, Rosenfeld MA, et al. Administration of an adenovirus
   containing the human CFTR cDNA to the respiratory tract of individuals with cystic
   fibrosis. Nature Genet 1994; 8:42–51.
5. Flotte TR, Afione SA, Conrad C, et al. Stable in vivo expression of the cystic fibrosis
   transmembrane conductance regulator with an adeno-associated virus vector. Proc
   Natl Acad Sci USA 1993; 90:10613–10617.
6. Flotte TR, Solow R, Owens RA, Afione S, Zeitlin PL, Carter BJ. Gene expression
   from adeno-associated virus vectors in airway epithelial cells. Am J Respir Cell Mol
   Biol 1992; 7:349–356.

7. Canonico AE, Plitman JD, Conary JT, Meyrick BO, Brigham KL. No lung toxicity after repeated aerosol or intravenous delivery of plasmid-cationic liposome complexes. J Appl Physiol 1994; 77:415–419.
8. Malone RW, Felgner PL, Verma IM. Cationic liposome-mediated RNA transfection. Biochemistry 1989; 86:6077–6081.
9. Drumm ML, Pope HA, Cliff WH, et al. Correction of the cystic fibrosis defect in vitro by retrovirus-mediated gene transfer. Cell 1990; 62:1227–1233.
10. Epstein SL. Regulatory concerns in human gene therapy. Human Gene Ther 1991; 2:243–249.
11. Vanin EF, Kaloss M, Broscius, Nienhuis AW. Characterization of replication-competent retroviruses from nonhuman primates with virus-induced T-cell lymphomas and observations regarding the mechanism of oncogenesis. J Virol 1994; 68:4241–4250.
12. Gareis M, Harrer P, Bertling WM. Homologous recombination of exogenous DNA fragments with genomic DNA in somatic cells of mice. Cell Mol Biol 1991; 37:191–203.
13. Diaz RS, Sabino EC, Mayer A, Mosley JW, Busch MP, Transfusion Safety Study Group. Dual human immunodeficiency virus type 1 infection and recombination in a dually exposed transfusion recipient. J Virol 1995; 69:3273–3281.
14. Cornetta K, Moen RC, Culver K, et al. Amphotropic murine leukemia retrovirus is not an acute pathogen for primates. Human Gene Ther 1990; 1:15–30.
15. Blaese RM, Culver KW, Miller AD, et al. T lymphocyte-directed gene therapy for ADA-SCID: Initial trial results after 4 years. Science 1995; 270:475–480.
16. Bordignon C, Notarangelo LD, Nobili N, et al. Gene therapy in peripheral blood lymphocytes and bone marrow for ADA-immunodeficient patients. Science 1995; 270:470–475.
17. Donahue RE, Kessler SW, Bodine D, et al. Helper virus induced T cell lymphoma in nonhuman primates after retroviral mediated gene transfer. J Exp Med 1992; 176:1125–1135.
18. Farhood H, Gao X, Son K, et al. Cationic liposomes for direct gene transfer in therapy of cancer and other diseases. Ann NY Acad Sci 1994; 716:23–34.
19. Gao X, Huang L. A novel cationic liposome reagent for efficient transfection of mammalian cells. Biochem Biophys Res Commun 1991; 179:280–285.
20. Raina AK, Adams JR. Gonad-specific virus of corn earworm. Nature 1995; 374:770.
21. Tsukamoto M, Ochiya T, Yoshida S, Sugimura T, Terada M. Gene transfer and expression in progeny after intravenous DNA injection into pregnant mice. Nature Genet 1995; 9:243–248.
22. Felgner JH, Kumar R, Sridhar CN, et al. Enhanced gene delivery and mechanism studies with a novel series of cationic lipid formulations. J Biol Chem 1994; 269:2550–2561.
23. Ledley FD. After gene therapy: issues in long-term clinical follow-up and care. Adv Genet 1995; 32:1–16.

18

Some Issues Affecting Progress Toward Human Gene Therapy and Potential Application to the Lung

THEODORE FRIEDMANN

University of California at San Diego
San Diego, California

I. Introduction

The concept of gene therapy for human disease has become very popular during the past decade or so. Its history has been reviewed extensively in both the biomedical and popular literatures (1–5), and the rationale for the use of genetic material as a therapeutic tool so intuitively understood and so widely accepted that gene therapy has become a central conceptual force in modern medicine. As a result of elegant basic research and results from many in vitro proof of principal studies, expectations for quick clinical success have been very high. These expectations have included application to many kinds of human disease, including inborn errors of metabolism, neoplasms, cardiovascular disease, disorders of the central nervous system, infectious diseases such as AIDS, and, of particular interest in this volume, diseases of the lung—in particular, cystic fibrosis (CF) (6). The potential for genetic approaches to therapy has been recognized and strengthened not only by traditional funding mechanisms but also by a number of timely and targeted funding opportunities from agencies such as the National Institutes of Health and from disease foundations such as the Cystic Fibrosis Foundation and others. However, for a variety of reasons, the level of expectation for easy clinical success was too high, the result first and foremost of the true and obvious clinical potential for dire diseases but also, less laudably, to an enthusiastic and, at times, overzealous science, to overactive public relations efforts at many institution, most notably the National Institutes of Health, and to a uncritical and sensational popular press (1).

II. Lessons From Recent Clinical Studies

During the past year, we have seen some of those hopes and expectations seemingly dashed by a series of publications presenting the first careful and rigorous reports of initial clinical results from some of the most highly touted disease models, including muscular dystrophy, ADA deficiency, familial hypercholesterolemia, and CF. In all cases, the studies have reported results that fall far short of a demonstration of therapeutic efficacy. At best, the studies with ADA deficiency have demonstrated prolonged survival of some genetically modified T cells in patients grafted with autologous peripheral blood cells infected with a retrovirus vector expressing the wild-type ADA cDNA as well as evidence for some prolonged gene expression in vivo (7,8). Although studies of the restoration of CFTR function in the airway epithelium of patients with CF after gene transfer with liposomes vectors have demonstrated only transient and low level of CFTR reconstitution (9), other studies of CFTR gene transfer with an adenovirus vector have shown no detectable reconstitution (10).

These results have been greeted by an outpouring of dismay from the general popular press and a feeling in some biomedical circles that there is less potential for human gene therapy than had been previously presented. At the same time, there has been a call by the director of the National Institutes of Health, Dr. Harold Varmus, for a re-evaluation of the federal review process for the approval of human gene therapy studies, an examination of the state of the technology underlying current clinical studies and an understanding of the overinflated expectations that the field had generated. The reports from the NIH advisory committees have been presented to the director (1) and paint a picture of reasonable and generally effective review procedures but of overstated preclinical results, inadequate tools for stable and therapeutically effective gene transfer, and the need for more rigorously designed clinical studies.

III. Technical Needs

The successful application of gene transfer technology to human disease will require major advances in our understanding of disease pathogenesis, in the availability of appropriate gene transfer tools and techniques and in our understanding of the actors that permit effective cell and tissue grafting.

A. Retrovirus Vectors

Currently available tools for gene transfer into human cells and tissues, including viral and nonviral reagents, are largely inadequate for the difficult job of transferring a therapeutic gene efficiently and safely into defective human cells in a tissue-specific way to ensure effective genetic modification of only specific target cells and an effective attack on a specific genetic function. For gene delivery to the lung via the airway, current retrovirus vectors, most of which are derived from Moloney leukemia virus, are not optimal because of the low mitotic index of normal airway epithelial cells, by their low titers, and by their relative instability in vivo. However, with sufficient improvements in vector design, delivery to the lung might be feasible, since it has been shown that a Moloney-based retrovirus is reasonably stable to aerosolization (11). Retrovirus vector-mediated gene transfer to the lung via the circulation would seem to represent a far greater problem, since in the absence of effective cell-specific targeting, the low-titer preparations would be further markedly diluted in the circulation and made susceptible to virus destruction by complement-mediated mechanisms. This class of retrovirus vectors, which represent the most commonly used gene transfer vectors in clinical gene therapy studies, would therefore be made far more useful in general if they can be improved by the following:

1. Major increases in vector titers to allow more efficient ex vivo applications and to permit in vivo gene delivery. Significant advances have been made toward this goal through the development of retroviruses pseudotyped with the G protein of vesicular stomatitis virus (VSV) (12–14), but still further increases are likely to be needed for truly efficient in vivo gene delivery.
2. Introduction, presumably into the viral envelope protein, of ligands or antibodies to cell surface receptors or other membrane components to permit cell-specific targeting. A number of studies have shown that the replacement of the Moloney retroviral env protein with other proteins such as the VSV-G protein, influenza virus hemagglutinin (15), or the env of Gibbon ape leukemia virus (16,17) modifies the host range of the resulting virus vectors. Furthermore, introduction of epitopes such as the erythropoietin protein (18) or single-chain antibodies into the viral env (19–21) also is able to specify cell tropism.
3. Modification of vectors to permit infection of quiescent cells such as normal airway epithelial cells, neurons, macrophages, and hepatocytes. The beginnings of this technology are beginning to appear

through the inclusion into Moloney-based vectors of the elements from the HIV virus that permit that retrovirus to infect nonreplicating cells, the nuclear localization signal of the HIV matrix (MA) protein, and the function of the HIV accessory protein Vpr (22–26).
4. Specifying the site of retrovirus integration into the host cell to reduce the potential for insertional mutagenesis. It has been demonstrated that the presence of sequence-specific transcription factors can confer integration site specificity to the Ty elements of yeast and that integrase-protein interactions also occur during integration of the HIV provirus (27). It seems likely that addition of appropriate sequence-specific elements to a retrovirus vector can help to direct the integration event in a similar manner.

B. Adenovirus Vectors

Adenovirus vectors have attained a favorite niche in gene therapy approaches to CF. They can be produced to very high titers, they infect nonreplicating or slowly replicating cells such as those of the airway epithelium, and they are relatively, but only *relatively*, noncytopathic. The first class of adenovirus vectors, of the sort used in the initial clinical CF studies, were designed to be replication-defective and incapable of viral protein synthesis through elimination of E1 viral functions (28,29), but it has become clear that these agents remained capable of some degree of replication and sufficient viral protein synthesis to produce powerful inflammatory and host immune responses. More recent modifications in vector design, including the inclusion of additional deletions in other adenoviral genes such as E2 and E4 (28,30), have led to vectors that are less immunogenic because of their reduced capacity for viral protein synthesis. In vivo gene transfer with these agents has been reported to lead to prolonged transgene expression and a reduced inflammatory response, making them potentially more attractive than the initial adenovirus vectors for human clinical application in cystic fibrosis and many other disease models.

C. Other Virus Vectors

A number of additional viral vector systems are becoming available for therapeutic gene transfer in a variety of disease models, including disorders of the lung. Of particular interest have been the adeno-associated vectors (AAV). The vectors, like adenovirus, can infect relatively quiescent cells and have the additional feature of being able to integrate into the host cell genome, to some extent at least into a preferred region of a single human chromosome, No. 19 (32). As in the case of retrovirus vectors, this capacity would be useful for an in vivo gene delivery scheme by reducing the potential for insertional mutagenesis. AAV vectors have also been shown to infect epithelial cells of the airway quite efficiently. However, recent studies have shown that infection of cells with very highly purified AAV vector is markedly less efficient than infection with AAV in the presence of helper adenovirus (33), a feature that in principle makes this agent somewhat less appealing. However, the development of the newer classes of adenovirus vectors described above might provide helper adenoviruses that are sufficiently nonimmunogenic and noncytopathic to serve as effective helpers in clinical human applications. Nevertheless, early versions of the AAV vector have been applied to the cystic fibrosis model. AAV vectors have been found to express the CFTR cDNA efficiently (34) and to lead to prolonged CFTR expression after instillation of a CFTR-expressing AAV vector into the lungs of rabbits (35).

D. Liposome Vectors

Liposomes have been used in several early in vivo clinical gene therapy applications, including studies of tumor immunotherapy (36–38) and, most notably, in studies of CFTR transfer into the airway epithelium (9,39–43). Studies with early versions of liposome vectors in the airway have demonstrated transient and low levels of CFTR reconstitution. However, the great advances that have been made in the past several years in the efficiency of gene transfer with cationic lipid-DNA conjugates (44,45) provide some optimism that future studies in the airway delivery of genes to the airway epithelium will show better levels of gene transfer and CFTR reconstitution and correction of the chloride transport defect in cystic fibrosis.

E. Receptor-Mediated Uptake

Great progress has been made over the past several years in the development of techniques for the introduction of foreign and potentially therapeutic genetic material into cells by specific interaction of DNA-virus or other conjugates to cell-specific receptors and other membrane components (46–50). In addition to being useful as approach to neoplasms and other disorders, it is possible that this approach might be useful in the case of pulmonary disease such as cystic fibrosis. Indeed, transgene delivery to respiratory epithelial cells has been reported by targeting these receptors (31). This approach may serve as a prototype either for the airway delivery or even the hematogenous delivery of foreign genes to cells of the respiratory epithelium.

IV. Conclusion

Gene therapy is in its earliest stages of development. While some of the concepts are established and accepted, the technology for the safe, efficient, functional, and therapeutic delivery of foreign genetic information to any tissue in the human body, including the airway, is not yet adequate. But the progress is rapid, and the next several years will almost assuredly see the emergence of markedly improved tools and techniques for therapeutic gene delivery for the treatment of human disease.

References

1. Friedmann T. Gene therapy—an immature genie but certainly out of the bottle. Nature Med. 1996; 2:144–147.
2. Brenner MK. Human somatic gene therapy: progress and problems. J Intern Med 1995; 237:229–239.
3. Mulligan RC. The basic science of gene therapy. Science 1993; 260:926–932.
4. Anderson WF. Human gene therapy. Science 1992; 256:808–813.
5. Verma IM. Gene therapy. Sci Am 1990; 263:68–72.
6. Alton EWFW, Geddes DM. Gene therapy for respiratory diseases: potential applications and difficulties. Thorax 1995; 50:484–486.
7. Bordignon C, Notarangelo LD, Nobili N, et al. Gene therapy in peripheral blood lymphocytes and bone marrow for ADA-immunodeficient patients. Science 1995; 270:470–475.
8. Blaese RM, Culver KW, Miller AD, et al. T lymphocyte-directed gene therapy for ADA-SCID: initial trial results after 4 years. Science 1995; 270:475–480.
9. Caplen NJ, Alton EWFW, Middleton PG, et al. Liposome-mediated *CFTR* gene transfer to the nasal epithelium of patients with cystic fibrosis. Nature Med 1995; 1:39–46.
10. Knowles MR, Hohneker KW, Zhou Z, et al. A controlled study of adenoviral-mediated gene transfer in the nasal epithelium of patients with cystic fibrosis. N Engl J Med 1995; 333:823–831.
11. Stanley C, Rosenberg MB, Friedmann T. Gene transfer into rat airway epithelial cells using retroviral vectors. Somat Cell Mol Genet 1991; 17:185–190.
12. Friedmann T. Pseudotyped retroviral vectors for studies of human gene therapy. Nature Med 1995; 1:275–277.
13. Burns JC, Matsubara T, Lozinski G, et al. Pantropic retroviral vector-mediated gene transfer, integration, and expression in cultured newt limb cells. Dev Biol 1994; 165:285–289.
14. Miyanohara A, Yee JK, Bouic K, LaPorte P, Friedmann T. Efficient in vivo transduction of the neonatal mouse liver with pseudotyped retroviral vectors. Gene Ther 1994. In press.
15. Dong J, Roth MG, Hunter EA. A chimeric avian retrovirus containing the influenza virus hemagglutinin gene has an expanded host range. J Virol 1992; 66:7374–7382.
16. Wilson C, Reitz MS, Okayama H, Eiden MV. Formation of infectious hybrid virions with gibbon ape leukemia virus and human T-cell leukemia virus retroviral envelope glycoproteins and the gag proteins of Moloney murine leukemia virus. J Virol 1989; 63:2374–2378.

17. Kim T, Leibfried-Rutledge ML, First NL. Gene transfer in bovine blastocysts using replication-defective retroviral vectors packaged with Gibbon ape leukemia virus envelopes. Mol Reprod Dev 1993; 35:105–113.
18. Kasahara N, Dozy A, Kan YW. Targeting retroviral vectors to specific cells. Science 1995; 269:417.
19. Chu T-HT, Dornburg R. Retroviral vector particles displaying the antigen-binding site of an antibody enable cell-type-specific gene transfer. J Virol 1995; 69:2659–2663.
20. Cosset F-L, Morling FJ, Takeuchi Y, Weiss RA, Collins MKL, Russell SJ. Retroviral retargeting by envelopes expressing an N-terminal binding domain. J Virol 1995; 69:6313–6322.
21. Russell SJ, Hawkins RE, Winter G. Retroviral vectors displaying functional antibody fragments. Nucleic Acids Res 1993; 21:1081–1085.
22. Heinzinger NK, Bukrinsky MI, Haggerty SA, et al. The Vpr protein of human immunodeficiency virus type 1 influences nuclear localization of viral nucleic acids in nondividing cells. Proc Natl Acad Sci USA 1994; 91:7311–7315.
23. Lewis P, Hensel M, Emerman M. Human immunodeficiency virus infection of cells arrested in the cell cycle. EMBO J 1992; 11:3053–3058.
24. Bukrinsky MI, Haggerty S, Dempsey MP, et al. A nuclear localization signal within the HIV-1 matrix protein that governs infection of non-dividing cells. Nature 1993; 365:666–669.
25. Gallay P, Swingler S, Aiken C, Trono D. HIV-1 infection of nondividing cells: C-terminal tyrosine phosphorylation of the viral matrix protein is a key regulator. Cell 1995; 80:379–388.
26. Von Schwedler U, Kornbluth RS, Trono D. The nuclear localization signal of the matrix protein of human immunodeficiency virus type 1 allows the establishment of infection in macrophages and quiescent T lymphocytes. Proc Natl Acad Sci USA 1994; 91:6992–6996.
27. Bushman FD. Tethering human immunodeficiency virus 1 integrase to a DNA site directs integration to nearby sequences. Proc Natl Acad Sci USA 1994; 91:9233–9237.
28. Smith AE. Viral vectors in gene therapy. Annu Rev Microbiol 1995; 49:807–838.
29. Bramson JL, Graham FL, Gauldie J. The use of adenoviral vectors for gene therapy and gene transfer in vivo. Curr Opin Biotechnol 1995; 6:590–595.
30. Engelhardt JF, Ye X, Doranz B, Wilson JM. Ablation of E2A in recombinant adenoviruses improves transgene persistence and decreases inflammatory response in mouse liver. Proc Natl Acad Sci USA 91:6196–6200.
31. Ferkol T, Kaetzel CS, Davis PB. Gene transfer into respiratory epithelial cells by targeting polymeric immunoglobulin receptors. J Clin Invest 1993; 92:2394–2400.
32. Giraud C, Winocour E, Berns KI. Site-specific integration by adeno-associated virus is directed by a cellular DNA sequence. Proc Natl Acad Sci USA 1994; 91:10039–10043.
33. Fisher KJ, Gao GP, Weitzman MD, DeMatteo R, Burda JF, Wilson JM. Transduction with recombinant adeno-associated virus for gene therapy is limited by leading-strand synthesis. J Virol 1996; 70:520–532.

34. Flotte TR, Afione SA, Solow R, et al. Expression of the cystic fibrosis transmembrane conductance regulator from a novel adeno-associated promoter. J Biol Chem 1993; 268:3781–3790.
35. Flotte TR, Afione SA, Conrad C, et al. Stable in vivo expression of the cystic fibrosis transmembrane conductance regulator with an adeno-associated virus vector. Proc Natl Acad Sci USA 1993; 90:10613–10617.
36. Lew D, Parker SE, Latimer T, et al. Cancer gene therapy using plasmid DNA: pharmacokinetic study of DNA following injection in mice. Human Gene Ther 1995; 6:553–564.
37. Nabel GJ, Felgner PL. Direct gene transfer for immunotherapy and immunization. Tibtech 1993; 11:211–215.
38. Plautz GE, Yang Z, Wu B, Gao X, Huang L, Nabel GJ. Immunotherapy of malignancy by in vivo gene transfer into tumors. Proc Natl Acad Sci USA 1993; 90:4645–4649.
39. Colledge WH, Evans MJ. Cystic fibrosis gene therapy. Br Med Bull 1995; 51:82–90.
40. Sorscher EJ, Logan JJ, Frizzell RA, et al. Gene therapy for cystic fibrosis using cationic liposome-mediate gene transfer: a phase I trial of safety and efficacy in the nasal airway. Human Gene Ther 1994; 5:1259–1277.
41. Hyde SC, Gill DR, Higgins CF, et al. Correction of the ion transport defect in cystic fibrosis transgenic mice by gene therapy. Nature 1993; 362:250–255.
42. Alton EWFW, Middleton PG, Caplen NJ, et al. Non-invasive liposome-mediated gene delivery can correct the ion transport defect in cystic fibrosis mutant mice. Nature Genet 1993; 5:135–142.
43. Caplen NJ, Middleton PG, Kinrade E, et al. DNA-liposome-mediated transfection of epithelial cells in vivo and in vitro. Pediatr Pulmonol 1993; 9(suppl):243.
44. Koopmann J, Maintz D, Schild S, et al. Multiple polymorphisms, but no mutations, in the WAF1/CIP1 gene in human brain tumours. Br J Cancer 1995; 72:1230–1233.
45. Felgner JH, Kumar R, Sridhar CN, et al. Enhanced gene delivery and mechanism studies with a novel series of cationic formulations. J Biol Chem 1994; 269:2550–2561.
46. Schwarzenberger P, Spence SE, Gooya JM, et al. Targeted gene transfer to human hematopoietic progenitor cell lines through the c-kit receptor. Blood 1996; 87:472–478.
47. von Rüden T, Stingl L, Cotten M, Wagner E, Zatloukal K. Generation of high-titer retroviral vectors following receptor-mediated, adenovirus-augmented transfection. BioTechniques 1995; 18:484–489.
48. Deshane J, Siegal GP, Alvarez RD, et al. Targeted tumor killing via an intracellular antibody against erbB-2. J Clin Invest 1995; 96;2980–2989.
49. Zatloukal K, Schmidt W, Cotten M, Wagner E, Stingl G, Birnstiel ML. Somatic gene therapy for cancer: the utility of transferrinfection in generating "tumor vaccines." 1993; 135:199–207.
50. Conry RM, LoBuglio AF, Kantor J, et al. Immune response to a carcinoembryonic antigen polynucleotide vaccine. Cancer Res 1994; 54:1164–1168.

AUTHOR INDEX

Underlined numbers give the page on which the complete reference is listed.

Subject Index